NMR 30

Basic Principles and Progress

Solid-State NMR I
Methods

Guest-Editor: B. Blümich

With contributions by
B. Blümich, P. Blümler, B.F. Chmelka,
G. Fleischer, F. Fujara, A.-R. Grimmer,
F. Lauprêtre, D. Raftery

With 141 Figures and 10 Tables

Springer-Verlag
Berlin Heidelberg NewYork
London Paris Tokyo
Hong Kong Barcelona Budapest

ISBN-13 : 978-3-642-78485-9 e-ISBN-13 : 978-3-642-78483-5
DOI: 10.1007/978-3-642-78483-5

Softcover reprint of the hardcover 1st edition 1994

Library of Congress Catalog Card Number 93-9522

Typesetting: Thomson Press (India) Ltd, New Delhi

51/3020 - 5 4 3 2 1 0 - Printed on acid-free paper

Preface

Solid-State NMR is a branch of Nuclear Magnetic Resonance which is presently experiencing a phase of strongly increasing popularity. The most striking evidence is the large number of contributions from Solid-State Resonance on NMR meetings, approaching that of liquid-state Resonance. Important progress can be observed in three areas: Methodical developments, applications to inorganic matter, and applications to organic matter. These developments are intended to be captured in three volumes of this series, each of them being devoted to more or less one of these areas.

The present volume on Solid-State NMR I is devoted largely to methodical aspects. The first chapter provides an introduction which reviews some of the basic features of Solid-State spectroscopy. The second chapter demonstrates methods and application of high-resolution ^{13}C-NMR for the study of local dynamics in polymers. Chapter three reviews xenon NMR as a tool for the characterization of complex chemical systems and host phases. Chapter four covers translational diffusion NMR in analogy to generalized incoherent scattering experiments, a topic which promises to become of great value also in Solid-State research. The last chapter reviews the developments of methods for acquisition of NMR information in solids with spatial resolution.

Of course, the topics chosen in all three Solid-State NMR volumes by no means cover the entire area of Solid-State NMR, but it is hoped that they treat an attractive cross-section of today's research. Particular thanks goes to the authors for their pleasant cooperation and, most importantly, for writing the contributions. Springer-Verlag has been very helpful in its assistance and editorial supervision.

Aachen, August 1993

B.Blümich
R. Kosfeld

Guest-Editor

Prof. Dr. Bernhard Blümich

Lehrstuhl für Makromolekulare Chemie,

RTWH Aachen, Worringer Weg 1,

D-52056 Aachen, FRG

Table of Contents

Table of Contents of Volume 31

Solid-State NMR II - Inorganic Matter

Introduction to Solid-State NMR

Arnd-Rüdiger Grimmer[1] and Bernhard Blümich[2]

[1] Projektgruppe Festkörper-NMR in KAI e.V., Rudower Chaussee 5, Haus 4.1, D-12489 Berlin-Adlershof, FRG
[2] Lehrstuhl für Makromolekulare Chemie, RWTH Aachen, Worringerweg 1, D-52074 Aachen, FRG

Table of Contents

NMR Basic Principles and Progress, Vol. 30

Solid-state NMR spectra generally reflect the sum of several independent interactions such as Zeeman, dipolar and quadrupolar coupling as well as magnetic shielding effects. To obtain the desired information, separation of these effects and their decoding is essential. Within the first part, as a basis of the successful separation of the NMR effects, the common and also the different (tensorial) properties of each interaction are discussed in detail. The second part mainly deals with the experimental design of solid-state high-resolution NMR experiments to suppress certain interactions and with wide-line experiments. Finally the principles of NMR imaging and NMR microscopy are described as an alternative method for investigating solids by NMR.

1 Introduction

Solids have always assumed a substantial role in development of human culture, from the very beginning in the Stone Age—named after predominantly siliceous solids—until the present period of sophisticated materials. Depending on their wide range of structure, solids exhibit specific mechanical, thermal, electric, magnetic, optic, and, last but not least, biological properties. The knowledge of these structures in the widest sense is an essential condition for their proper economic and ecological use. Among the physical methods contributing to this knowledge, solid-state NMR is now established as a technique with widespread applications.

The first NMR experiments were carried out in 1945 on solid paraffin [1a] as well as on liquid water [1b]. Immediately followed by investigations of dipolar interactions in solids [2], which created the basis of *wide-line* NMR. Due to restrictions imposed by effects of the dipole–dipole couplings, however, these early experiments were of limited value, with investigations of dynamic processes in polymers being one possible exception. The same is valid for quadrupolar effects [3], although studies of inorganic glasses [4] should be mentioned as a positive example for wide-line NMR.

A little later, the discovery of chemical shift [5, 6] and indirect dipolar coupling [7] marked the starting point of dynamic growth of the still young technique. Due to the more or less convenient accessibility of these parameters in liquids most of the work following has been devoted to liquid-state NMR. During the 1950s and the 1960s the condition of solid-state NMR was defined by the generally accepted statement [8, 9] that chemical shift is not detectable in solids. Contrasting this pessimism, singular examples of wide-line measurements of chemical shifts in solids (Fig. 1) seemed to bear the message of an imminent vigorous development.

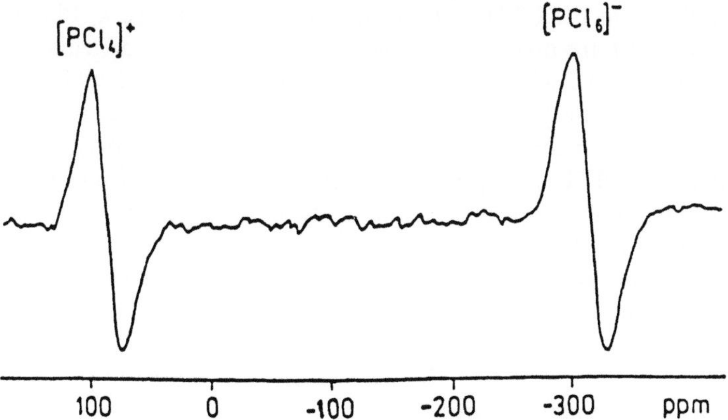

Fig. 1. ^{31}P CW NMR spectrum of PCl_5 at 24 MHz (static sample, first derivative spectrum). The two 1:1 signals reflect the structure as ionic PCl_4PCl_6 [26]

Widespread applications of solid-state NMR became possible at the end of the 1960s after the impressive discovery that the different, useful or troublesome, interactions in solids could be manipulated with more or less ease [10–13]. Eventually, the last decade has witnessed the genesis of highly sophisticated variants of solid-state NMR such as multidimensional NMR [14, 15] and imaging [16–18].

It is the aim of the present chapter to give an introductory review of basic solid-state NMR with special emphasis on their common aspects and their practical advantages. More detailed information including theoretical aspects will be found in other chapters of this volume as well as in relevant textbooks [19–25].

2 Principles of NMR

NMR spectroscopy is a physical method for direct investigation of nuclear energy levels. The origin of these levels goes back to the angular momentum **P** of the nucleus. Therefore, the NMR effect is based on a nuclear property. Consequently the discussion of the fundamental aspects of NMR requires the use of quantum mechanics. The density matrix [27] is the appropriate concept for the description of the motion of an ensemble of interacting magnetic moments. Nevertheless many aspects of NMR can be described in terms of the classical vector model of magnetization precessing in a magnetic field.

2.1 The NMR Phenomenon

The majority of nuclei possesses an angular momentum **P**, which is proportional to a quantity called *spin* **I** (ZUM1). The proportionality constant is Planck's constant $\hbar$, and **P** and **I** are quantum mechanical operators. In particular, the eigenvalue of $\mathbf{I}^2$ is $I(I+1)$, where I is the spin quantum number which can assume integral and half-integral values. Often the spin quantum number itself is referred to as spin, for instance ^{1}H, ^{13}C, ^{29}Si, and ^{31}P are spin-1/2 nuclei with $I = 1/2$, and ^{2}H and ^{6}Li are spin-1 nuclei with $I = 1$. A list of nuclei with spin relevant to materials science is given in Table 1 together with other information pertinent to NMR spectroscopy.

The angular momentum and thus the spin is proportional to the magnetic moment μ, of the nucleus,

$$\mu = \gamma \hbar \mathbf{I} \tag{1}$$

This equation defines the *magneto-gyric ratio* γ.

The macroscopic thermodynamic equilibrium magnetization M_0 is formed by the sum of projections of all nuclear magnetic moments along the axis of

Table 1. Nuclei and their NMR properties

isotope	spin	nat. abundance (%)	quadrupole moment[a]	rel. sensitivity[b]	ν_0 (MHz) at 2.3488 T	chemical shift range (ppm)	chemical shift reference
^1H	1/2	99.98	—	1.0	100.000	20 to −1	$SiMe_4$
^2H	1	1.5×10^{-2}	0.002875	9.65×10^{-3}	15.351	2 to 0	—
^7Li	3/2	92.58	−0.040	0.29	38.863	50 to −10	1MLiCl
^{11}B	3/2	80.42	0.040	0.17	32.084	100 to −120	BF_3OEt_3
^{13}C	1/2	1.108	—	1.59×10^{-2}	25.144	240 to −10	$SiMe_4$
^{15}N	1/2	0.37	—	1.04×10^{-3}	10.133	1200 to −500	$MeNO_2$
^{17}O	5/2	3.7×10^{-2}	−0.026	2.91×10^{-2}	13.557	1400 to −100	H_2O
^{19}F	1/2	100	—	0.83	94.077	800 to −450	$CFCC_3$
^{23}Na	3/2	100	0.108	9.25×10^{-2}	26.451	10 to −60	1MNaCl
^{27}Al	5/2	100	0.150	0.21	26.057	200 to −200	$[Al(H_2O)_6]^{3+}$
^{29}Si	1/2	4.7	—	7.84×10^{-3}	19.865	100 to −400	$SiMe_4$
^{31}P	1/2	100	—	6.63×10^{-2}	40.481	230 to −200	H_3PO4

[a]electric quadrupole moment Q in multiplies of $|e|\,10^{-24}\,cm^2$.
[b]at constant field and equal number of nuclei

the magnetic field $\boldsymbol{B}_0$. Its value is expressed by the Curie law as a function of the magnetic field and the temperature. The magnetization $\boldsymbol{M}$ is a macroscopic magnetic dipole moment.

2.1.1 Zeeman Interaction

The potential energy of a magnetization $\boldsymbol{M}$ in a magnetic field $\boldsymbol{B}_0$ depends on the angle θ between the dipole moment and the field (Fig. 2a)·

$$E = -\boldsymbol{M}\boldsymbol{B}_0 = -|\boldsymbol{M}|\,|\boldsymbol{B}_0|\cos\theta = M_z B_0. \tag{2}$$

Because in NMR the orientation of the field $\boldsymbol{B}_0 = (0, 0, B_0)$ defines the z axis of the laboratory coordinate frame of reference, $M_z = |\boldsymbol{M}|\cos\theta$ is the projection of the magnetization vector on to the direction of the magnetic field.

The quantum mechanical operator corresponding to the energy is the *Hamilton operator*. The potential energy of a single magnetic moment in a magnetic field is given in analogy to the last part of Eq. (2) by:

$$\mathbf{H}^Z = -\gamma \mathbf{I}_Z B_0. \tag{3}$$

It is called *Zeeman interaction*. For this reason the upper index Z is used. Here and in the following the NMR notation of writing the Hamilton operators in rad/s is used. Energy units are obtained by multiplication with h. The interaction of magnetic moments with a magnetic field gives rise to a splitting of energy levels.

The energy levels E_m are defined as the eigenvalues of the Hamilton operator:

$$E_m = -\gamma \hbar m B_0. \tag{4}$$

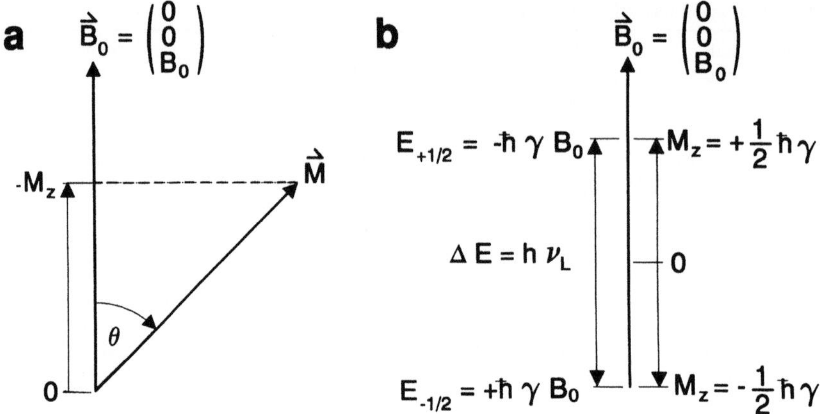

Fig. 2a,b. Classical and quantum mechanical energies of magnetic dipoles in a magnetic field. **a**. The potential energy of the macroscopic magnetization $\boldsymbol{M}$ in a magnetic field $\boldsymbol{B}_0$ is the product of the magnetic field with the projection of the magnetization on to the axis of the field. It depends on the angle θ between the magnetization and the field. **b**. For a quantum mechanical magnetic moment with spin $I = 1/2$ there are two stable states in a magnetic field. One has its projection parallel, the other antiparallel to the direction of the field. Both states differ in energy

Here m is the *magnetic quantum number*. It can assume the values:

$$-I \leqq m \leqq +I. \tag{5}$$

Thus a nuclear spin with quantum number I can be in one of $2I + 1$ stable positions in a magnetic field. Nuclei like ^{1}H and ^{13}C with spins $I = 1/2$ have two eigenstates. These are referred to as spin-up and spin-down, depending on whether the z component of the magnetic moment is parallel or antiparallel to the magnetic field (Fig. 2b). Nuclei like ^{2}H with spins $I = 1$, have three

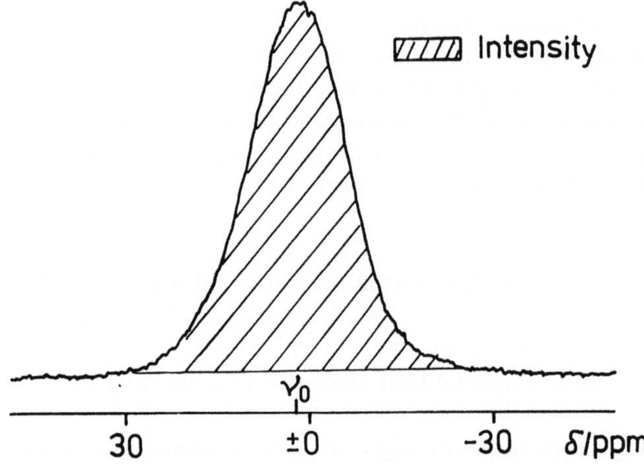

Fig. 3. Position and intensity of the NMR signal

eigenstates. The energy difference ΔE between neighboring energy levels is absorbed or emitted by a nuclear spin when it reorients and moves from one energy level to the next. This energy difference determines the NMR frequency ν_0:

$$\Delta E = |E_m - E_{m-1}| = \hbar\gamma B_0 = -\hbar\omega_0 = -h\nu_0. \tag{6}$$

It defines the position ν_0 of the (Zeeman) signal in the NMR spectrum (Fig. 3). In Table 1 the NMR frequencies ν_0 of different nuclei are given in MHz for a magnetic field of 2.3488 T.

2.1.2 Population Differences

Figure 2 also serves to illustrate the second important NMR parameter after the position in the spectrum. This is the intensity of the signal. As mentioned before, the polarization is the sum of all components of the nuclear magnetic moments parallel to the applied field. But from the quantum mechanics above it is known, that in thermodynamic equilibrium all magnetic moments are found in one of the energy eigenstates E_m having one of the $2I + 1$ allowed projections along the z axis. Thus the nuclear magnetic polarization is determined by the differences in population of the energy levels. The relative number n_{m-1}/n_m of spins in these states is given by the *Boltzmann distribution*:

$$n_{m-1}/n_m = \exp\{-\hbar\omega_0/k_B T). \tag{7}$$

From this the population difference $\Delta n = n_{m-1} - n_m$ is calculated. For ^1H at room temperature and a frequency of 100 MHz the exponent is given by $\hbar\omega_0/k_B T = 1.6 \times 10^{-5}$. At this temperature $k_B T \ll \hbar\omega_0$, so that the exponential in Eq. (7) can be expanded and the expansion truncated after the second term (high temperature approximation),

$$\Delta n = n_m - n_{m-1} \approx N_0\hbar\omega_0/(2k_B T). \tag{8}$$

Given $N_0 = 10^{20}$ spins in a sample, only 1.6×10^{15} of them make up for the nuclear magnetization which determines the integral of the signal. For this reason NMR spectroscopy is a method which is insensitive compared to infrared and optical spectroscopy with respect to the amount of sample needed. But this disadvantage is offset by the unsurpassed manifold and specificity of information accessible by NMR. One such advantage is illustrated in Fig. 3. The intensity (area, not amplitude of the signal) of the NMR signal depends linearly on Δn and therefore on N_0, i.e. the NMR spectrum yields straightforward information on the relative number of the spins under investigation.

According to Eqs. (6) and (8) each NMR signal provides specific information on the quality (sort of the chemical element) and the quantity (content of the chemical element) of the corresponding nucleus via γ and N_0, respectively. In comparison to other methods of analysis, the expenses for such a detailed analysis by NMR may be too high, but in many cases the presence or absence of an NMR signal can discriminate between different structural variants with high accuracy. Examples are the proof of small amounts of hydrogen in bismuth ultraphosphates [29] or of aluminium in silica-rich zeolites [28], where the sensitivity problem has been overcome by the accumulation of large numbers of scans.

The Zeeman NMR signal discussed so far does not discriminate between nuclei of the same kind, which are in different environments. Thus this interaction by itself is of no interest for practical applications of NMR. Nevertheless this case is of fundamental importance at least in terms of an absolute standard for theoretical calculations of the magnetic shielding, e.g. the absolute shielding of bare nuclei related to the shielding of the standard compound used by the experimenter [30].

2.1.3 Relaxation

Relaxation is the process by which the initial state of thermodynamic equilibrium is restored. There are two elementary kinds of relaxation.

Energy relaxation or spin-lattice relaxation is denoted with the symbol T_1. It characterizes the time needed to establish longitudinal thermodynamic equilibrium magnetization after the sample has initially been exposed to the magnetic field, or after manipulation of the magnetization by rf pulses. To this end energy must be exchanged between the nuclear spins and the lattice according to Eq. (2).

Once coherently oscillating magnetization components have been generated by pulses, the coherence will eventually be lost so that the amplitude of the oscillation decays to zero by interference of signals with different phases. The accumulation of phase differences is the result of different local magnetic fields to which different nuclei are exposed. If these fields fluctuate statistically, the magnetization components cannot be refocussed by echoes, and the coherence of the precession phases is lost irreversibly. The characteristic time for such a process is the phase or spin-spin relaxation time T_2.

Relaxation theory is concerned with the formal description of the processes achieving relaxation. In most cases, these are rotational and translational molecular motions which lead to fluctuating local fields at the sites of the nuclei. The origin of these fields is in one or more of the spin interactions treated below. Most important are the dipole–dipole interaction, the anisotropy of the chemical shift, and the quadrupole interaction. The analysis of relaxation times can provide important information about molecular motion at different time scales. These are defined by the resonance frequency, twice the resonance frequency and zero frequency in most cases, but can also be correlated to the amplitude of an applied rf field, the rotation frequency in magic angle spinning experiments, or the cycle time in multipulse experiments. A more detailed analysis of relaxation is beyond the scope of this introduction but can be found in many classic textbooks [22, 25, 31–33].

2.2 *Spin Interactions in the Solid State*

In condensed matter each nucleus is exposed to the influence of other nuclei which modify the local magnetic field, to electric field gradients, and to the coupling to the surroundings or the lattice. Similar to the magnetic field B_0 each coupling independently effects a specific shift of the energy levels. The differences between energy levels are the measurable quantities. Formally they are given by the eigenvalues of the total Hamiltonian operator $\mathbf{H}^{\text{total}}$.

2.2.1 *Types of Interaction*

The total Hamiltonian operator $\mathbf{H}^{\text{total}}$ (in frequency units rad/s) is the sum of operators for each individual interaction:

$$\begin{array}{lllllll} \mathbf{H}^{\text{total}} = & \mathbf{H}^{Z} + & \mathbf{H}^{Q} + & \mathbf{H}^{C} + & \mathbf{H}^{\sigma} + & \mathbf{H}^{K} & \mathbf{H}^{J} + \\ & \gg 100 \times 10^{6} & \gg 1 \times 10^{6} & \gg 5 \times 10^{4} & \gg 2 \times 10^{4} & \gg 1 \times 10^{4} & \gg 1 \times 10^{3} \end{array} \tag{9}$$

The frequencies given in s^{-1} under each operator indicate the order of magnitude of the interaction. Depending on the chemical and physical state of the nucleus under consideration as well as on $\boldsymbol{B}_0$ they can vary within a wide range of values. The numbers given are representative for ^{27}Al. The largest interaction next to the dominating Zeeman interaction $\mathbf{H}^{Z}$ is the quadrupole interaction $\mathbf{H}^{Q}$, followed by the direct dipole-dipole coupling $\mathbf{H}^{C}$, the magnetic shielding interaction $\mathbf{H}^{\sigma}$, the knight shift $\mathbf{H}^{K}$, and the comparatively small indirect coupling $\mathbf{H}^{J}$. Further interactions like the spin-rotation interaction are of minor importance for the NMR spectrum, but need to be taken into account for relaxation. These and the Knight shift are neglected in the following.

The magnetic shielding interaction $\mathbf{H}^{\sigma}$, also referred to as chemical shift δ, provides unique information about the chemical environment of the nucleus. Therefore the determination of $\mathbf{H}^{\sigma}$ is the most popular aim of solid-state NMR spectroscopy. To this end the interaction must be isolated from the others. This can be done by (i) elimination or (ii) separation of $\mathbf{H}^{i}$ (i = C, J and Q). There is no need for the separation of $\mathbf{H}^{Z}$, because for a given nucleus $\mathbf{H}^{Z}$ is independent of the environment and consequently acts as a constant additive term. Chemical shift data are almost universally reported in terms of a frequency shift from a reference compound. This subtraction cancels the effect of $\mathbf{H}^{Z}$. The constraint to separate or eliminate $\mathbf{H}^{i}$ is the central problem for the determination of $\mathbf{H}^{\sigma}$ in solids. In the case of nuclei with $I = 1/2\,\mathbf{H}^{Q}$ influences the spectra only in special cases [34], but the remaining couplings $\mathbf{H}^{C}$ and $\mathbf{H}^{J}$ are almost always present and require special techniques for experiments and data analysis.

In isotropic *liquids*, molecules and ions reorient rapidly on the NMR time scale. As a consequence the direct dipole-dipole and the quadrupole interactions are cancelled by motional averaging. With $\mathbf{H}^{D} = 0 = \mathbf{H}^{Q}$, Eq. (9) is reduced for liquids to:

$$\mathbf{H}^{C} = \mathbf{H}^{Z} + \mathbf{H}^{\sigma} + \mathbf{H}^{J} \tag{10}$$

Consequently in liquids the measurement of chemical shift δ and indirect coupling J is significantly easier than in solids. The advantage given by Eq. (10) is the basis of the efficiency of *high-resolution NMR* in the liquid phase.

2.2.2 General Formalism: Magnetic Shielding

The individual components of the total Hamiltonian, Eq. (9), represent the respective interactions of spins with their coupling partners. The spin **I** as well as its partners, the applied magnetic field $\boldsymbol{B}_0$, the local field $\boldsymbol{B}_{\text{local}}$ another spin **I** of the same nuclear species (homonuclear coupling) or a spin **S** of a different nuclear species (heteronuclear coupling) are *vectors*. Thus they are quantities with magnitude and orientation. The interaction of two vectors is described by a (second rank) *tensor*. The tensor description of interaction Hamiltonians applies to all spin interactions. But it is more instructive to explain it by example of the magnetic shielding or chemical shift tensor σ.

The Magnetic Shielding Tensor

Interactions denote the influenc of one coupling partner upon the other. The action of the magnetic field $\boldsymbol{B}_0$ generates the (Zeeman) magnetization and *vice versa* $\boldsymbol{B}_0$ experiences a small perturbation by the spin system: When the sample is placed into the external field $\boldsymbol{B}_0$ it becomes magnetized and the applied field is modified by an induced field. This modification arises from a perturbation of the electron motion around the nuclei. The induced field $\boldsymbol{B}_{\text{ind}}$ is directly proportional to the applied field $\boldsymbol{B}_0$,

$$\boldsymbol{B}_{\text{ind}} = -\sigma \boldsymbol{B}_0, \tag{11}$$

where σ is the proportionality constant. The local field then becomes

$$\boldsymbol{B}_{\text{local}} = (1 - \sigma)\boldsymbol{B}_0,$$

The proportionality constant σ correlates two vectors, which are not necessarily in parallel.

Equation (11) is analyzed by writing the components of $\boldsymbol{B}_{\text{ind}}$ in cartesian coordinates x, y, z as

$$\boldsymbol{B}_{\text{ind}} = B_{\text{ind,x}}\boldsymbol{i} + B_{\text{ind,y}}\boldsymbol{j} + B_{0z}\boldsymbol{k}. \tag{12}$$

They can be expressed as homogeneous linear functions of the components of the second vector $\boldsymbol{B}_0$

$$\boldsymbol{B}_0 = B_{0x}\boldsymbol{i} + B_{0y}\boldsymbol{j} + B_{0z}\boldsymbol{k}, \tag{13}$$

which gives rise to them,

$$B_{\text{ind},x} = -(\sigma_{xx}B_{0x} + \sigma_{xy}B_{0y} + \sigma_{xz}B_{0z}),$$
$$B_{\text{ind},y} = -(\sigma_{yx}B_{0x} + \sigma_{yy}B_{0y} + \sigma_{yz}B_{0z}),$$
$$B_{\text{ind},z} = -(\sigma_{zx}B_{0x} + \sigma_{zy}B_{0y} + \sigma_{zz}B_{0z}), \tag{14}$$

Equivalently this can be expressed in terms of a product

$$\begin{bmatrix} B_{\text{ind},x} \\ B_{\text{ind},y} \\ B_{\text{ind},z} \end{bmatrix} = \begin{bmatrix} \sigma_{xx} & \sigma_{xy} & \sigma_{xz} \\ \sigma_{yx} & \sigma_{yy} & \sigma_{yz} \\ \sigma_{zx} & \sigma_{zy} & \sigma_{zz} \end{bmatrix} \begin{bmatrix} B_{0x} \\ B_{0y} \\ B_{0z} \end{bmatrix}. \tag{15}$$

of the applied magnetic field vector with the matrix of the nine coefficients σ_{rs}, where $r, s = x, y, z$. Equation (15) is the equation of transformation of the vector $\boldsymbol{B}_0$ into the vector $\boldsymbol{B}_{\text{ind}}$. It is written in short form as Eq. (11), where the coefficient matrix in Eq. (15) defines the chemical shielding tensor σ.

The Chemical Shift

The observable of the NMR experiment is the frequency ν. Therefore magnetic shielding data are generally given as the frequency shift related to a reference compound with σ_{ref} defining the origin of the σ scale. The logical choice of origin would be an isolated bare nucleus, but this is impractical for experimental calibration of shielding scales.

An alternative scale is the chemical shift scale. Chemical shift data δ are usually reported as the frequency shift from the reference compound giving a signal at $\nu_{\text{ref}} = 0$ normalized by the resonance frequency of the reference compound in parts per million (ppm). The corresponding normalization is used for shielding data.

The relation between the two scales is [35]

$$\delta_{\text{obs}} = (\sigma_{\text{ref}} - \sigma_{\text{obs}})/(1 - \sigma_{\text{ref}}), \tag{16}$$

where δ is the chemical shift generally used in liquid-state NMR. With the realistic assumption

$$\sigma_{\text{ref}} \ll 1 \tag{17}$$

Eq. (3) can be rewritten as

$$\delta_{\text{obs}} \approx \sigma_{\text{ref}} - \sigma_{\text{obs}}. \tag{18}$$

The non-logical but usual definition for the reference compound

$$\sigma_{\text{ref}} \equiv 0 \tag{19}$$

gives

$$\delta_{\text{obs}} \approx -\sigma_{\text{obs}}. \tag{20}$$

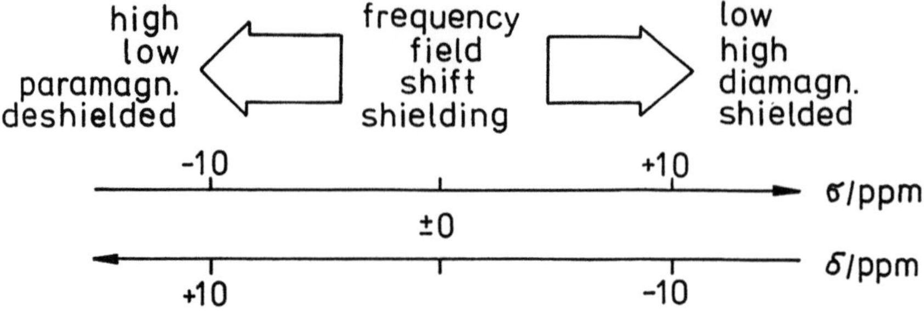

Fig. 4. Comparison of δ and σ scale and qualitative notations for resonance frequencies

This inversion of sign corresponds to different directions of the σ scale and the δ scale, which are used in NMR for measuring magnetic shielding and chemical shift respectively (Fig. 4).

Decomposition of the Coupling Tensor

A general second rank tensor σ_{rs} with nine independent components σ_{rs} can be decomposed into an antisymmetric and a symmetric part. To first-order NMR spectra are determined only by the symmetric part [20], for which $\sigma_{rs} = \sigma_{sr}$ is valid. The antisymmetric part plays a role in relaxation and can be observed only in special cases [36, 37]. Therefore, all coupling tensors are considered to be symmetric in the following. Thus they can be expressed by only six independent components. These six components can be interpreted in a geometric fashion. Three of them are used to define a 3D ellipsoid and the other three define the orientation of the ellipsoid in the laboratory frame, where the z axis is parallel to the applied magnetic field $\boldsymbol{B}_0$ (Fig. 5). The eccentricity and size of the ellipsoid are determined by the eigenvalues σ_{11}, σ_{22}, and σ_{33}, of the tensor, that is by the values on the diagonal after diagonalization ($\sigma_{rs} = 0$ for $r \neq s$):

$$\sigma = \begin{bmatrix} \sigma_{11} & 0 & 0 \\ 0 & \sigma_{22} & 0 \\ 0 & 0 & \sigma_{33} \end{bmatrix} \tag{21}$$

The cartesian coordinate frame in which the ellipsoid is defined is called *the principal axes frame*. The angles α, β and γ in Fig. 4 which define its orientation in the laboratory frame are the Euler angles. Recently alternative pictorial representations of a tensor have been developed [38].

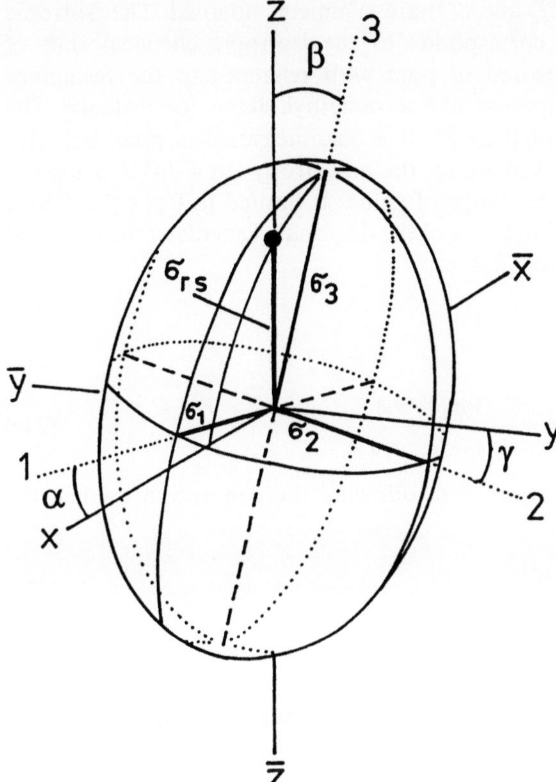

Fig. 5. Magnetic shielding tensor represented by an ellipsoid (second rank surface) in the laboratory system (x, y, z) and the sample system $(1, 2, 3)$

Symmetry Adapted Parameters

A symmetric tensor σ can be separated into an isotropic part σ, a traceless axially symmetric part σ^{ax}, and a traceless nonaxial part σ^{nonax},

$$\begin{bmatrix} \sigma_{11} & 0 & 0 \\ 0 & \sigma_{22} & 0 \\ 0 & 0 & \sigma_{33} \end{bmatrix} = \bar{\sigma} \begin{bmatrix} 1 & & \\ & 1 & \\ & & 1 \end{bmatrix} + \delta^* \begin{bmatrix} -1/2 & & \\ & -1/2 & \\ & & 1 \end{bmatrix} + \eta\delta^* \begin{bmatrix} +1/2 & & \\ & -1/2 & \\ & & 0 \end{bmatrix} \quad (22)$$

Equation (22) is written in cartesian coordinates. In many cases, a representation in spherical coordinates is appropriate, because the tensorial properties of all interactions lead to an angular dependence of the observed resonance frequency

$$\nu = \nu_0 - \nu_L = \nu_{zz} = \nu_0\{\sigma + \delta^*/2[3\cos^2\beta - 1 - \eta\sin^2\beta\cos(2\alpha)]\} \quad (23)$$

The coefficients of Eqs. (22) and (23) are symmetry adapted. The isotropic part is denoted by σ. This corresponds to the isotropic chemical shift of liquid-state NMR. It is measured in ppm with reference to the resonance frequency of a standard compound like tetramethyl silane, for instance. The largest principal value is denoted by δ^*. It is also measured in ppm, but with reference to σ). Instead of δ^* sometimes the anisotropy $\Delta\sigma = 3\delta^*/2$ is used. η is the asymmetry parameter. Its range of values is limited to $0 \leqq \eta \leqq 1$. These parameters can be determined in terms of the diagonal elements of the chemical shift tensor in the principal axes frame,

$$\bar{\sigma} = \tfrac{1}{3}\operatorname{tr}\sigma = \tfrac{1}{3}(\sigma_{11} + \sigma_{22} + \sigma_{33}) \tag{24a}$$

$$\delta^* = \tfrac{2}{3}\Delta\sigma = \sigma_{33} - \bar{\sigma} = \tfrac{2}{3}[\sigma_{33} - \tfrac{1}{2}(\sigma_{11} + \sigma_{22})] \tag{24b}$$

$$\eta = \frac{\sigma_2 - \sigma_1}{\delta^*} = \frac{\sigma_2 - \sigma_1}{\sigma_3 - \bar{\sigma}} = \frac{3(\sigma_2 - \sigma_1)}{2\sigma_3 - (\sigma_2 + \sigma_1)} \tag{24c}$$

where the principal values are ordered following the convention used in the monographs [19–21, 24, 39],

$$|\sigma_{33} - \bar{\sigma}| \geqq |\sigma_{11} - \bar{\sigma}| \geqq |\sigma_{22} - \bar{\sigma}|. \tag{25}$$

Other authors use the convention [40, 41]

$$\sigma_{33} > \sigma_{22} > \sigma_{11}. \tag{26}$$

Unfortunately a commonly accepted definition has not been established.

2.2.3 Magnetic Shielding: Experimental Aspects

Angular Dependence of the Magnetic Shielding

In a strong applied magnetic field $\boldsymbol{B}_0(|\boldsymbol{B}_0| \gg |\boldsymbol{B}_{\text{ind}}|)$ the one and only measurable quantity of the shielding tensor σ is the σ_{zz} component in the laboratory frame

$$\sigma_{zz} = \cos^2\theta_{1z}\sigma_{11} + \cos^2\theta_{2z}\sigma_{22} + \cos^2\theta_{3z}\sigma_{33}. \tag{27}$$

It can be expressed in terms of a linear combination of the three principal values σ_{ii}, where the coefficients are the squares of the corresponding directional cosines. Alternatively σ_{zz} can be expressed in terms of the Euler angles α and β (cf. Fig. 5). Depending on the particular choice of description (Eqs. (22) and (24), respectively) Eq. (27) changes to:

$$\sigma_{zz} = \sigma_{11}\sin^2\alpha\cos^2\beta + \sigma_{22}\sin^2\alpha\sin^2\beta + \sigma_{33}\cos^2\alpha \tag{28a}$$

or equivalently

$$\sigma_{zz} = \bar{\sigma} + \Delta\sigma/3[(3\cos^2\beta - 1) + \eta\sin^2\beta\cos^2\alpha]. \tag{28b}$$

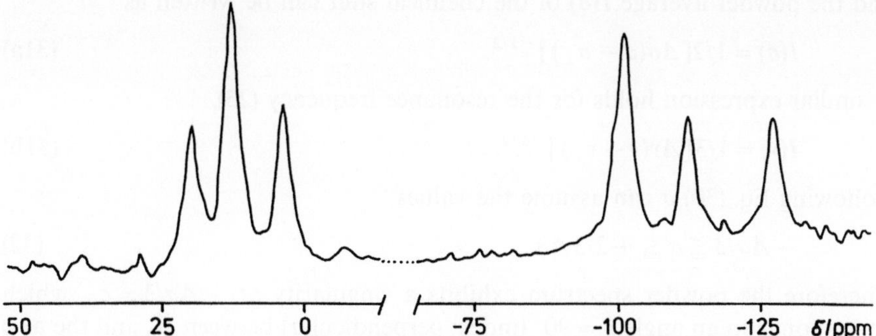

Fig. 6. ^{29}Si NMR spectrum (12 MHz) of a single crystal of $[(CH_3)_3Si]_8[Si_8O_{20}]$. Due to the coincidence of 2 signals at -100 and $+12$ ppm resp. only six lines (intensity ratio 1:2:1:2:1:1) are visible [42]

The use of Eq. (28b) for calculation of the resonance frequency $\nu_L = -\gamma B$ local from the local field (cf. Eq. 11) immediately leads to the angular dependent offset frequency $\nu = \nu_0 - \nu_L = \nu_0 \sigma_{zz} = \nu_{zz}$ in Eq. (23). It applies to the magnetization from nuclei in a *single crystal* at positions which are related by translation only (crystallographically and magnetically equivalent sites). Reorientation of the crystal changes the angles α and β, and thus the resonance frequency. Therefore the resonance frequency can be used as a protractor to measure angles of molecular orientation.

Systematic measurements of σ_{zz} for different orientations of the crystal in the magnetic field (x, y, z frame) provide the principal values σ_{ii} as well as the orientation of the principal axes system (1, 2, 3 frame, in Fig. 5) in terms of the Euler angles. Thus all six independent values of the symmetric part of the second rank coupling tensor are accessible. N magnetically non-equivalent sites within an unit cell give rise to N signals corresponding to N different magnetic shielding tensors. This is illustrated in Fig. 6 for $N = 8$ by the ^{29}Si NMR spectrum of a single crystal of $[(CH_3)_3Si]_8[Si_8O_{20}]$ [42].

Powder Average

In *polycrystalline* materials as well as in *non-crystalline* samples all orientations of σ arise with equal probability, i.e. there is a random distribution of α and β in Eq. (41). Consequently the values of the resonance frequency are distributed in the manner defined by the angular dependence. Eq. (31a), which leads to characteristic wide-line spectra, also called *powder patterns*.

For an axially symmetric tensor with

$$\sigma_\perp = \sigma_{11} = \sigma_{22} \neq \sigma_{33} = \sigma_\parallel \tag{29}$$

the asymmetry parameter $\eta = 0$ vanishes, and Eq. (28b) reduces to

$$\sigma = \sigma_{zz} - \bar{\sigma} = \Delta\sigma/3(3\cos^2\beta - 1), \tag{30}$$

and the powder average $I(\sigma)$ of the chemical shift can be written as

$$I(\sigma)=1/2[\Delta\sigma(\sigma-\sigma_{\perp})]^{-1/2}. \tag{31a}$$

A similar expression holds for the resonance frequency (23),

$$I(\nu)=1/3[\Delta\nu(\nu-\nu_{\perp})]^{-1/2}. \tag{31b}$$

Following Eq. (30) σ can assume the values

$$-\Delta\sigma/3\leqq\sigma\leqq+2/3\,\Delta\sigma. \tag{32}$$

Therefore the powder spectrum exhibits a singularity at $-\Delta\sigma/3=\sigma_{\perp}$ which corresponds to an angle $\beta=90°$ (index: perpendicular) between $\boldsymbol{B}_0$ and the axis 33 of the shielding tensor. The cut-off at $+2/3\ \Delta\sigma$ corresponds to an angle

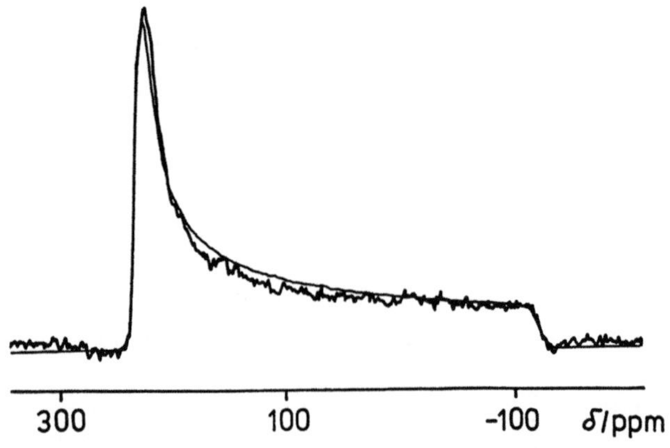

Fig. 7. ^{31}P NMR spectrum of P_4O_6 (static, $-60°$ C) and the calculated spectrum with $\sigma=-113$ ppm, $\Delta\sigma=+350$ ppm, or $\sigma_{\perp}=+120$ ppm and $\sigma_{\parallel}=-230$ ppm, respectively, and a dipolar broadening of 1000 Hz [42]

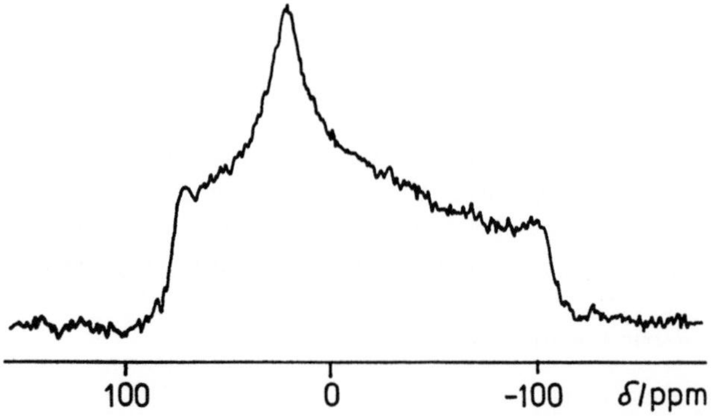

Fig. 8. ^{31}P NMR spectrum of $Ba(Et)_2PO_4$ (static, proton decoupled) with $\sigma_{11}=-81$ ppm, $\sigma_{22}=-23$ ppm, $\sigma_{33}=+104$ ppm and $\sigma=\pm 0$ ppm, $\Delta\sigma=+156$ ppm, and $h=0.56$, respectively. The dipolar broadening amounts to $\gg 600$ Hz (adapted from [40])

$\beta = 0°$ (index: parallel) between $\boldsymbol{B}_0$ and the 33 axis. This is illustrated in Fig. 7 by an experimental powder spectrum of the nonaxial ^{31}P magnetic shielding tensor in P_4O_6 and the corresponding signal intensity calculated from Eq. (31a).

In the general case $\eta \neq 0$, the line shape cannot be expressed by a closed algebraic formula. Figure 8 shows a typical spectrum of a non-axial shielding tensor in $Ba(Et)_2PO_4$. From the high and low frequency cut-offs and the form the singularity in the centre the three principal values σ_{ii} can be derived. In the third case ($\eta = 0$, $\Delta\sigma = 0$) there is no orientation dependence, and singularity and cut-offs coincide, giving a symmetric line as shown in Fig. 3.

Magic Angle Spinning

When rewriting Eq. (27) as

$$\sigma_{zz} = 1/3[3\cos^2\theta_{1z}\sigma_{11} + 3\cos^2\theta_{2z}\sigma_{22} + 3\cos^2\theta_{3z}\sigma_{33}] \tag{33a}$$

the isotropic and the anisotropic parts can be separated,

$$\sigma_{zz} = \bar{\sigma} + \frac{1}{3}\sum_{i}^{3}\sigma_{ii}(3\cos^2\theta_{iz} - 1). \tag{33b}$$

Separating the first, isotropic term of Eq. (33b) from the two following, anisotropic terms is the fundamental idea underlying the *m*agic-*a*ngle *s*pinning (MAS) technique [10, 25, 43] which is applied routinely in many fields of solid-state NMR.

For MAS NMR experiments, the powder sample is placed into a gas driven turbine which is supported by gas bearings. The rotation axis of which is inclined

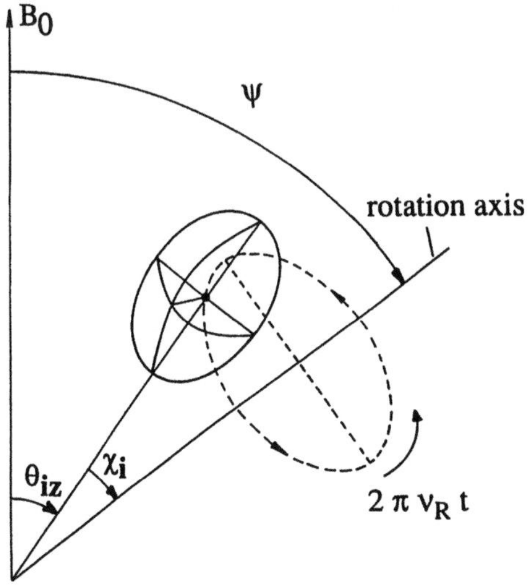

Fig. 9. Magic-angle spinning: Definition of angles

by an angle ρ against $\boldsymbol{B}_0$ (Fig. 9). The time averaged value of the angle dependence in Eq. (33b) is given by

$$\langle 3\cos^2\theta_{iz} - 1\rangle = 1/2(3\cos^2\psi - 1)(3\cos^2\chi_i - 1) \tag{34}$$

with χ_i as the constant angle between the principal tensor axis i and the axis of rotation, within the range of

$$\pi \leqq \chi_i \leqq 2\pi. \tag{35}$$

Adjusting the angle ψ to the *magic angle*

$$\psi_{\text{magic}} = \arccos 3^{-1/2} = 54.74° \tag{36}$$

gives the time-averaged value

$$\langle 3\cos^2\theta_{iz} - 1\rangle = 0, \tag{37}$$

independent of the orientation of the tensor given by χ_i, and Eq. (30) is truncated to

$$\sigma = \sigma_{zz} - \sigma = 0. \tag{38}$$

The comparison of Eqs. (38) and (32) shows that the anisotropic parts of the magnetic shielding tensor are eliminated so that high spectral resolution can

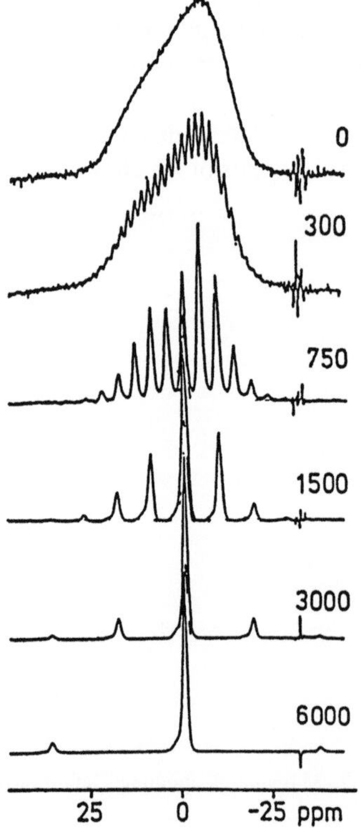

Fig. 10. ^{31}P NMR spectra of $WO_3 \cdot P_2O_5$ at various spinning frequencies ν_R [44]

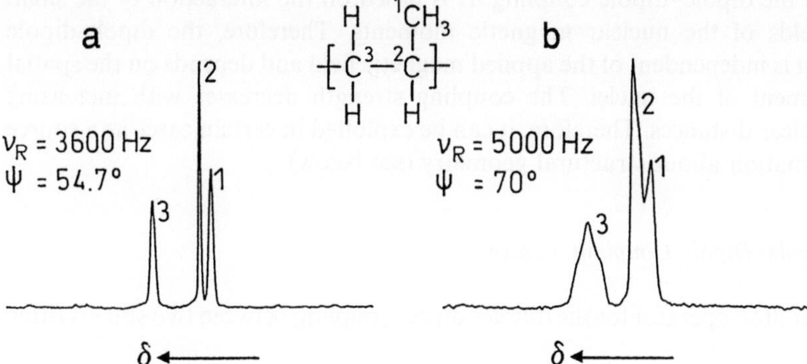

Fig. 11a,b. ^{13}C NMR spectra of isotactic polypropylene (proton decoupled) with MAS at $\nu_R = 3600$ Hz (**a**) and OMAS at $\nu_R = 5000$ Hz (**b**) [45]

be achieved. However, the line narrowing is paid for by a loss of information about the anisotropy of shielding tensor. But this information can partly be regained in two ways by compromising spectral resolution. Equation (38) is strictly valid for high rotation rates ($\nu_R \gg \Delta\nu$) only. One way consists of reducing the rotation rate. At medium rates ($0 < \nu_R < \Delta\nu$) the isotropic signal at ν_L is flanked on both sides by rotational sidebands at multiples of the spinning frequency ν_R. At slow rotation rates the sideband envelope approaches the shape of the static powder spectrum. The stepwise disintegration of the static spectrum with increasing rotation rates is illustrated in Fig. 10. From an analysis of sideband intensities, information about the chemical shift tensor can be obtained [40]. Experiments at slow rotation rates are particularly useful for investigations of amorphous solids and of powder samples with crystallographically non-equivalent sites. The second way of regaining the information about the chemical shielding anisotropy is fast sample spinning at a known angle ψ slightly different from the magic angle (OMAS: *o*ff-*m*agic-*a*ngle *s*pinning) (Fig. 11). Then the anisotropic part of the interaction is preserved (cf. Eq. 34 above and Eqs. 65 to 68 below) but scaled down so that overlap between wide line shapes can be decreased.

It should be noted, that rotation rates of several tens of kilohertz produce centrifugal accelerations of up to 10^7 m/s^2, i.e. a million times the acceleration caused by the earth's gravitational field. Another technical problem can be heating at extremely high spinning rates by friction or by eddy currents induced in conducting samples. Thus, even today, MAS may be a demanding technique.

2.2.4 Dipole–Dipole Coupling

The formalism developed by the example of the magnetic shielding applies to all other second rank spin interactions as well. The dominant interaction in solids is the dipole–dipole coupling. It is based on the interaction of the small local fields of the nuclear magnetic moments. Therefore, the dipole-dipole coupling is independent of the applied magnetic field and depends on the spatial arrangement of the nuclei. The coupling strength decreases with increasing internuclear distances. Therefore, it can be exploited in certain cases as a source of information about structural geometry (see below).

The Dipole–Dipole Coupling Tensor

The Hamilton operator for the dipole–dipole coupling between two spins written as

$$\mathbf{H}^{\mathrm{c}} = \mathbf{I}^{\mathrm{i}}\mathbf{C}\mathbf{I}^{\mathrm{j}}, \tag{39}$$

where $\mathbf{I}^{\mathrm{i}}$ and $\mathbf{I}^{\mathrm{j}}$ are the spin vector operators of the nuclei i and j. Because the coupling tensor $\mathbf{C}$ transforms two equivalent vectors, it is symmetric.

There are two mechanisms, by which spins can be coupled via the dipole fields of their magnetic moments. The first one is the *direct* coupling. It acts through space in a straightforward fashion. Here the trace of the coupling tensor is zero, so that it vanishes under motional averaging. The second mechanism is called *indirect*, because it proceeds via the bonding electrons. There the trace of the coupling tensor, differs from zero, so that the isotropic value is observed under motional narrowing in liquids. It is also referred to as *scalar* or J coupling, because it is customarily given the symbol J. Both mechanisms are accommodated in Eq. (39) if the coupling tensor $\mathbf{C}$ is split into two parts,

$$\mathbf{C} = 2\mathbf{D} + \mathbf{J}, \tag{40}$$

where $\mathbf{D}$ is the tensor of the direct and $\mathbf{J}$ the tensor of the indirect dipolar coupling. Analogous to Eq. (22) the symmetric tensor $\mathbf{D}$ can be separated into three terms. The first (isotropic) and the third (nonaxial) term are zero. Therefore $\mathbf{D}$ is given by:

$$\mathbf{D} = \begin{bmatrix} D_1 & & \\ & D_2 & \\ & & D_3 \end{bmatrix} = \tfrac{2}{3}\Delta D \begin{bmatrix} +1/2 & & \\ & +1/2 & \\ & & -1 \end{bmatrix} \tag{41}$$

Thus this axially symmetric tensor shows remarkable properties:

$$D_\perp = D_{11} = D_{22} \neq D_{33} = D_\parallel, \tag{42a}$$

$$D = 0, \quad \Delta D \neq 0, \quad \eta^{\mathrm{D}} = 0. \tag{42b}$$

The anisotropy ΔD is given by:

$$2/3\Delta D = 2/3(D_\parallel - D_\perp) = R, \tag{43}$$

where R is direct dipolar coupling constant,

$$R = \mu_0/(4\pi\gamma_i\gamma_j\hbar r_{ij}^{-3}). \tag{44}$$

Under the assumption of no motion, this parameter can be calculated in a straightforward fashion from structural data.

The properties of the tensor **J** are far more complex. Because like the magnetic shielding tensor its trace is nonzero, and its full description requires three values, cf. Eq. (22),

$$J_{11} \neq J_{22} \neq J_{33}, \tag{45a}$$

$$J \neq 0, \quad \Delta J \neq 0, \quad \eta^{J} \neq 0. \tag{45b}$$

As a consequence of the equivalence of the two Hamiltonians $\mathbf{H}^{C}$ and $\mathbf{H}^{J}$ in an NMR experiment with nonspinning samples, both contributions cannot be separated.

Angular Dependence of the Dipole–Dipole Coupling

In a single crystal of a heteronuclear two-sin system ($I = 1/2, S = 1/2$) the position of the signal(s) is given by:

$$\nu_{zz}^{C}(\pm\tfrac{1}{2}) = \{C \pm \tfrac{1}{2}C + \Delta C/3[(3\cos^2\beta - 1) + \eta^{C\,\sin 2}\beta\cos 2\alpha)]\}, \tag{46}$$

with $m = 1/2$ and $m = -1/2$ for each of the two transitions, respectively. The spectrum shows two signals, the splitting of which equals the orientation dependent coupling constant C_{zz},

$$C_{zz} = f(\alpha, \beta) = \nu_{zz}^{C}(+\tfrac{1}{2}) - \nu_{zz}^{C}(-\tfrac{1}{2}). \tag{47}$$

In the case of the dipole–dipole coupling with ^{1}H nuclei as one of the coupling partners the J coupling can usually be neglected, because

$$2\mathbf{D} \gg \mathbf{J}. \tag{48}$$

Then, by using Eqs. (42), (43), and (48), Eq. (46) simplifies to the classical Pake equation of a homonuclear two-spin system. For single crystals

$$\nu_{zz}^{C}(\pm\tfrac{1}{2})^{\text{hetero}} = \nu_0 \pm \frac{R}{2}(3\cos^2\beta - 1). \tag{49a}$$

Due to additional flip-flop transitions which arise in homonuclear two-spin systems Eq. (49) is modified by a factor $\tfrac{3}{2}$,

$$\nu_{zz}^{C}(\pm 1/2)^{\text{homo}} = \nu_0 \pm 3/4R(3\cos^2\beta - 1). \tag{49b}$$

Powder Average

In polycrystalline samples the line shape for an axially symmetric coupling tensor **C** is given in analogy to Eq. (31b) by

$$I^{C}(\nu)(\pm\tfrac{1}{2}) = \pm\tfrac{1}{2}[\Delta C(\nu - C_{\perp})]^{-1/2}. \tag{50}$$

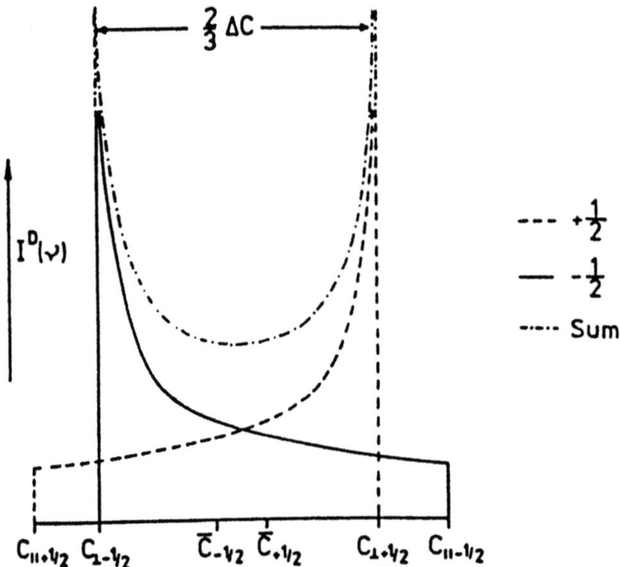

Fig. 12. Heteronuclear two-spin systems: Powder pattern arising from dipolar coupling

Equation (50) is illustrated in Fig. 12 by the calculated spectrum of a heteronuclear two-spin systems. It consists of two subspectra with axial symmetry but inverse signs, corresponding to the two transitions with $m = \pm\frac{1}{2}$. Their sum gives the doublet with characteristic singularities and shoulders.

It should be mentioned that by Eq. (50) only the parameter

$$\Delta C = \pm \Delta D + \Delta J$$

is observable. Especially in the case of dipolar coupling of heavy nuclei like ^{31}P-^{31}P (homonuclear) or ^{31}P-^{19}F (heteronuclear) ΔJ is in the order of

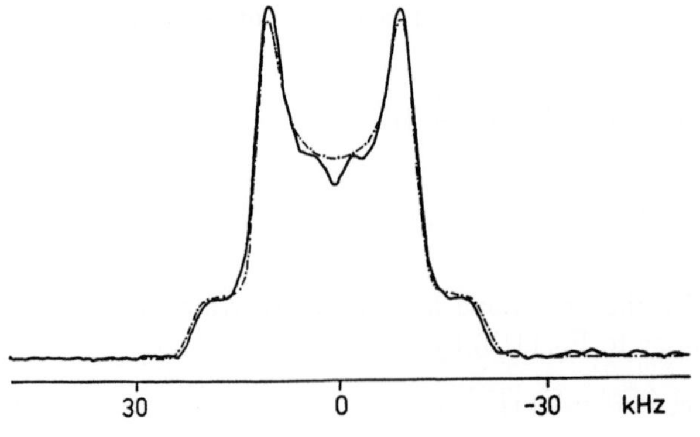

Fig. 13. ^1H NMR spectrum of Dellaite $Ca_6(OH)_2(Si_2O_7)$. The homonuclear dipolar splitting of 22 kHz corresponds to $R = 2.2$ Å [48]

magnitude of several kHz [46]. This leads to erroneous results for internuclear distances (Eq. 44) if only the direct dipole–dipole interaction is taken into account for data evaluation [47]. On the other hand, by knowledge of internuclear distances from diffraction data it is possible to determine the anisotropy ΔJ of the J coupling, though with an ambiguity in sign.

With Eqs. (40) and (48), Eq. (50) changes to

$$I^{D}(\nu)(\pm\tfrac{1}{2}) = \pm[\Delta D(\nu - D_{\perp})], \tag{51}$$

which is known as the *Pake equation* for polycrystalline samples. An example of a Pake spectrum is shown in Fig. 13 with the ^{1}H NMR spectrum of Dellaite $Ca_6(OH)_2(Si_2O_7)$.

Magic-Angle Spinning

The effect of the MAS technique on the dipolar interactions is the elimination of the anisotropic terms of **C** corresponding to Eq. (50). It is here, where the differences between **D** and **J** become most obvious. Following Eq. (42b) the direct dipolar coupling will be completely eliminated at sufficiently high rotation rates, and the remaining observable parameter is the isotropic part J of the indirect coupling,

$$C_{zz}^{MAS} = J. \tag{52}$$

For a heteronuclear two-spin system a 1:1 doublet with a splitting equal to J is observed. With increasing number N of the coupling nuclei the values of the total magnetic quantum number m increase as well as the possible geometric arrangements. The least complicated arrangement of spins next to the two-spin systems consists of three spins at the corners of an equilateral triangle [49], followed by $2+1$ spins at the corners of an isosceles triangle [50–52].

In the most general case with $N \gg 1$ all the N subspectra merge to form an unspecific line profile typical for multiple couplings among protons. This *dipolar broadening* is of the order of up to 120 kHz for protons, but can be much smaller for other nuclei at larger distances and lower gyromagnetic ratios. Then it is possible to *spin out* this interaction by MAS. As an example Fig. 14 shows the ^{31}P MAS NMR spectrum of $Na_5P_3O_{10}\cdot 6H_2O$. It should be mentioned that in rare cases the direct dipolar coupling gives rise to interesting "cross singular" line shape effects in polycrystalline materials [53].

2.2.5 *Quadrupolar Coupling*

Nuclei with spin quantum number $I > 1/2$ exhibit an electric quadrupole moment Q (Table 1) in addition to the magnetic dipole moment μ, which reflects the deviation from spherical symmetry of the electric charge distribution at the site of the nucleus. It is worthwhile mentioning that nuclei with $I = 1/2$ can

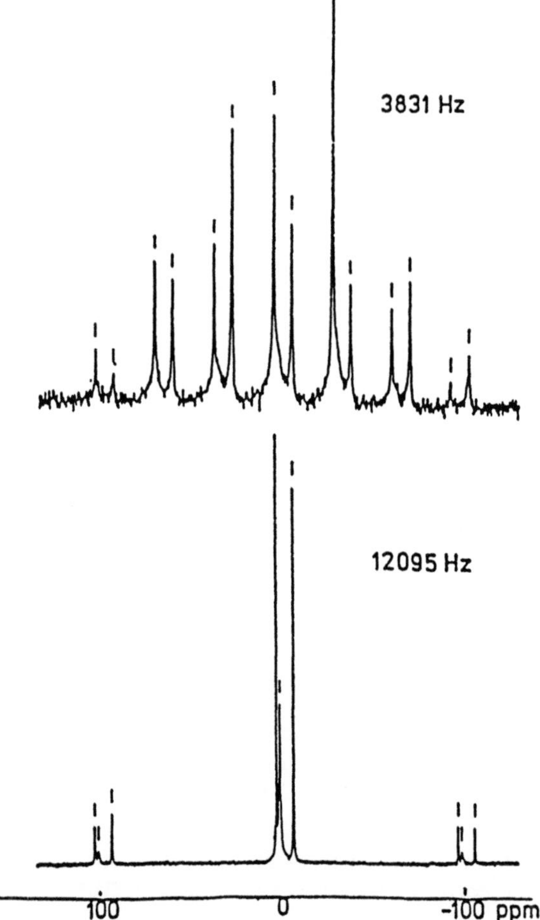

Fig. 14. ^{31}P MAS NMR spectra of $Na_5P_3O_{10} \cdot 6H_2O$ at low (*top*) and high (*bottom*) rotation rates. Due to the strong dipolar coupling of one of the three different PO_4 units to five neighbouring H_2O molecules the corresponding signal is broadened and only visible at high rotation rates [54]

also possess a quadrupole moment if they are in an excited state. For example ^{19}F as a typical spin 1/2 nucleus possesses a spin $I = 5/2$ in he triplet state [55].

Because the quadrupole moment Q couples to the electric field gradient (EFG) established by the electrons surrounding the nucleus, the (electric) quadrupole interaction is highly sensitive to the electronic structure. As a consequence of this structural sensitivity the quadrupole coupling fluctuates within wide limits, Eq. (9), in contrast to the magnetic shielding and to dipolar effects. It can even exceed the Zeeman coupling. Two limiting cases can be distinguished,

$$\mathbf{H}^Z \gg \mathbf{H}^Q, \quad \text{and} \quad \mathbf{H}^Z \ll \mathbf{H}^Q. \tag{53a, b}$$

The existence of very strong quadrupolar interaction leads to pure nuclear quadrupolar resonance (NQR) spectroscopy. For solid-state NMR only the first case is lucid (high field conditions), and with increasing strength of the quadrupole interaction the spectral analysis becomes more and more complicated [34]. Following Eq. (3), $\mathbf{H}^{Z}$ increases with $\boldsymbol{B}_0$, whereas $\mathbf{H}^{Q}$ is to the first order independent of $\boldsymbol{B}_0$ and in second order inversely proportional to $\boldsymbol{B}_0$. Therefore the use of cryomagnets with high $\boldsymbol{B}_0$ fields is a prerequisite for solid-state NMR of quadrupolar nuclei so far as second order quadrupolar interactions mask the chemical shift effects. If the experimental aim is the exact determination of the quadrupolar shift (e.g. DOR experiments) measurements at lower $\boldsymbol{B}_0$ field are advantageous because of the large dispersion of the different EFG tensors.

The Quadrupole Coupling Tensor

Like the other interaction Hamiltonians, the quadrupole-coupling Hamiltonian is written as

$$\mathbf{H}^{Q} = \mathbf{IQI}. \tag{54}$$

The orientation dependence of the interaction is given by the quadrupole coupling tensor

$$\mathbf{Q} = \mathbf{Q}^{(1)} + \frac{1}{n_0}\mathbf{Q}^{(2)} + \frac{1}{n_0^2}\mathbf{Q}^{(3)} + \cdots, \tag{55}$$

which is expanded into a series according to the order n of the interaction.

The symmetric tensor $\mathbf{Q}$ is separated into three parts, cf. Eq. (22). The isotropic part Q is zero,

$$Q = 0. \tag{56}$$

The coefficients of the two remaining parts are given by

$$\tfrac{2}{3}\Delta Q = \mathrm{eq} \tag{57a}$$

and

$$\eta^{Q}. \tag{57b}$$

For historical reasons $\frac{2}{3}\Delta Q$ is called the *electric field gradient eq* and not the *quadrupolar anisotropy*. In practical applications this component of the quadrupole coupling tensor is expressed in terms of the quadrupolar coupling constant Q

$$Q = \frac{eq\,eQ}{h} \tag{58}$$

or in terms of the *quadrupolar coupling frequency*

$$\nu_Q = \frac{3e^2Qq}{2I(2I-1)h}. \tag{59}$$

Because the quadrupole coupling tensor is traceless and symmetric, it is defined by five independent parameters.

Angular Dependence of the Quadrupole Coupling

In the case of *first-order* quadrupole interaction the quadrupolar coupling for a single crystal is given in analogy to Eq. (28b) by

$$Q_{ZZ}^{(1)} = \Delta Q/3[(3\cos^2\beta - 1) + \eta^Q \sin^2\beta \cos 2\alpha]. \tag{60}$$

The $Q_{ZZ}^{(1)}$ component is observed via the frequency

$$\nu_{ZZ,m\to(m-1)}^{(1)} = \nu_0 - Q_{ZZ}(m - \tfrac{1}{2}) \tag{61}$$

Combination of Eqs. (60) and (61) yields an expression similar to Eq. (23),

$$\nu_{m\to(m-1)}^{(1)} = \nu_0 - \nu_Q \tfrac{1}{2}[(3\cos^2\beta - 1) + \eta^Q \sin^2\beta \cos 2\alpha](m - \tfrac{1}{2}). \tag{62}$$

Following Eq. (62) all 2I line positions except for the central transition ($m = 1/2$ to $m = -1/2$) depend on the orientation of the quadrupole coupling tensor in the magnetic field.

Powder Average

In powder samples the values of the transition frequencies in Eq. (62) are distributed corresponding to the distribution of the orientations of the quadrupole coupling tensors. Instead of separate lines, profiles similar to those for the dipole–dipole and the magnetic shielding interaction are observed. As an example

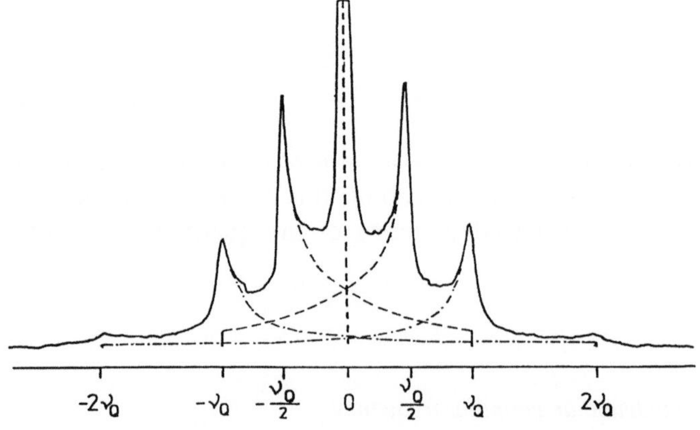

Fig. 15. First order quadrupolar inteactions: ^{27}Al NMR powder spectrum of $AlCl_3 \cdot 6H_2O$ showing the central transition and the satellite transitions for the first order quadrupole interaction of a spin with $I = 5/2$ and $\eta = 0$ [56]

Fig. 15 shows the line shape for nuclei with $I = 3/2$ and $\eta^Q = 0$. As a consequence of the factor $(m - 1/2)$ in Eq. (62), the line shapes of the two satellite transitions from $+3/2$ to $+1/2$ and from $-1/2$ to $-3/2$ differ in sign and are symmetrically displaced with respect to ν_0. The central transition from $+1/2$ to $-1/2$ is unaffected by the anisotropy. Thus the prominent feature of first order quadrupolar spectra of half-integral nuclei is a dominant, narrow line in the center.

Magic-Angle Spinning

Under MAS the anisotropy of first-order quadrupole interaction is averaged, and one obtains

$$\nu_{ZZ}^{(1)} = \nu_0. \tag{63}$$

If the spinning speed exceeded the anisotropy of the spectrum just one narrow signal sould be observed at ν_0. However, in practice this is not the case and a central line is flanked by a comb of spinning sidebands. But because the signal intensity is concentrated from the whole area under the wide line shape into a limited number of narrow signals, a significant gain in sensitivity is achieved by MAS in this case [57].

Higher-Order Quadrupole Coupling

In the rather frequent case of a strong quadrupole interaction Eq. (62) has to be extended to include higher terms. Then the central line is shifted with respect

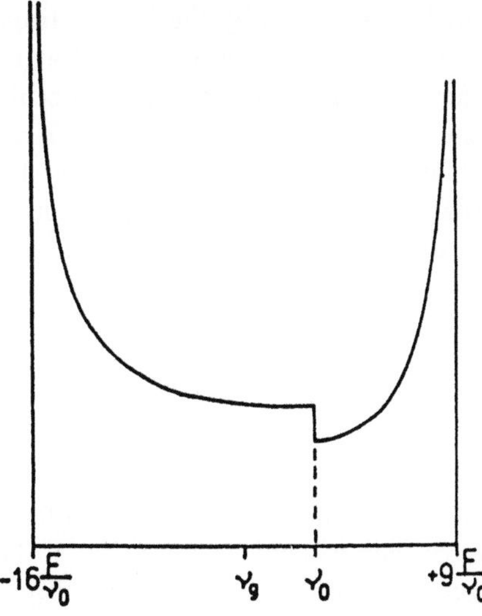

Fig. 16. Powder spectrum of the central transition for the second order quadrupole interaction of a nucleus with half integral spin

to ν_0 and becomes dependent on the orientation. A typical asymmetric line shape of the central line is shown in Fig. 16. Usually the satellite transitions are too broad to be visible in such powder spectra.

For second-order quadrupole coupling the NMR frequency depends on geometric factors like $\sin\theta$, $\cos\theta$, $\sin^2\theta$, and $\cos^2\theta$ terms. For this reason, classical MAS cannot be used to eliminate second-order line broadening. Rather the sample must be spun around more than one axis [43, 58]. This can be achieved either by simultaneous rotation around two axis (DOR: *d*ouble *r*otation) [59] or by successive rotation around different axes in subsequent time periods (DAS: *d*ynamic-*a*ngle *s*pinning) [60]. By these techniques the first (see above) and the second-order quadrupole interactions are averaged, so that high resolution spectra of many quadrupolar nuclei like ^{11}B, ^{27}Al, ^{17}O, and ^{23}Na can be obtained.

3 NMR Spectroscopy

Important goals of NMR spectroscopy are the determination of molecular structure, molecular order, and molecular mobility. The treatment of the first topic requires high resolution so that individual chemical shifts are revealed free of overlap from other interactions as well as the anisotropy of the magnetic shielding. To this end magic angle spinning and multipulse techniques are used. The second topic relies entirely on the evaluation of the anisotropy of a given interaction. The influence of other interactions must be eliminated either by experiment or by chemistry in terms of isotope substitution. Here slow magic angle spinning and wideline techniques are used. The third topic covers a wide range of techniques, depending on the characteristic time scale of the motion. The time scale accessible by NMR is limited on one end by the fast motional cut-off of the spectral densities for unrestricted segmental motion, and at the other end by the length of the spin-lattice relaxation time. Thus molecular dynamics within a range 10^{-12} s to some 100 s can be investigated. High-resolution NMR is often used to investigate fast molecular motion, and wide-line NMR is used for studies of slow molecular motion. In this introduction the focus is entirely on techniques for slow motion.

3.1 High-Resolution NMR

The crucial problem in solid-state NMR is the separation of the different interactions. There are two principal approaches to achieve this goal: 1. The calculation of the total Hamiltonian by summing up the components in Eq. (9)

and subsequently comparison with the experimental. 2. The suppression or reduction of the disturbing interactions by suitable NMR techniques. Of course, this can also be achieved by isotope dilution. This particular procedure is applied with great success in ^{2}H NMR for studies of molecular motion (cf. Sec. 3.2).

The first approach corresponds to the philosophy of the recently published computer program ANTIOPE [61] for simulating a variety of real and hypothetical NMR experiments. The second approach is the philosophy behind experimental high-resolution solid-state NMR. Usually this term is used as a synonym for *line-narrowing* techniques as a counterpart to *wide-line* techniques. Independent of the type of NMR experiment, the prevailing interactions, and the available apparatus, high-resolution solid-state NMR can be defined as a class of methods for studying chemical shifts in solids.

The fundamental line-narrowing technique is sample spinning including MAS, DAS, and DOR variants because it effectively suppresses all anisotropic interactions except for those with anisotropies larger than the spinning speed. In such cases other decoupling methods have to be applied. In the following the general aspects of various experimental decoupling methods are discussed.

3.1.1 Magic-Angle Spinning

The effect of mechanical sample rotation at the magic angle has been described above by example of the chemical shielding anisotropy, cf. Eq. (34), the dipole–dipole, and the quadrupole interaction. Other than for uncovering the isotropic shifts, MAS can be used to obtain information about macroscopic molecular order in partially oriented samples, such as in fibers or streched films by ^{13}C NMR. To this end the information about the anisotropy is needed. It can be obtained from the spinning sidebands, which arise under slow sample rotation for samples lacking molecular order. The information about molecular order is obtained from an analysis of the experimental sideband spectra in terms of simulated spectra. Therefore the formal description of MAS NMR is outlined in the following.

Time Dependence of the Resonance Frequency

For a mathematical description of MAS, different coordinate frames have to be considered (Fig. 17). The z_s axis of the sample frame is defined parallel to the preferential direction of the sample. Thus the coupling tensor attached to a molecule appears non-diagonal within this frame. The same frame can also be used to specify the orientation of a crystallite in the sample. Its orientation in the spinner fixed frame is specified by the Euler angles α, β, and γ. The orientation of the spinner fixed frame in the laboratory frame is determined by the polar coordinates Ψ and ξ. In the spinner fixed frame the magnetic field $\boldsymbol{B}_0$ precesses around the rotor axis at an angle Ψ, which is adjusted to 54.7° for MAS.

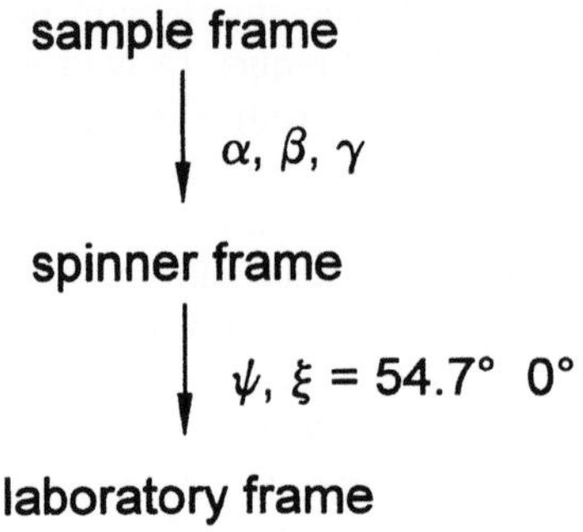

sample frame

↓ α, β, γ

spinner frame

↓ $\psi, \xi = 54.7°$ 0°

laboratory frame

Fig. 17. Euler angles used to define the relationships between coordinate frames in sample spinning

The phase, which the spinner accumulates during sample rotation with constant speed ν_R is given by

$$\xi(t) = \xi_0 + 2\pi\nu_R t. \tag{64}$$

If the preferential order axis $\boldsymbol{n}$ of the sample is not aligned with the spinner axis, then the sample orientation becomes time dependent via the angle $\xi(t)$. In this way also the resonance frequency ν becomes dependent on time. But it can be split into a time-independent and a time-dependent part [25], cf. Eq. (33b),

$$\nu(t) = \nu' + \nu''(t). \tag{65}$$

The time-independent part is given by

$$\begin{aligned}\nu' = & -\nu_0\{\bar{\sigma} + (1/2)(3\cos^2\Psi - 1)[(1/2)(3\cos^2\beta - 1)(\sigma_{33} - \bar{\sigma}) \\ & + (1/2)(\sigma_{11} - \sigma_{22})\sin^2\beta\cos(2\alpha)]\},\end{aligned} \tag{66}$$

and the time-dependent part is

$$\begin{aligned}\nu''(t) = & A_2\cos(2\gamma + 4\pi\nu_R t) + B_2\sin(2\gamma + 4\pi\nu_R t) + A_1\cos(\gamma + 2\pi\nu_R t) \\ & + B_1\sin(\gamma + 2\pi\nu_R t).\end{aligned} \tag{67}$$

The quantity $\bar{\sigma}$ is the isotropic chemical shift Eq. (32a). The coefficients A_1, B_1, A_2, and B_2 are given at the magic angle by [25]

$$\begin{aligned}A_1 = & -(2\sqrt{2}\nu_0/3)\{\sin(2\beta)[(\sigma_{11} + \sigma_{22} + 2\sigma_{33})/4 - \cos(2\alpha)(\sigma_{22} - \sigma_{11})/4 \\ & + \sigma_{12}\sin(2\alpha)] + \cos(2\beta)[\sigma_{13}\cos\alpha + \sigma_{23}\sin\alpha]\},\end{aligned} \tag{68a}$$

$$\begin{aligned}B_1 = & -(2\sqrt{2}\nu_0/3)\{\sin\beta[(\sigma_{22} - \sigma_{11})\sin(2\alpha)/2 + \sigma_{12}\cos(2\alpha)] \\ & - \cos\beta[\sigma_{13}\sin\alpha - \sigma_{23}\cos\alpha]\},\end{aligned} \tag{68b}$$

$$\begin{aligned}A_2 = & -(\nu_0/3)(-\sigma_{11} - \sigma_{22} + 2\sigma_{33})\sin^2\beta/2 - (\sigma_{22} - \sigma_{11})\cos(2\alpha) \\ & \cdot(1 + \cos^2\beta)/2 + \sigma_{12}\sin(2\alpha)(1 + \cos^2\beta) - \sin(2\beta) \\ & \cdot[\sigma_{13}\cos\alpha + \sigma_{23}\sin\alpha]\},\end{aligned} \tag{68c}$$

$$\begin{aligned}B_2 = & -(\nu_0/3)\cos\beta[(\sigma_{22} - \sigma_{11})\sin(2\alpha) + 2\sigma_{12}\cos(2\alpha)] \\ & + 2\sin\beta[\sigma_{13}\sin\alpha - \sigma_{23}\cos\alpha]\}.\end{aligned} \tag{68d}$$

Because the principal axes of the coupling tensor are not aligned with the axes of the sample frame, all elements of the coupling tensor appear in Eq. (68). Though the calculation explicitly deals with the chemical shielding, similar expressions are valid for other anisotropic second rank interactions like the dipole–dipole coupling between two spins and the first order quadrupole interaction. At the magic angle $\Psi_{\text{magic}} = \arccos(3^{-1/2}) = 54.7°$ the time-independent part, Eq. (66), of the resonance frequency becomes independent of the orientation angles α and β. Deviations from the magic angle lead to powder spectra the widths of which are scaled by the factor $(3\cos^2\Psi - 1)/2$ compared to the spectra of non-spinning samples, cf. Eq. (34) and Fig. 11b.

The Free-Induction Decay under Magic-Angle Spinning

For calculation of the FID under MAS the precession phase ζ of an arbitrary magnetization component corresponding to a fixed orientation of the coupling tensor in the rotor frame is written as the time integral over the frequency $v(t)$,

$$\zeta(t_1, t_2) = 2\pi \int_{t_1}^{t_2} v(t)\mathrm{d}t = \theta(t_2) - \theta(t_1). \tag{69}$$

The integral is determined by the difference in angles $\theta(t_2)$ and $\theta(t_1)$, which the magnetization vector forms with the x axis of the coordinate frame at times t_1 and t_2, respectively. Then the signal $g(t)$ of the free-induction decay of this component is given by

$$\begin{aligned} g(t) &= \exp\{-\mathrm{i}2\pi v_0 \bar{\sigma} t\} \exp\left\{2\pi \int_0^t v''(t')\mathrm{d}t'\right\} \\ &= \exp\{-\mathrm{i}2\pi v_0 \bar{\sigma} t\} \exp\{\mathrm{i}\Theta(t)\} \exp\{-\mathrm{i}\Theta(0)\}. \end{aligned} \tag{70}$$

Because $v''(t)$ depends on the orientation of the principal axis frame of the coupling tensor relative to the magnetic field, each magnetization vector in the sample follows an individual phase trajectory in the transverse plane, which is described by (69) [62]. Therefore the total magnetization dephases rapidly. The time-dependent part $v''(t)$ of the resonance frequency, however, exclusively depends on terms which oscillate periodically with v_R and $2v_R$. For this reason the integral in (69) vanishes for each magnetization vector after an integral multiple of the rotor period $T_R = 1/v_R$, and rotary echoes are observed at these times in the FID of a powder sample (cf. Fig. 18b).

The f Functions

The formal description of MAS time-domain signals is facilitated by use of the so-called f functions [25, 63]. For spinning at the magic angle $\Psi_{\text{magic}} = 54.7°$ the FID, Eq. (70) of a given magnetization component is rewritten with the help

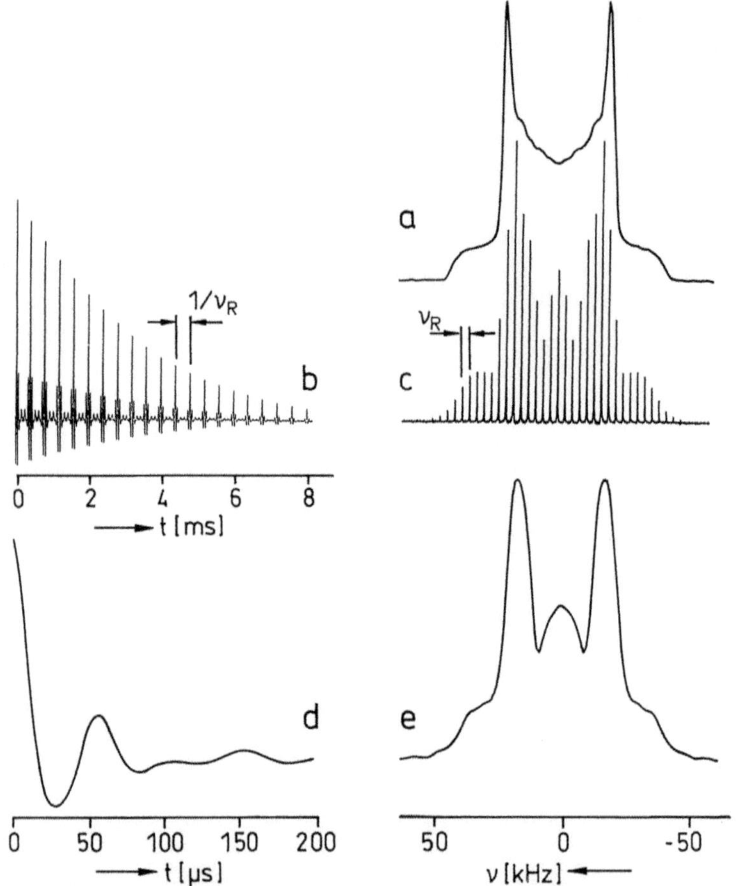

Fig. 18a–e. ^{2}H MAS NMR of dimethylsulphone-d$_6$ at a rotor frequency of $\nu_R = 2500$ Hz. **a.** Solid-echo spectrum of the non-spinning sample. **b.** On-resonance FID ($\bar{\sigma} = 0$) with rotary echoes. **c.** MAS sideband spectrum. **d.** Decay of a rotary echo signal. **e.** Fourier transform of a rotary echo [64]

of Eqs. (64) and (65) as

$$g(t) = \exp\{-\mathrm{i}2\pi\nu_0\bar{\sigma}t\} f(\gamma + \xi(t)) f^*(\gamma + \xi_0), \tag{71}$$

where * denotes the complex conjugate,

$$f(\gamma) = \exp\{\mathrm{i}[A'_2 \sin(2\gamma) - B'_2 \cos(2\gamma) + A'_1 \sin\gamma - B'_1 \cos\gamma]\}, \tag{72}$$

and

$$A'_1 = A_1/(2\pi\nu_R), \quad B'_1 = B_1/(2\pi\nu_R), \tag{73a}$$

$$A'_2 = A_2/(4\pi\nu_R), \quad B'_2 = B_2/(4\pi\nu_R). \tag{73b}$$

For clarity, the arbitrary but fixed spinner phase ξ_0 and the rotation angle $2\pi\nu_R t$ are not always carried along explicitly in the f functions.

Partially Ordered Samples

For calculation of the free-induction decay signal of all magnetization components in the sample, (70) is rewritten by use of the delta function

$$\delta(\vartheta-\gamma-2\pi\nu_{R}t)=(2\pi)^{-1}\sum_{N=-\infty}^{+\infty}\exp\{\pm iN(\vartheta-\gamma-2\pi\nu_{R}t)\} \tag{74}$$

to yield

$$\begin{aligned}g(t)&=\exp\{-i2\pi\nu_0\bar{\sigma}t\}f^*(\gamma)\int_{t_1}^{2\pi}\delta(\vartheta-\gamma-2\pi\nu_{R}t)f(\vartheta)d\vartheta\\&=(2\pi)^{-1}\exp\{-i2\pi\nu_0\bar{\sigma}t)\sum_{N=-\infty}^{+\infty}\exp\{iN2\pi\nu_{R}t\}\exp\{iN\gamma\}f^*(\gamma)\\&\quad\int_0^{2\pi}\exp\{-iN\vartheta)\}f(\vartheta)d\vartheta.\end{aligned} \tag{75}$$

The signal of the total magnetization derives from (75) by summation of all magnetization components. This can be achieved by integration over all orientations α,β,γ of the sample frame in the rotor frame. For partially ordered samples the integration must be weighted by the *orientational distribution function* $P(\alpha,\beta,\gamma)$ in the sample frame,

$$\begin{aligned}s(t)&=\int_0^{2\pi}\int_0^{\pi}\int_0^{2\pi}P(\alpha,\beta,\gamma)\exp\{-(1/T_2-i2\pi\nu_0\bar{\sigma})t\}g(t)d\gamma\sin\beta\, d\beta\, d\alpha\\&=(2\pi)^{-1}\exp\{-(1/T_2+i2\pi\nu_0\bar{\sigma})t\}\sum_N\int_0^{2\pi}\int_0^{\pi}\int_0^{2\pi}P(\alpha,\beta,\gamma)\exp(iN\gamma\}f^*(\gamma)\\&\quad\cdot d\gamma\sin\beta\, d\beta\, d\alpha\int_0^{2\pi}\exp\{-iN\vartheta\}f(\vartheta)d\vartheta.\end{aligned} \tag{76}$$

By use of Eq. (76) the FID signal can be calculated for oriented samples under MAS conditions, and MAS spectra are obtained by subsequent Fourier transformation. The finite width of center lines and sideband signals can be accounted for by multiplication of the FID with the exponential line-broadening function $\exp\{-(t/T_2)\}$. For practical applications the distribution function $P(\alpha,\beta,\gamma)$ can be expressed in terms of the molecular orientation in the sample frame and the orientation of the coupling tensor in the molecular frame (cf. Fig. 17).

For an isotropic orientational distribution $P(\alpha,\beta,\gamma)=(8\pi^2)^{-1}$, so that

$$s(t)=(2\pi)^{-2}\exp\{-(1/T_2+i2\pi\nu_0\bar{\sigma})t\} \tag{77}$$

where

$$F_N=\int_0^{2\pi}\exp\{-iN\vartheta\}f(\vartheta)\}d\vartheta. \tag{78}$$

Thus by Fourier transformation of (4.3.17) a sideband spectrum is obtained with signals at frequencies $2\pi\nu_0\bar{\sigma} + N2\pi\nu_R$ with intensities $F_N^* F_N$.

The MAS signals are illustrated in Fig. 18 with experimental ^2H data of dimethylsulphone [64]. The FID (b) consists of series of rotary echoes (d) which are formed every rotor period $T_R = 1/\nu_R$. The Fourier transform (c) of the FID is the sideband MAS spectrum, consisting of narrow lines separated by the rotor frequency ν_R. The maxima of the rotary echoes in (b) follow the evolution of the isotropic mean. If the MAS FID is sampled at the echo maxima only, no sidebands will appear in the MAS spectrum. A similar situation is encountered for fast spinning, when the rotary echoes are no longer resolved. The envelope (e) of the sideband spectrum is the Fourier transform of the rotary echo decay (c). For slow spinning speeds it approaches the shape of the wide-line spectrum (a) of the nonspinning sample.

Sideband Suppression

Rotational sideband signals contain important information about anisotropies of spin interactions. Nevertheless, they can be a nuisance in crowded high-

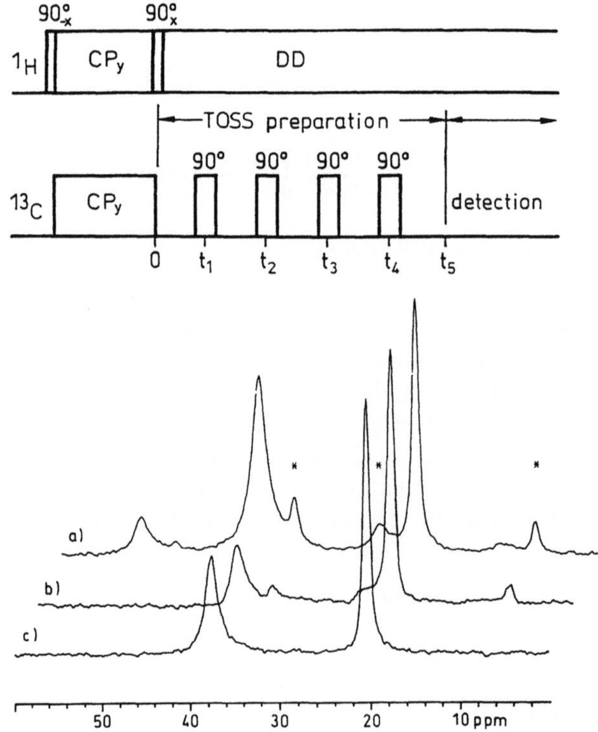

Fig. 19a–c. TOSS preparation in MAS NMR for suppression of spinning sidebands. *Top*: Pulse sequence. *Bottom* [150]: ^{13}C MAS spectra of succinic acid without TOSS **(a)**, with conventional TOSS **(b)**, and with phase-cycled TOSS **(c)**. The spinning sidebands are marked by an *asterisk*

resolution spectra in cases where the isotropic chemical shift needs to be identified. Sideband signals do not arise, if the spinner frequency ν_R is larger than the anisotropy Δ of the interaction. This can generally be achieved with current spinning frequencies of $\nu_R = 5$ to 20 kHz for the anisotropic shielding of aliphatic carbons, but only in special cases or at low magnetic fields for aromatic carbons. By use of certain pulse sequences, however, the magnetization can be prepared before data acquisition in such a way, that the sideband signals are modulated or even suppressed.

The pulse sequence used most often for this purpose is the *TOSS* sequence (*t*otal *s*uppression *o*f *s*pinning *s*idebands (Fig. 19) [66]. Before the start of data acquisition at $t = t_5$ the spin system is prepared by four 180° pulses at well-defined times t_1, t_2, t_3, and t_4. The timing of the pulses is chosen in a way so that the magnetization components giving rise to the sidebands in the MAS spectrum interfere to zero during data acquisition, and only the signals at the isotropic chemical shifts remain. The TOSS sequence works in this way only for isotropic samples. For macroscopically ordered samples, sidebands which are modulated by the spinner phase are still observed. This can be exploited for instance for separation of signals from the isotropic and the anisotropic parts of the interaction.

3.1.2 Heteronuclear High-Power Decoupling

High-power decoupling denotes the suppression of the heteronuclear dipole-dipole coupling **C** between *I* and *S* spins. This technique is routinely applied during acquisition of *S* magnetization from rare spins like ^{13}C in the presence of abundant *I* spins like ^1H. The most effective method for solid-state applications requires application of a strong continuous rf field of amplitude $\nu_{1I} = -\gamma B_{1I}/2\pi$ at the *I* spin frequency ν_{0I} (*DD* in Fig. 19).

The decoupling efficiency depends on two factors [23, 25]: 1) The amplitude ν_{1I} of the decoupling field in comparison with the strength of the heteronuclear dipole–dipole interaction. 2) The modulation of the heteronuclear dipole–dipole coupling by flip-flop transitions in the system of the abundant *I* spins, which communicate by the homonuclear dipole–dipole interaction. In addition to this the influence of thermal motion has to be considered [25].

A measure for the decoupling efficiency is the line width in the spectrum of the *S* spins. It grows with the square of the rf frequency offset from the resonance of the *I* spins. The amplitude of the decoupler field has to be chosen large compared to the strength of the heteronuclear dipole–dipole interaction. For decoupling of ^{13}C and ^1H this means, that $|\gamma B_{1H}| > 2\pi$ 100 kHz should be chosen. In practice this requires transmitter power in excess of 1 kW for coil diameters of 5 to 10 mm, and problems with sample heating can be encountered. Continuous high-power decoupling not only averages the heteronuclear direct dipole–dipole coupling to zero, but also the heteronuclear indirect coupling (*J* coupling).

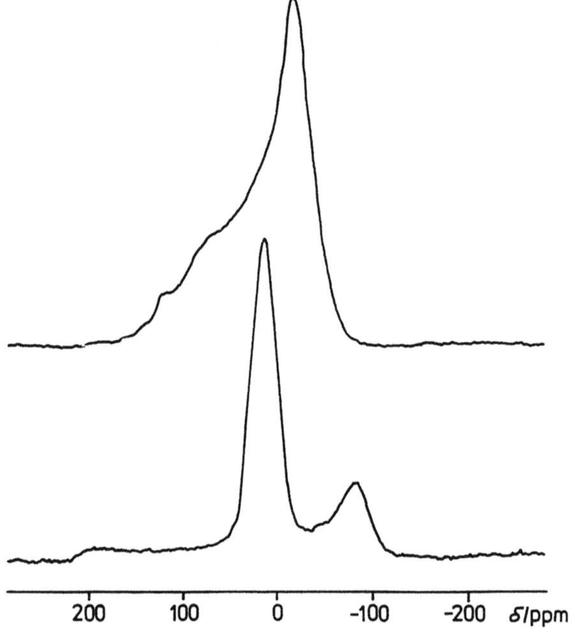

Fig. 20. ^{31}P MAS spectra of Na_2HPO_3 without (*top*) and with (*bottom*) ^{1}H decoupling [42]

Although dipolar decoupling is typical for investigating chemical shift of rare nuclei like ^{13}C and ^{29}Si it can be useful for suppressing dipolar decoupling between two types of abundant spins like ^{1}H or ^{31}P, too. Figure 20 demonstrates this by proton decoupled ^{31}P spectra of Na_2HPO_3. The ^{1}H decoupling eliminates the (weak) intermolecular dipolar ^{1}H–^{1}H interaction but does not change the (strong) intramolecular dipolar ^{1}H–^{31}P interaction.

3.1.3 *Multipulse Decoupling*

In cases of strong (>25 kHz) homonuclear dipole–dipole interactions MAS spinning rates ν_R and decoupler amplitudes $\nu_{11} = -\gamma B_{11}/2\pi$ are too weak for efficient decoupling. However, in this case, multipulse techniques can be employed, which, depending on the pulse sequence, selectively average different interactions [20, 21, 25, 65].

An important application of multipulses techniques is the elimination of the homonuclear dipole–dipole interaction for determination of the chemical shift in single crystals. Other applications are isolation of the heteronuclear dipole–dipole interaction between an abundant and a rare spin, isotropic magnetization transfer independent of chemical shielding effects [67] instead of cross-polarization, filtering of magnetization according to chemical [68] and morphological criteria [69] for studies of magnetization spread by spin diffusion, and multipulse imaging of rigid solids [70–73]. Multipulse excitation is experimentally demanding. It requires high-power pulses with extraordinary phase and amplitude stability.

The idea underlying multipulse decoupling is revealed when rewriting the dipolar coupling Hamiltonian, Eq. (39), in the high field approximation in terms of the product of a spin operator part and a geometric part,

$$\mathbf{H}^{c} = \mathbf{I}_i \mathbf{I}_j \tfrac{2}{3} \Delta C (3\cos^2\beta - 1). \tag{79}$$

By magic-angle spinning, the geometric factor is modified. Multipulse technique aims at forming the time average of the spin operator factor.

A multipulse sequence consists of a cycle of rf pulses, which is repetitively applied to the spin system. This effect can be seen as a way to enforce a motion in spin space which in other cases is inscribed on the molecules by thermal motion or MAS. Different multipulse sequences vary in the number and phases of the pulses per cycle. In this way the interactions can be made to cancel in the average Hamilton operator not only in zeroth order, but also in first and higher orders. Also the chemical shielding can be scaled by different factors R_σ, and effects of finite pulse lengths t_p, of B_1 inhomogeneities as well as of cross terms between different interactions can be compensated for to different degrees of accuracy. Further improvement of the averaging can be achieved by *second averaging*. This denotes introduction of a second axis of averaging, which is orthogonal to the first [20, 21, 25, 65, 74].

Some properties of important multipulse sequences are collected in Table 2. For homonuclear decoupling the MREV8 [75] sequence is used most often in practice. The WIM24 sequence [67] is applied for polarization transfer by isotropic mixing. It not only averages the homonuclear dipole–dipole coupling to zero but also the chemical shielding ($R_\sigma = 0$) [72, 76].

The use of multipulse excitation for high-resolution solid-state NMR of ^{1}H is illustrated in Fig. 21 by an MREV8 spectrum of a calcium formiate crystal acquired with a minimum pulse separation of $t = 4\,\mu s$ and a width of the 90° pulses of $t_p = 0.75\,\mu s$ [79]. The line splittings are revealed only under decoupling of the homonuclear dipole-dipole interaction. The sample shape plays an important role, as it influences the homogeneity of the magnetic field by differences in the magnetic susceptibilities of sample and surrounding matter.

Table 2. Multipulse sequences

cycle	Ref.	length	R_σ	$\mathbf{H}_D^{A(0)}$	$\mathbf{H}_D^{A(1)}$	$\mathbf{H}_D^{A(2)}$	t_n	compensation of B_1 inhom.
WAHUHA	[12]	6τ	$1/\sqrt{3}$	0	0	0	no	no
MREV8	[78] [75]	12τ	$\sqrt{2}/3$	0	0	0	yes	yes
BR24	[77]	36τ	$2/3\sqrt{3}$	0	0	0	yes	yes
WIM24	[67]	24τ	0	0	0	—	0	—

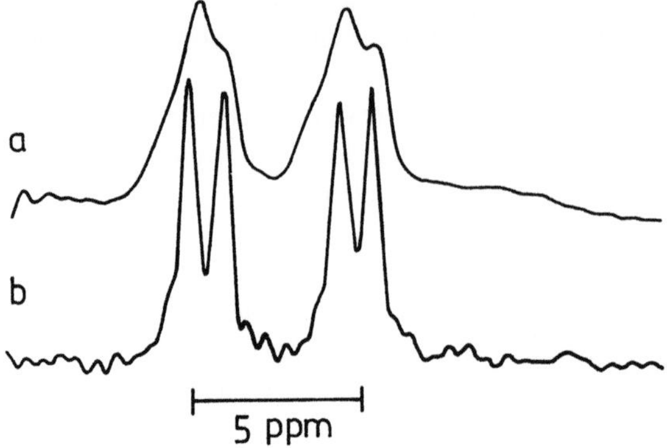

Fig. 21. ^1H-MREV8 spectra of a calcium formiate crystal for a pulse separation of $\tau = 4\,\mu s$ and a pulse width of $t_p = 0.75\,\mu s$. **a.** Crystal sphere glued to the tip of a glass rod. **b.** Rotational ellipsoid inside a long glass tube without a glass rod [79]

3.1.4 Combined Rotation and Multipulse Spectroscopy: CRAMPS

In the case of strong dipole–dipole coupling efficient decoupling is obtained by a combination of both, MAS and multipulse decoupling. By use of the *C*ombination of *R*otation *a*nd *M*ultiple *P*ulse *s*equences (*CRAMPS*) [81–83] the chemical shift dispersion can be uncovered even for nuclei like ^1H [84] and ^{19}F in organic as well as inorganic [85] solids. Figure 22 shows the ^1H CRAMPS spectrum of $Ca_2[(HSiO_4)(OH)]$. The suppression of the proton–proton dipolar coupling and the ^1H magnetic shielding anisotropy results in two signals with intensity ratio 1:1 at the isotropic shifts of the two chemically different protons (acid and base, respectively).

3.1.5 Magnetic Dilution: "Chemical" Decoupling

Other than MAS and multipulse sequences a third method for reduction of the dipole–dipole interaction, cf. Eq. (44), consists of physically reducing the internuclear distance r_{ij}. This can be achieved by magnetic dilution in terms of substitution of nuclei or groups of nuclei by isotopes or structurally similar groups of nuclei of smaller or zero nuclear magnetic moment. Examples are

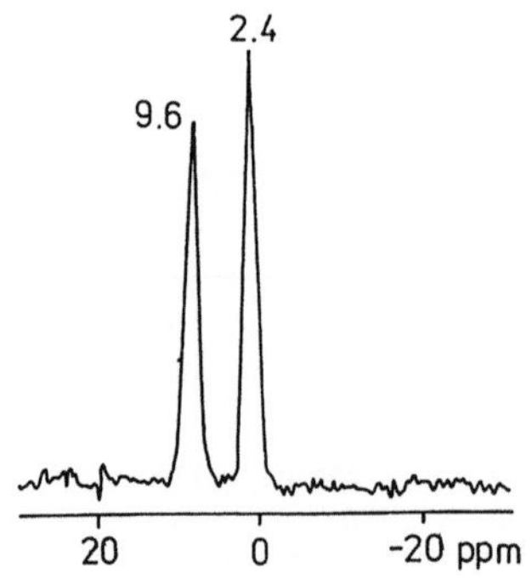

Fig. 22. ^1H NMR spectra of $Ca_2[(HSiO_4)(OH)]$. *Top*: Single pulse excitation of the static sample. *Middle*: Multipulse (WAHUHA) excitation of static sample. Two signals with overlapping magnetic shielding anisotropies can be identified. *Bottom*: CRAMPS spectrum at $\nu_R = 4\,kHz$. The isotropic signal at $\delta = -2.6\,ppm$ corresponds to the OH^- unit, the signal at $\delta = 9.4\,ppm$ is caused by the acidic protons [86]

the substitution of ^1H by ^2H [87–92] and the substitution of $^{31}PO_3F^{2-}$ by SO_4^{2-} [93]. The substitution of entire groups of molecules is often connected with more or less significant structural changes and therefore this form of "chemical" decoupling cannot compete with real physical decoupling methods described above. Isotope labelling on the other hand has become a widely accepted technique, in particular the site-specific substitution of ^1H by ^2H for investigations of molecular order and dynamics. Examples of the use of deuteron NMR are given in Sec. 3.2.

3.1.6 *Cross-Polarization*

When observing rare nuclei like ^{13}C, a significant gain in signal-to-noise ratio can be achieved if magnetization is transferred to the rare spins S from abundant spins I like ^1H [11]. The gain can be twofold: A higher polarization of the S spins is achieved, and the often longer relaxation times of the S spins are circumvented in magnetization build-up.

The Hartmann-Hahn Condition

Cross-polarization is achieved in a double resonance experiment (Fig. 23a). Transverse magnetization of the I spins is generated by a 90° pulse at frequency ν_{0I}. However, the transmitter is not turned off afterwards, only the rf phase is shifted by 90°. Thus the $\boldsymbol{B}_{1I}$ field is now applied parallel to the I magnetization. In the frame rotating with frequency ν_{0I} around the z axis, the $\boldsymbol{B}_{1I}$ field is the dominant magnetic field which the I spins experience, if its amplitude ν_{1I} is larger than the other interactions of the I spin. The I magnetization is locked by it, so that this technique is called *spin locking*. A similar protocol is followed for the S spins. A second rf field $\boldsymbol{B}_{1S}$ is applied at frequency ν_{0S}, coincident with the lock field for the I spins. If the magnitudes B_{1S} and B_{1I} of both rf fields are matched by the *Hartmann-Hahn condition* [80],

$$\gamma_S B_{1S} = \gamma_I B_{1I}, \tag{80}$$

each spin species precesses with the same frequency $\nu_1 = -\gamma B_1/2\pi$ around the axis of its rf field in its own rotating frame. But because both rotating frames share the same z axis, there will be an oscillation of local I and S magnetization components along the z axis with the same frequency ν_1 (Fig. 23, c). By this frequency match magnetization can be exchanged between both spin species. But because only the I spins were polarized to begin with, magnetization is transferred from the I spins to the S spins.

Selectivity in Cross-Polarization

The efficiency of cross-polarization is determined by the size of the dipole–dipole interaction between I and S spins, and by the relaxation times $T_{1\rho I}$ and $T_{1\rho S}$

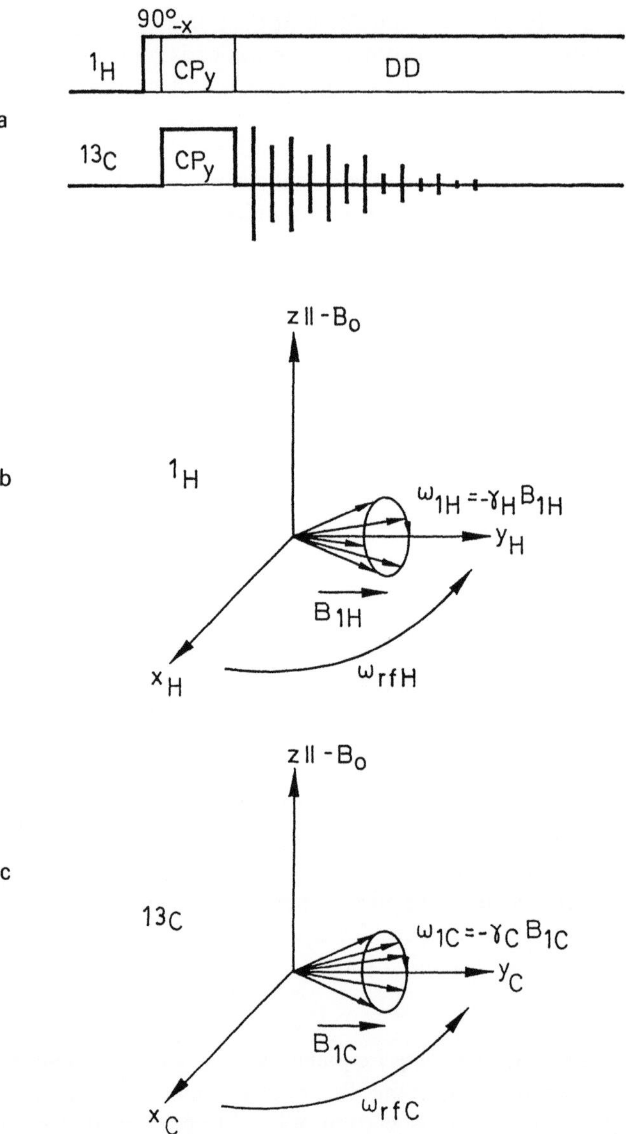

Fig. 23a–c. Illustration of cross-polarization (CP): **a.** Timing diagram of rf excitation for cross-polarization of ^{1}H magnetization to ^{13}C. DD denotes heteronuclear dipolar decoupling. **b.** Precession of ^{1}H magnetization components during cross-polarization with a frequency ν_{1H} around the magnetic field $\boldsymbol{B}_{1H}$ in the coordinate system, which rotates with frequency ν_{0H}. **c.** Precession of ^{13}C magnetization components during cross-polarization with a frequency ν_{1C} around the magnetic field $\boldsymbol{B}_{1C}$ in the coordinate system, which rotates with frequency ν_{0C} [94]

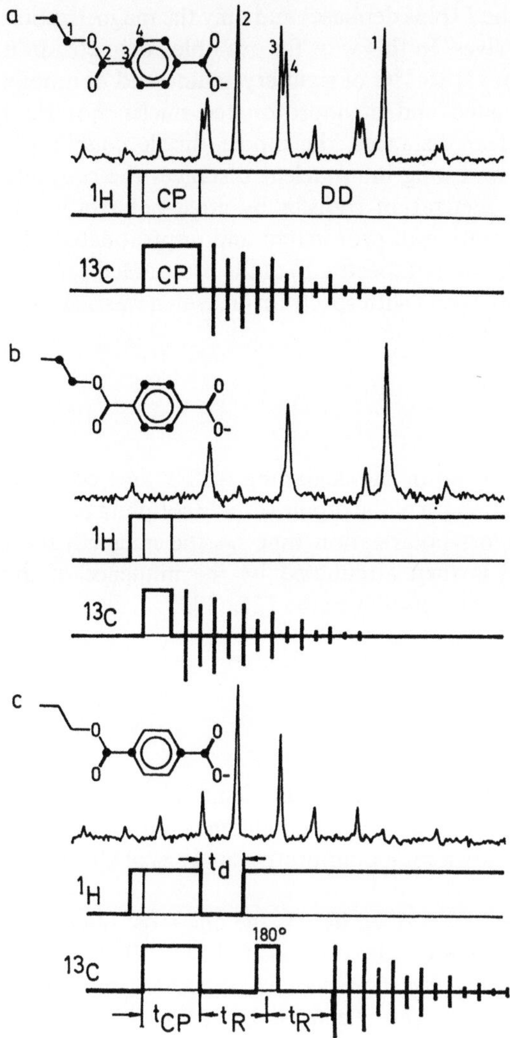

Fig. 24a,b. Selectivity of cross-polarization demonstrated for polyethelene therephthalate at room temperature and a spinning frequency of $\nu_R = 4090$ Hz. **a.** Polarization of protonated and unprotonated ^{13}C nuclei with $t_{CP} = 5$ ms. **b.** Polarization of protonated ^{13}C nuclei for $t_{CP} = 50$ µs. **c.** Polarization of unprotonated ^{13}C nuclei for $t_{CP} = 5$ ms and subsequent dephasing for $t_D = 100$ µs under the influence of the ^1H-^{13}C dipole–dipole coupling [94]

of the spin-locked I and S magnetizations. $T_{1\rho}$ is the *relaxation time in the rotating frame*. It is an energy relaxation time similar to T_1. By variation of the contact time t_{CP}, that is the length of the spin-lock time, local differences in the dipole–dipole couplings and in the $T_{1\rho}$ relaxation times can be exploited to selectively polarize different chemical and morphological structures. For short contact times only the strongly coupled nuclei are polarized, for long contact times also the weakly coupled ones are.

During data acquisition the dipole–dipole interaction between I and S spins is made ineffective by heteronuclear *d*ipolar *d*ecoupling (DD) (Sec. 3.1.2). However, if the decoupler is turned off for a short time t_D, the magnetization of the S spins strongly coupled to the I spins dephases and only the magnetization of the weakly coupled S spins survives. In this way, for example, magnetization can selectively be transferred from ^1H to ^{13}C of rigid crystalline and of mobile amorphous regions, or of protonated and of unprotonated nuclei. For rigid segments and for short internuclear distances, the dipole–dipole coupling is strong, and for mobile segments and long internuclear distances the coupling is weak. The use of selective magnetization transfer by cross-polarization is illustrated in Fig. 24 for the polarization of protonated and unprotonated ^{13}C in polyethylene therephthalate. The sum of spectra (b) of the protonated carbons and (c) of the unprotonated carbons agrees with spectrum (a), which was acquired with polarization transfer to all carbons.

Cross-Polarization Efficiency

The S magnetization $M(t_{CP})$ is built up in the beginning with a time constant T_{CH} which is characteristic for the strength of the dipole–dipole coupling between I and S spins. With increasing cross-polarization time t_{CP} the magnetization passes through a maximum and is then attenuated by the influence of the relaxation times $T_{1\rho H}$ and $T_{1\rho C}$ in the rotating frame [25, 31],

$$M(t_{CP}) = (M_0/\lambda)[1 - \exp\{-\lambda t_{CP}/T_{CH}\}]\exp\{-t_{CP}/T_{1\rho H}\}, \tag{81}$$

where

$$\lambda = 1 + T_{CH}/T_{1\rho C} - T_{CH}/T_{1\rho H}. \tag{82}$$

While $T_{1\rho H}$ is identical throughout the sample as a result of multiple homonuclear dipole–dipole couplings among ^1H, $T_{1\rho C}$ and T_{CH} are different for different parts of the molecule. They need to be known for a quantitative analysis of CP signals [95].

The cross-polarization technique described here is the one most often used in practice. However, other techniques can be used as well [11]. One consists of simultaneous application of multipulse sequences, cf. Sect. 3.1.3, on both, I and S nuclei. Then different interactions can selectively be averaged out during polarization transfer. An example is cross-polarization by isotropic mixing. Here the homonuclear dipole–dipole interaction and the chemical shift are scaled to zero under the influence of a multipulse sequence, so that the frequency dependence of the cross-polarization efficiency is reduced [66, 67].

3.2 *Wide-Line NMR*

Until MAS, multipulse, and Fourier techniques became available on a commercial basis, *wide-line NMR* was the most frequently used method of solid state NMR spectroscopy. Originally this meant the measurement and evaluation of structureless, wide-line solid-state NMR spectra in terms of frequency averages or *moments* [22, 31]. Today, however, the line shapes of the anisotropic chemical shielding, the dipole–dipole, and the quadrupole interaction can be isolated in addition to isotope enrichment, by homo- and heteronuclear decoupling, so that a single interaction dominates the shape of the wide-line spectrum. Then a *line-shape analysis* reveals far more detailed information about molecular oder and mobility than a simple moment analysis.

For ^1H in organic solids, a wide-line-NMR line-shape analysis is the exception. The strong dipole–dipole interaction among the protons cannot be reduced enough by multipulse sequences, and the chemical shift range of ^1H is so small, that the powder spectra of the anisotropic chemical shielding, which are centered at different chemical shifts strongly overlap. This overlap also applies to ^{13}C spectra. But here the homonuclear dipole–dipole coupling is often negligible because of the low natural abundance (1%) of ^{13}C, and the heteronuclear dipole–dipole coupling can effectively by eliminated by high-power ^1H decoupling (cf. Sec. 3.1.2). The overlap of wide-line resonances can be reduced by use of 2D techniques [15] or by fast sample rotation at an angle different from the magic angle [96]. The most straight forward approach to wide-line spectra of a single interaction is also the most labor-intensive one. It requires site-selective isotope enrichment. The most well-known representative of this approach is ^2H spectroscopy [88–92]. Here the quadrupole coupling dominates all other interactions, and wide-line spectra can be measured by relatively simple techniques on modern solid-state NMR spectrometers.

Because ^2H posesses a spin with $I = 1$, the deuteron resonance for a single molecular orientation exhibits two lines, which correspond to the two transitions with $\Delta m = \pm 1$ in the energy level diagram. The quadrupole coupling constant of ^2H is small enough to be treated in first order. For aliphatic deuterons the asymmetry parameter η^Q is close to zero, so that often axially symmetric powder patterns are observed for rigid molecular segements, which are similar in shape to the Pake pattern of Fig. 13.

3.2.1 Molecular Order

Molecular order is determined conventionally by X-ray scattering and in some other cases by neutron scattering. Highly ordered structures lead to sharp reflexes, and weakly ordered ones to washed-out reflexes. Scattering methods are therefore primarily used for analysis of high order. For the characterization of molecular order by solid-state NMR lineshape analysis [97] [92] such a restriction does not apply. But compared to scattering techniques the strength of NMR methods is the analysis of weak order. It can arise in many materials, for instance, during processing by straining, plastic deformation, and molding.

The Orientational Distribution Function

Molecular order is described by the *orientational distribution function* $P(\theta)$ [98]. This is the probability density of finding a preferential direction $\boldsymbol{n}$ in the sample

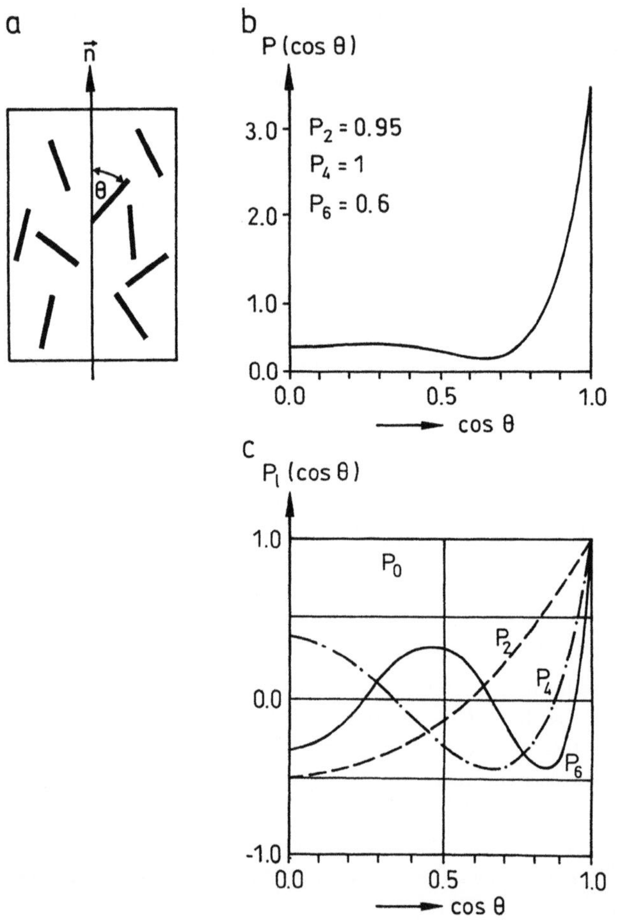

Fig. 25a–c. The orientational distribution function. **a.** Definition of the orientation angle θ. **b.** Angular distribution as a function of $\cos\theta$. **c.** Legendre polynomials

under an angle θ in a molecule-fixed coordinate frame (Fig. 25a). For simplicity macroscopically uniaxial samples with cylindrically symmetric molecules are considered. Then one angle is sufficient to characterize the orientational distribution function. In practice not the angle θ itself, but its cosine is used as the variable, and for weak order the distribution function is expanded into Legendre polynomials $P_l(\cos\theta)$,

$$P(\cos\theta) = \sum_{l=0}^{\infty} \chi_l P_l(\cos\theta). \tag{83}$$

The expansion coefficients χ_l are related to the moments $P_l = \langle P_l(\cos\theta)\rangle$ of the distribution function according to

$$\chi_l = (2l+1)\langle P_l(\cos\theta)\rangle/(8\pi^2), \tag{84}$$

where $\langle \cdots \rangle$ denotes the average weighted by the orientational distribution function. For an isotropic distribution $\langle P_0 \rangle = 1$ and $\langle P_l \rangle = 0$ for all $l > 0$. For a completely ordered sample $\langle P_l \rangle = 1$. $\langle P_0 \rangle$ quantifies the isotropic part of the orientational distribution function. $\langle P_2 \rangle$ is the *second moment* or the *degree of order*. With increasing order higher values of $\langle P_l \rangle$ become significant. Because the expansion treats molecular oder as a perturbation of the isotropic state, it is convergent for weak order. To achieve convergence at high order the orientational distribution function may be expanded for instance into planar or conical distributions [92].

Many methods for determination of molecular order can only measure $\langle P_2 \rangle$ [98]. Among them are measurements of the infrared dichroism and of the refractive index. $\langle P_2 \rangle$ and $\langle P_4 \rangle$ can be determined by Raman fluorescence depolarization and from ^{1}H wideline NMR spectra. However, X-ray scattering and line-shape analysis of resonances from isolated NMR interactions can also provide higher moments.

Determination of the Orientational Distribution Function by Wide-Line NMR

The orientational distribution function $P(\theta)$ enters the shape of the wideline spectrum $S(\nu)$ in a slightly hidden way. The angular dependence of the resonance frequency is given by Eq. (31a) via the orientation of the magnetic field in the principal axes system of the coupling tensor, while the orientational distribution function specifies the distribution of the preferential direction $\boldsymbol{n}$ in a molecule-fixed coordinate frame (Fig. 25a). Figure 26 shows the relationship between the different coordinate frames and the definition of the relative orientation angles.

The orientational distribution function $P'(\cos\beta)$ of axially symmetric coupling tensors ($\eta = 0$) can be read directly from the NMR spectrum $S(\nu)$. Its moments χ_l' can be transformed to the moments χ_l of the orientational distribution function

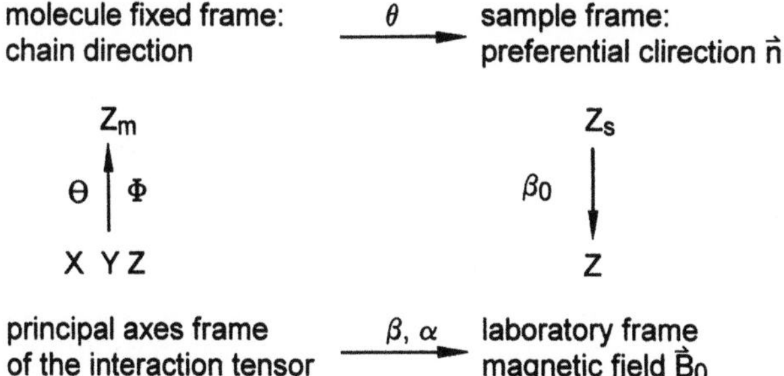

Fig. 26. Relationships between coordinate systems for the description of molecular order. The orientation of the laboratory coordinate system in the principal axes system of the coupling tensor determines the angular dependence of the resonance frequency. The orientations of the preferential sample direction in a molecule-fixed coordinate frame determines the orientational distribution function

$P(\cos\theta)$ of the molecules by using the known orientation angles Θ, Φ of the principal axes frame in the molecule-fixed coordinate frame, and the orientation angle β_0 of the sample frame in the laboratory frame. By convention, the preferential axis $\boldsymbol{n}$ of the sample is parallel to the z_s axis of the sample frame.

The wide-line NMR spectrum $S(\nu)$ can be written as a convolution of the resonance frequency with the orientational distribution $P'(\cos\beta)$ [99],

$$S(\nu') = \int_0^{\pi} \delta(\nu' - \nu(\cos\beta))P(\cos\beta)\sin\beta\,\mathrm{d}\beta, \tag{85}$$

where $\nu(\beta)$ is the angular-dependent part of the resonance frequency, Eq. (23), for an axially symmetric coupling tensor ($\eta = 0$), $\nu = \nu_0(\Delta/2)(3\cos^2\beta - 1)$. Because of the dependence on $\cos^2\beta$ only angles β can be discriminated which are found in the first quadrant between 0 and $\pi/2$,

$$S(\nu') = \delta(\nu' - \nu_0(\Delta/2)(3\cos^2\beta - 1))(1/2)[P'(\cos\beta) + P'(-\cos\beta)]\,\mathrm{d}\cos\beta \tag{86}$$

Thus only the even moments $\langle P_1' \rangle, 1 = 0, 2, 4, \ldots$, of the orientational distribution function P' $(\cos\beta)$ can be determined by NMR. Using an expansion similar to Eq. (83) one arrives immediately at

$$S(\nu') = \sum_{1=0,2,4,\ldots}^{\infty} \chi_1' S_1(\nu'). \tag{87}$$

This equation defines the Legendre subspectra $S_1(\nu')$,

$$S_1(\nu') = \int_0^1 \delta(\nu' - \nu_0(\Delta/2)(3\cos^2\beta - 1))P_1(\cos\beta)\mathrm{d}\cos\beta. \tag{88}$$

The expansion coefficients χ_1 and therefore also the moments $\langle P_1 \rangle$ cf. Eq. (85), of the desired orientational distribution function $P(\cos\theta)$ are related to the expansion coefficients χ_1' of Eq. (87) by the orientation angles Θ and β_0 (Fig. 26) [97]:

$$\chi_1' = (2l+1)/(8\pi^2)\langle P_l(\cos\theta)\rangle P_1(\cos\Theta)P_1(\cos\beta_0). \tag{89}$$

Thus from the expansion coefficients χ_1' of a Legendre sub-spectral analysis according to Eq. (87) the moments $\langle P_1(\cos\theta)\rangle$ of the orientational distribution function can be determined.

The validity of Eq. (89) is restricted to the symmetries mentioned above (cylindrical molecules, macroscopically uniaxial samples, $\eta = 0$). For many samples these are fulfilled when using ^{2}H NMR. In ^{13}C wide-line NMR the anisotropy of the magnetic shielding is used. Here the angular resolution is lower, and the calculation has to be extended to include $\eta > 0$ [97].

As an example for a Legendre subspectral analysis, orientation-dependent NMR spectra of a ^{13}C-labelled liquid-crystalline polymer film are shown in Fig. 27 [100]. The mesogenic sidechain of this polymer contains azobenzene as a dye, which changes its conformation from *trans* to *cis* upon irradiation with

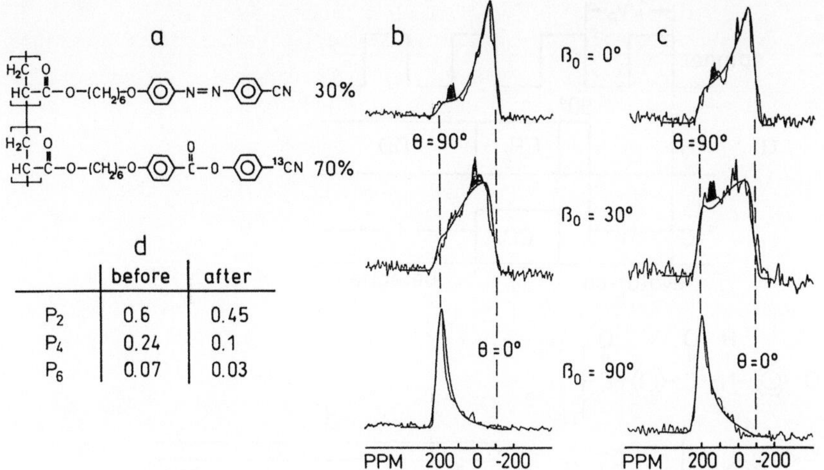

Fig. 27a–d. Orientation dependent ^{13}C NMR spectra of the liquid-crystalline side-chain polymer **a, b** before and **c** after irradiation with light. The *shaded regions* are ^{13}C resonances from nuclei in natural abundance. The angle β_0 measures the orientation of the optical axis of the film relative to the magnetic field $\boldsymbol{B}_0$. **d.** Moments of the orientational-distribution function before and after irradiation [100]

light. This transition is associated with a change in order of the liquid-crystalline matrix. The resonance of the ^{13}C-labelled cyano group in the phenyl benzoate of the mesogen has been analyzed in terms of Legendre subspectra. Switching of the dye by irradiation with light reduces the Legendre moments P_2, P_4, and P_6. This indicates a reduction in molecular order after switching.

Determination of the Orientational-Distribution Function by MAS NMR

In combination with MAS the Legendre subspectral analysis is used successfully for the determination of molecular order in partially ordered polymers [85]. If a macroscopically ordered sample is rotated, the preferential axis $\boldsymbol{n}$ of which is not aligned with the spinning axis, the phase of the sideband spectrum is modulated by the phase ξ_0 of the spinner (cf. Sec. 3.1.1). This modulation can be interrogated experimentally by synchronization of the data acquisition with the spinner phase in a 2D-experiment (Fig. 28a) [85, 101]. To this end the spinner phase is incremented in typically 16 steps through one rotor period in the evolution time t_1,

$$\xi_0 = 2\pi \nu_R t_1. \tag{90}$$

A 2D spectrum is obtained by 2D Fourier transformation over the spinner phase and the acquisition time t_2. It exhibits spinning sideband signals in both dimensions. Those in the additional dimension ω_1 are characteristic of the molecular order. No sidebands appear in this direction if the sample is isotropic.

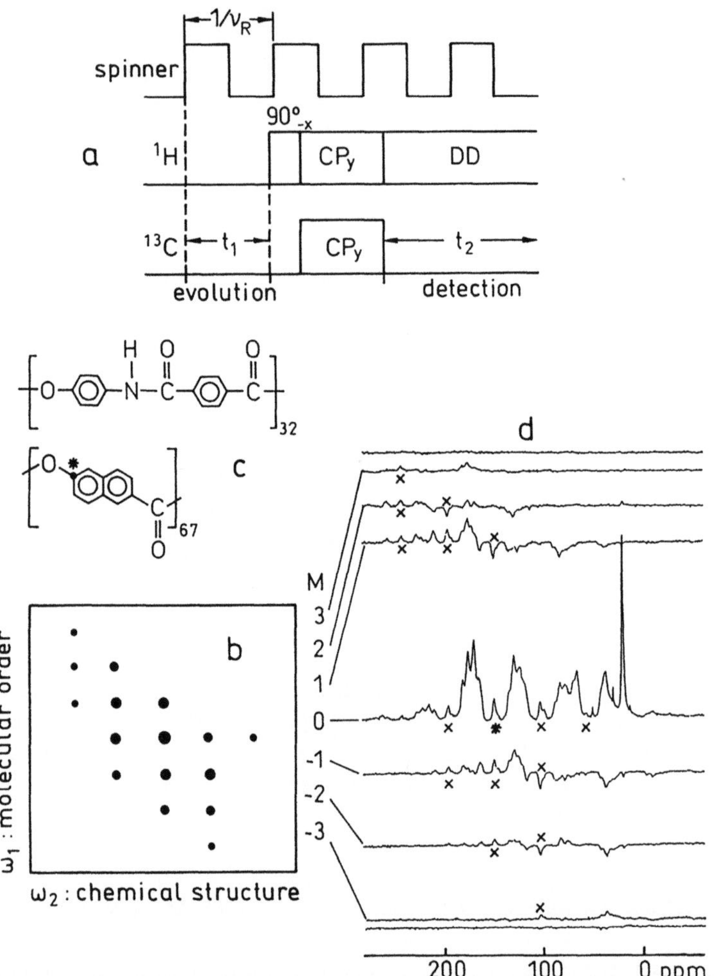

Fig. 28a–d. 2D MAS NMR with rotor synchronization for the detection of molecular order. **a.** Spinner signal and rf excitation. $T_R = 1/\nu_R$ denotes the rotor period. **b.** Arrangement of signals in the 2D sideband spectrum for the ^{13}C nucleus marked in the chemical formula **c.** of the high-performance fiber VECTRA B900. **d.** 1D cross-sections through the 2D sideband spectrum. Rotor frequency and angle between fiber and spinner axes were set to $\nu_R/2\pi = 3500$ Hz and 45°, respectively. The corresponding orientational distribution function is depicted in Fig. 25b [45]

The 2D sideband signals can be analyzed to obtain the orientational-distribution function.

An example is depicted in Fig. 28 with ^{13}C data of the high modulus fiber *Vectra B900*. The sideband signals of the quaternary carbon marked in the formula (c) and in the cross-sections (d) of the spectrum are well resolved and can be used for quantitative analysis. Given the orientation of the chemical shielding tensor in the molecule fixed frame and the orientation of the fiber axis in rotor frame, the orientational distribution function can be calculated from the experimental 2D spectrum. In analogy to the procedure outlined above, an analysis in terms of 2D Legendre subspectra can be followed for weak molecular order. The resulting distribution function is depicted in Fig. 25. It is well approximated by the Legendre moments P_2, P_4, and P_6.

3.2.2 Molecular Reorientation

Wide-line spectra can provide detailed information about type and time scale of slow molecular motion. Because of the angular dependence of the resonance frequency the information is restricted to reorientational processes. Molecular translation cannot yet be sensed by NMR on a molecular distance scale, but on a larger scale in the range of 0.1 μm up to about 10 μm by measuring the particle diffusion in magnetic field gradients [74, 102–104].

The Correlation Time

Molecular motions appear incoherent and are described by a stochastic process $n(t)$. One important quantity to characterize stationary stochastic processes is the auto-correlation function $a(\sigma)$,

$$a(\sigma) = \lim_{T\to\infty} \frac{1}{2T} \int_{-T}^{T} n(t)n(t-\sigma)\mathrm{d}t \propto \exp\{-\sigma/\tau_c\}. \qquad (91)$$

In many cases the auto-correlation function is an exponential function with a time constant τ_c, which is called the *correlation time* of the process. In polymeric materials often distributions of correlation times are observed for molecular motions. Such a distribution can be interpreted in two ways. Either different molecules exhibit different correlation times during the time of observation (heterogeneous distribution) or a single molecule exhibits different correlation times in different observation intervals (homogeneous distribution).

Solid Echo Spectra

For reorientation with correlation times in the range of the inverse spectral width of the powder-spectrum temperature-dependent changes of the line shape are observed, which are characteristic of the motional process [88–91]. As an example, Fig. 29 shows ^{2}H NMR spectra for different motional mechanisms and different correlation times [89]. However, such wide-line spectra cannot readily be measured with pulse excitation, because the beginning of the FID will decay within the receiver deadtime, and only the end can be acquired. Then, after Fourier transformation, only the narrow horns of the singularities will be observed but not the broad shoulders (cf. Fig. 13).

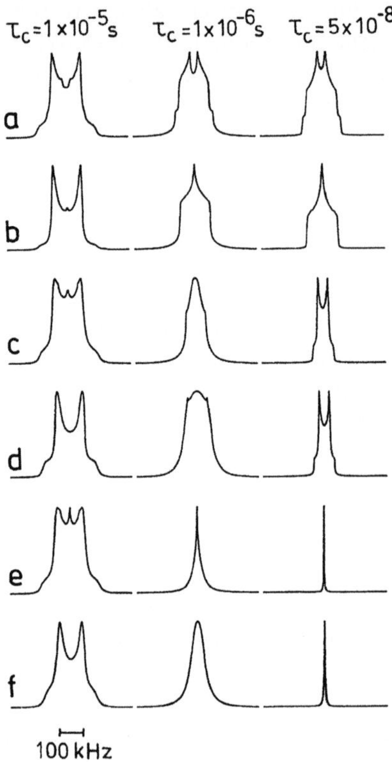

Fig. 29a–f. Deuteron wideline spectra for different motional mechanisms and correlation times τ_C. The angle between the axis of rotation and the C—^{2}H bond (principal axis 33 of the quadrupole coupling tensor) is denoted by θ. **a.** Two-fold jump with $\theta = 60°$. **b.** Two-fold jump with $\theta = 109°$. **c.** Three-fold jump with $\theta = 109°$ (rotation of a methyl group). **d.** Rotational diffusion on a cone with $\theta = 109°$. **e.** Tetrahedral jump. **f.** Isotropic rotational diffusion [89]

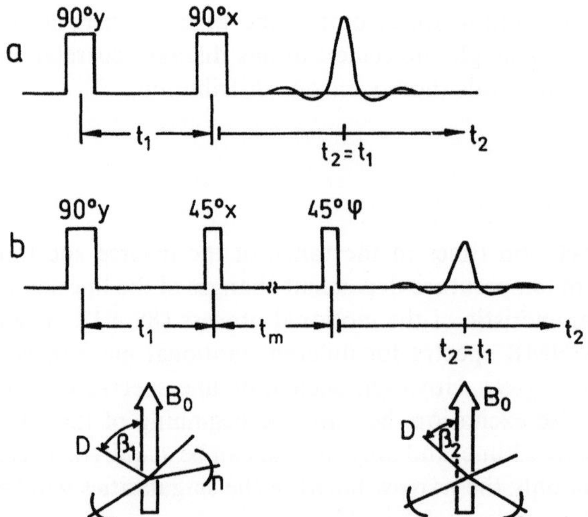

Fig. 30. Pulse sequences for magnetization echoes of spin-1 systems in the solid state. **a.** Quadrupole or solid echo. **b.** Jeener-Broekaert or alignment echo

To bypass receiver deadtime effects, wide-line spectra are derived in Fourier spectroscopy by Fourier transformation of the decay of an echo. By use of the Hahn and the stimulated echo wide-line spectra of ^{13}C and other spin-1/2 nuclei can be measured, for example, but not the spectra of dipolar coupled spins and of quadrupolar nuclei like ^2H. The magnetization of nuclei with spin $I = 1$ can be refocussed by the *quadrupole echo* or the *solid echo*, and by the *Jeener-Broekaert echo* or the *alignment echo* [31] (Fig. 30).

The Quadrupole and the Jeener-Broekaert Echo

A quadrupole echo is generated similar to a Hahn echo by two pulses which are separated by half the echo time t_1 (Fig. 30a). However, the second pulse is a 90° and not a 180° pulse, and it is shifted in phase by 90° with respect to the first. The quadrupole echo appears at a time $t_2 = t_1$ after the second pulse. The spectra shown in Fig. 29 have been simulated for the quadrupole echo technique with acquisition of the echo decay during $t_2 - t_1$. Clearly, the line shape strongly depends on type and times scale of the motion. In this way molecular reorientation with correlation times in the range of $10^{-8}\,\mathrm{s} < \tau_c < 10^{-4}\,\mathrm{s}$ can be characterized by ^2H NMR. Faster motion with correlation times in the range of $10^{-12}\,\mathrm{s} < \tau_c < 10^{-8}\,\mathrm{s}$ can be investigated by measurements of the spin–lattice relaxation time T_1. For investigations of slower processes with correlation times in the range of $10^{-4}\,\mathrm{s} < \tau_c < 10\,\mathrm{s}$ the Jeener-Broekaert echo can be employed [89, 91, 105].

The Jeener-Broekaert echo is stimulated with three pulses (Fig. 30b). The second and the third pulse are each a 45°-pulse, whereby the phase of the second pulse is shifted against that of the first pulse by 90°. In addition to double quantum coherences the second pulse generates a long-lived magnetization state along the z axis which is similar to z magnetization, but exhibits no net magnetization. This state is best explained by considering two spins with $I = 1/2$, which are coupled by the dipole–dipole interaction. In this state the z-projection of one spin is parallel to the z axis, that of the other is antiparallel. Thus, the spins produce no common polarization but an antiparallel order, that is they exhibit *alignment*. The alignment state decays with relaxation times T_{1D} and T_{1Q}, for *dipolar order* and *quadrupolar order*, respectively. The magnitude of these relaxation times is comparable to that of T_1.

At sufficiently long time after the second pulse the double-quantum coherences will have decayed, and only spin alignment remains. Then the memory of the magnetization to the phase of the first two pulses has been lost. Threfore the phase φ of the third pulse is arbitrary. It transforms the spin alignment into single quantum coherences, which refocus to form the Jeener-Broekaert echo after a time $t_2 = t_1$ following the third pulse. The time scale of the molecular motion which is interrogated by the alignment echo is determined by the separation t_m between the second and the third pulse. It is limited at long times by the alignment relaxation times T_{1D} or T_{1Q} and at short times by the separation t_1 of the first two pulses.

2D Exchange Spectroscopy

In the description of the echoes in Fig. 30, two time axes are introduced, because the echo is a fundamental phenomenon, upon which many two-dimensional NMR techniques are based, and the nomenclature used is that of 2D spectroscopy [15]. There, t_1 is the *evolution time*, t_m the *mixing time*, and t_2 the *detection time*. In fact, the Jeener-Broekaert echo can be executed as a 2D experiment by repeating the data acquisition with a systematic variation of the evolution time t_1. The data acquired as a function of $t_2 - t_1$ are written into a matrix, the rows of which are labelled by t_1. The acquired alignment signal is modulated by the sine component of the magnetization, which exists during the evolution time t_1 between the first and the second pulse. The cosine part can be measured with the stimulated echo $90°_X - t_1 - 90°_X - t_m - 90°_X - t_2$, which creates true longitudinal magnetization during the mixing time t_m. For generation of purely absorptive 2D spectra, both components have to be

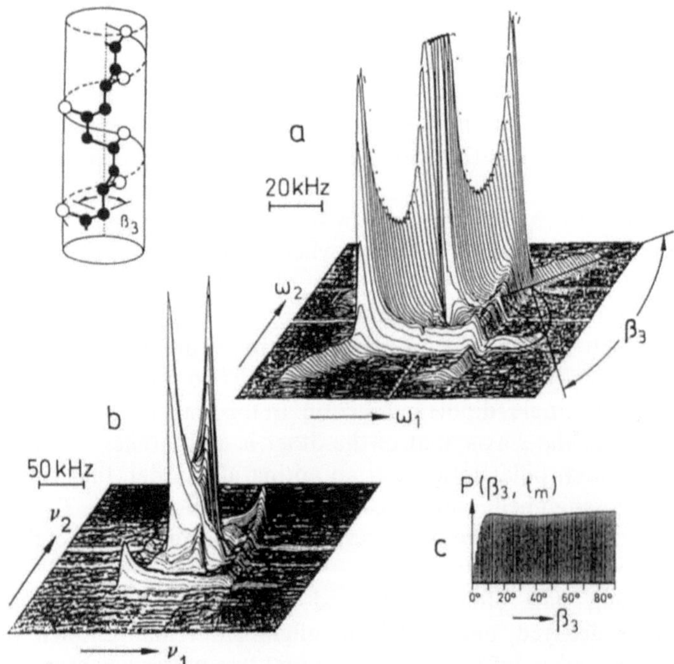

Fig. 31a–c. 2D-^{2}H exchange spectrum of polypropylene (PP). **a.** Methyl deuterated isotactic PP; $t_m = 150$ ms, $T = 380$ K. The helix molecules in the crystalline regions perform discrete reorientational jumps. The 33 principal axis of the axially symmetric quadrupole coupling tensor of the methylene groups is parallel to the axis of the C—C^{2}H$_3$ bond. The reorientation angle β_3 of this axis can be read directly from the 2D spectrum. **b.** Perdeuterated atactic PP; $t_m = 25$ ms, $T = 275$ K. The powder spectra of the deuterons at different chemical shifts overlap. The material is completely amorphous. The molecular segments undergo reorientation by isotropic rotational diffusion. **c.** Distribution of reorientation angles for atactic PP established within the mixing time t_m [45]

acquired, and the respective data matrices need to be combined during data processing [106]. 2D Fourier transformation along the echo decay axis $t_2 - t_1$ and the evolution time t_1 produces a 2D spectrum with frequency axes ν_1 and ν_2.

As an example, Fig. 31 depicts ^{2}H-2D spectra of polypropylene obtained in this way [107]. Signals are identified which spread over the 2D frequency plane. These signals characterize the reorientation of the molecules during the mixing time t_M. Because detailed balance requires that frequency components are exchanged by reorientation this type of 2D spectrum is referred to as a *2D exchange spectrum.*

Isotactic polypropylene is semicrystalline. In the crystalline regions the molecules form a 3_1 helix, that is, three monomer units form one turn of a helix. This helix performs rotational jumps around its own axis. After one jump, each monomer occupies the place of its next neighbor before the jump. Molecular reorientation by discrete jumps leads to characteristic ridges in the 2D plane, which assume the shape of an ellipse for axially symmetric coupling tensors. From the eccentricity of the ellipse the reorientation angle β_3 can be read off directly, by which the Z axis of the coupling tensor has been reorienting during the mixing time t_m (Fig. 31a). Diffusive small angle reorientation, on the other hand, leads to broad exchange signals in the 2D plane. Such motion is characteristic for amorphous polymers above the glass transition. This is illustrated for atactic polypropylene in Fig. 31. By numerical simulation of the 2D spectrum the distribution of reorientation angles (Fig. 31b) can be extracted from the measured data [99].

Statistical Theory

For simplicity it is assumed that the evolution time t_1 and the detection time t_2 are short compared to the mixing time t_m. Then it is a good approximation that the molecules reorient primarily during the mixing time. When also neglecting relaxation during the mixing time, the combined 2D time domain signal $s(t_1, t_m, t_2)$ measured with the stimulated and the Jeener-Broekaert echo can be written as ensemble average $\langle \cdots \rangle$ over all molecules in the sample,

$$\begin{aligned} s(t_1, t_m, t_2) &= \langle \exp\{i2\pi\nu_1 t_1\} \exp\{i2\pi\nu_2 t_2\} \rangle \\ &= \iint \exp\{i2\pi\nu_1(\Gamma_1)t_1\} \exp\{i2\pi\nu_2(\Gamma_2)t_2\} W(\Gamma_1, \Gamma_2; t_m) \mathrm{d}\Gamma_1 \mathrm{d}\Gamma_2. \end{aligned} \tag{92}$$

Here Γ_1 and Γ_2 denote the orientation angles (α, β) of the magnetic field $\boldsymbol{B}_0$ in the principal axes frame of the interaction tensor during the evolution time t_1 and the detection time t_2, respectively. The quantity which characterizes the reorientation process is the combined probability density

$$W(\Gamma_1, \Gamma_2; t_m) = W(\Gamma_1) P(\Gamma_1, \Gamma_2; t_m), \tag{93}$$

to find a frequency ν_1 in the interval $[\nu_1, \nu_1 + \mathrm{d}\nu_1]$ during the evolution time t_1 and a frequency ν_2 in the interval $[\nu_2, \nu_2 + \mathrm{d}\nu_2]$ during the detection time

t_2. It can be written as the product of the probability density $W(\Gamma_1)$ to find a molecule with orientation Γ_1 and the conditional probability density $P(\Gamma_1, \Gamma_2; t_m)$, that a molecule is found with orientation Γ_1 at the beginning of the mixing time t_m and with orientation Γ_2 at the end of the mixing time [22]. From the probability density $P(\Gamma_1, \Gamma_2; t_m)$ with initial and final orientations Γ_1 and Γ_2, the distribution $P(\Gamma_3; t_m)$ of reorientation angles Γ_3 can be calculated [99]. For axially symmetric coupling tensors this angle distribution is a function of the angle β_3 only, which specifies the reorientation of the angle between the 33 principal axis of the coupling tensor and the magnetic field. The angle distribution $P(\beta_3; t_m)$ is depicted in Fig. 31c for the rotational diffusion in atactic polypropylene corresponding to a mixing time of $t_m = 25$ ms and a temperature of $T = 275$ K. For a unique characterization of the motion the angle distribution must be determined for different mixing times and at different temperatures. The method is not restricted to ^{2}H NMR. It can also be applied to other nuclei like ^{31}P and ^{13}C [63, 108].

Computer Simulations

A detailed quantitative analysis of time scale and angle distribution requires numerical simulation of the spectra acquired with the solid and the alignment echoes. To this end, the time-domain signal is written as a vector. The components of which correspond to the different molecular orientations [22, 109],

$$s(t_1, t_m, t_2) = \mathbf{E}^t \exp\{(\Pi + i\Omega)t_2\} \exp\{\Pi t_m\} \exp\{(\Pi \pm i\Omega)t_1\} \mathbf{W}. \quad (94)$$

Here Π denotes the exchange matrix associated with the reorientation process, Ω is a diagonal matrix of the angle dependent resonance frequencies, $\mathbf{W}$ is a vector which determines the probability density to find a molecule at orientation Γ within the interval $[\Gamma, \Gamma + d\Gamma]$, and $\mathbf{E}^t$ is a row vector the elements of which are all equal to one.

Equation (94) is the starting point for numerical simulation of dynamic wide-line NMR spectra. With the Jeener-Broekaert echo the imaginary part of the exponential depending on t_1 is measured, with the stimulated echo the real part is measured. The line-shapes of solid-echo spectra follow from Eq. (94) with $t_m = 0$.

4. Imaging

NMR imaging is a noninvasive analytical technique, which is capable of producing images of arbitrarily oriented slices through optically nontransparent objects [18, 110–112]. The objects are placed in a strong magnetic field and investigated with radio-frequency (rf) waves. Biological tissue, plants, foodstuffs, and many synthetic nonconducting materials can be penetrated with rf waves, and the signal is attenuated only weakly by absorption and emission of rf energy at the resonance frequencies of the nuclear spins. In this sense the objects appear largely transparent to radio waves, while they are nontransparent to electromagnetic waves at optical frequencies, where the human eye is a sensitive detector, or exhibit different absorption properties at higher frequency irradiation, for instance, as provided by X-rays and electrons.

The most well known application of NMR imaging is in biomedicine where the method has become a valuable diagnostic tool complementing X-ray tomography [113–116], although the first reports of NMR imaging to medical [16] and materials [72] research were almost coincidental. Though the achievable spatial resolution was originally believed to be inferior to X-ray tomography, it was known at the start, that NMR provides different types of image contrast which can be fundamentally different from X-rays, [117, 118]. For instance, gray and white brain tissue can readily be discriminated by NMR imaging. For X-rays, the signal attenuation at the given frequency is the only source of contrast available. For this reason the use of contrast agents appears to be a necessity.

Philosophy

Only recently, nonclinical NMR imaging started to expand vigorously with steady progress in methodical developments and innovative applications. Major objections against the use of NMR imaging for solid materials were 1. the seeming lack of a need for a nondestructive method of analysis, and 2. The much larger line widths of solids compared to liquids, because the narrower the resonance, the better the achievable spatial resolution. The line width problem can be overcome by development of special imaging schemes which use line-narrowing techniques similar to those in solid-state NMR spectroscopy. The justification for relying on NMR imaging as an analytical tool in nonclinical research is its unsurpassed range of parameters for creating image contrast and localization of otherwise undetectable heterogeneities.

Contrast

The image contrast is determined not only by the density of the observed nucleus, but also by the numerous other parameters, which are measured in NMR spectroscopy to determine the molecular characteristics of condensed matter. These parameters include the relaxation times T_1 for energy dissipation and T_2 for dephasing of signal coherence, the self-diffusion constant, particle velocity components, the chemical shifts and indirect couplings of nuclear spins which are characteristic of molecular configuration and conformation, and the size and orientation dependence of different spin interactions which are effective in the solid state whenever the molecular motion is restricted to rates slower than the anisotropy Δ of the interaction.

By suitable NMR methods, the image contrast can be determined by the spin density weighted by various NMR parameters or it can be determined by just one parameter itself. In the first case one speaks of *parameter-weighted images*, in the second case of *parameter images*. Because the accessibility of these different types of contrast is the primary asset of NMR as an imaging method, an important goal for methodical developments of NMR imaging as a tool for materials analysis is the optimization and diversification of image contrast [119].

Equipment

The equipment typical for use in solid state imaging consists of a solid-state NMR spectrometer with a high-field magnet (e.g. 4 to 12 T, 80 mm bore), high-power rf amplifiers ($P = 1$ kW), one for ^{1}H and one for other nuclei, and an rf receiver. A computer controls the sequence of events during the experiment. Another one is used for data evaluation and display of the results. Particular imaging accessories are needed for generation and application of the magnetic field gradients necessary to achieve spatial resolution. They are control devices, linear audio amplifiers, and a set of gradient coils. For shaping of rf and gradient pulses special waveform memories are required.

Spatial Resolution

In NMR spectroscopy the applied magnetic field is made constant over the sample volume by shimming, a procedure following each change of sample in high-resolution NMR. For a homogeneous magnetic field the NMR frequencies are identical for each volume element of the sample, so that the signals from all parts of the sample add up in the spectrum, and the line width is determined by the sample and not by the homogeneity of the magnetic field.

To obtain spatial resolution, the applied magnetic field is deliberately made inhomogeneous by application of a field gradient. This is the spatial derivative of the field;

$$G_{\mathrm{kl}} = \partial B_{\mathrm{l}}/\partial x_{\mathrm{k}}. \tag{95}$$

A constant gradient denotes a linear variation of the field with space. By convention, the magnetic field is applied in the z direction of the laboratory frame. Then, in most cases, only three of the nine gradient tensor components are needed. They are concatenated to form the gradient vector $\mathbf{G} = (G_{\mathrm{xz}}, G_{\mathrm{yz}}, G_{\mathrm{zz}})^{\mathrm{t}}$. For ease of notation the second subscript z is often discarded.

In a gradient, the NMR frequency becomes a function of position $\mathbf{r} = (x, y, z)^{\mathrm{t}}$. in space,

$$\nu_{\mathrm{L}} = -\gamma|(1-\sigma)B_0 + \mathbf{Gr}|/2\pi. \tag{96}$$

Thus different locations in space can be discriminated by their NMR frequency as long as the local fields are not modified by some other interaction, for instance by the shielding. 2D-NMR imaging can be perceived as a particular form of 2D spectroscopy, where the frequency axes have been converted to space axes by application of magnetic field gradients.

In direct detection schemes, the achievable spatial resolution $1/\Delta x$ is determined by the width $\Delta \nu$ of resonance. Two points in space separated by Δx can only be discriminated as long as

$$|\Delta \nu| < |\gamma G_X \Delta_X|/2\pi. \tag{97}$$

In liquids the ^{1}H line widths are as narrow as 0.1 Hz. In solids, however, the line widths caused by the homonuclear dipole–dipole coupling among ^{1}H can be up to 100 kHz wide in rigid polymers, and are of the order of 100 Hz to 3 kHz for many elastomers, depending on the degree of motional narrowing. In ^{2}H NMR line widths of the order of 250 kHz are observed.

Strong Field Gradients

To increase the spatial resolution, either stronger magnetic field gradients need to be applied, the resonances need to be narrowed, or indirect detection schemes need to be employed. The first case is straight forward, but the number of spins contributing a signal within a given spectral region is reduced, so that the signal-to-noise ratio suffers accordingly, unless echoes are applied to recall the signal for coherent addition. Maximum gradients (of the order of 50 T/m) are obtained in the stray fields of superconducting magnets, and a method has been developed, by which morphological features of rigid solids can be revealed by mechanical reorientation of the sample in the stray field (STRAFI: stray field imaging) [120, 121].

Line Narrowing

The second approach to high spatial resolution in solids employs various methods of line narrowing, by which the signal decay is extended, so that smaller receiver bandwidths and higher Q values can be used. This is experimentally more demanding, but more successful in practice, because by employing line-narrowing techniques, the gradients can be weaker. Consequently, time-dependent gradients can be used, which provide greater freedom in tailoring the NMR method to the information requested. Line narrowing is achieved by creating a string of echoes [122, 123]. The Fourier transform of an individual echo dacay is an inhomogeneously broadened wide-line spectrum, that of the decay of the echo maxima provides a homogeneously broadened, narrow line. This is illustrated in Fig. 18 for a train of echoes obtained by magic-angle spinning of a deuterated solid sample [57, 124]. Various types of echo trains can be exploited to this end, leading to different imaging methods [125]. Examples are MAS imaging [126–128], multipulse imaging [70, 71, 73, 76, 126, 129–132], multisolid-echo imaging [125, 133, 134], rotary-echo imaging (magic-angle imaging) [135, 136], and magic-echo imaging [137–140]. For spectroscopic imaging, repetitive use of single echoes is sufficient [141–144].

NMR Microscopy

At present the isotropic spatial resolution of NMR imaging is limited to volumes larger than about $(5\,\mu m)^3$ in liquids and $(50\,\mu m)^3$ in rigid solids. Thus the name NMR microscopy appears misleading when comparing the spatial resolution achievable by NMR with that achievable by light microscopy, electron microscopy or even scanning tunnel or atomic force microscopy. Within the NMR community, however, it is well established and expresses the pride of resolving structures by NMR which are below the spatial resolution limit of the human eye [145–146].

Space Encoding

In the presence of a magnetic field gradient $\boldsymbol{G}$ the space information at position $\boldsymbol{r}$ is coded in the signal phase Φ of the transverse magnetization accumulated up to a time t. It is given by the time intergral over the resonance frequency Eq. (96),

$$\Phi(t) = \int_0^t 2\pi\nu_L(t)\mathrm{d}t = 2\pi\nu_0(1-\sigma)t - \gamma\boldsymbol{r}\int_0^t \boldsymbol{G}(t)\mathrm{d}t = 2\pi\nu_0(1-\sigma) + \boldsymbol{kr}. \tag{98}$$

This equation illustrates the fundamental fact, that the signal phase can be expressed in terms of the space vector $\boldsymbol{r}$ and its Fourier conjugate variable

$$\boldsymbol{k} = -\gamma\int_0^t \boldsymbol{G}(t)\mathrm{d}t. \tag{99}$$

As the phase of the precessing transverse magnetization changes with time under the influence of the magnetic field gradient, the magnitude of the $\boldsymbol{k}$ vector changes accordingly. The alignment of $\boldsymbol{k}$ is parallel to $\boldsymbol{G}$, and the sign of $\boldsymbol{k}$ depends on the time dependence of $\boldsymbol{G}$. Thus measurement of the NMR signal as a function of time t in the presence of the gradient $\boldsymbol{G}$ or as a function of $\boldsymbol{G}$ for fixed time intervals t provides the image information in $\boldsymbol{k}$ space or reciprocal space, and the actual image is retrieved by simple Fourier transformation of the $\boldsymbol{k}$-space signal [73 – 147]. NMR imaging methods therefore are designed in such a way, that the spacial information is acquired in the Fourier transform space.

The acquisition of spatial information in Fourier space is reminiscent of X-ray and neutron scattering, where the spatial information is obtained from inversion of the diffraction pattern [18, 128, 148, 149]. In fact the Patterson function is equivalent the Fourier transform of the magnitude square of the FID signal corresponding to the spatial auto-correlation function. For this reason scattering experiments provide information only on average displacements $\Delta \boldsymbol{r}$ but not on absolute space coordinates $\boldsymbol{r}$.

NMR imaging of solid materials is a rapidly expanding field of research. Various areas of application are being explored. Details on methods and applications are presented in the last chapter of this volume.

Acknowledgements. ARG is very grateful to A. P. Legrand and H. Zanni as well as to the director of the Ecole Superieure de Physique et Chimie Industrielles de la Ville de Paris, PG de Gennes, for the invitation as visiting professor giving the possibilility to finish the present paper. For critical reading and comments we thank D. Müller and for technical assistance I. Busch.

5 References

1a. Purcell EM, Torey HC, Pound ME (1946) Phys. Rev. 69: 37
1b. Bloch F, Hansen WW and Purcell ME (1946) Phys. Rev. 69: 127
2. Pake GE (1948) J. Chem. Phys. 16: 327
3. Dehmelt HG, Kruger H (1950) Naturwiss. 37: 111
4. Silver AH, Bray JB (1958) J. Chem. Phys. 29: 984
5. Proctor WG, Yu FC (1950) Phys. Rev 77: 717
6. Dickinson WC (1950) Phys. Rev. 77: 736
7. Gutowsky HS, McCall DW, Slichter CP (1953) J. Chem. Phys. 21: 279
8. Ebert I, Seifert G (1966) Kernresonanz im Festkörper. Leipzig, p 51
9. Zschunke A (1977) Kernmagnetishe Resonanzspektroskopie in der organischen Chemie, Berlin, p 70
10. Andrew ER (1961) Nuclear magnetic resonance. Oxford University Press, Oxford
11. Pines A, Gibby MG, Waugh JS (1972) J. Chem. Phys. 56: 1776, ibd. (1973) 59: 569
12. Waugh JS, Huber LM, Haeberlen U (1986) Phys. Rev. Lett. 20: 108
13. Ernst RR, Anderson WA (1966) Rev. Sci. Instr. 37: 93
14. Jeener J (1971) Lecture at the Ampere Summer School, Basko Polje
15. Ernst RR, Bodenhausen G, Wokaun A (1987) Principles of NMR in one and two dimensions, Clarendon, Oxford
16. Lauterbur PC (1973) Nature 242: 190
17. Mansfield P, Hahn EL (eds) (1990) NMR Imaging Royal Society. London
18. Callaghan PT (1991) Principles of nuclear magnetic resonance microscopy. Clarendon, Oxford
19. Fyfe CA (1983) Solid state NMR for Chemists Guelph
20. Haeberlen U (1976) High resolution NMR in solids: Selective averaging. Adv. Magn. Reson. Suppl. 1. Academic, New York
21. Gerstein BC, Dybowski C: R (1985) Transient techniques in NMR of solids. Academic, New York
22. Abragam A (1961) The principles of Nuclear Magnetism. Clarendon, Oxford
23. Engelhardt G, Michel D (1987) High Resolution Solid State NMR of Silicates and Zeolites. Wiley, New York
24. Harris RK (1987) NMR spectroscopy–a physicochemical view. Harlow
25. Mehring M (1980) Principles of high resolution NMR in solids, 2nd edn. Springer, Berlin Heidelberg, New York
26. Wieker W, Grimmer A-R (1966) Z. Naturf, 21b: 1103

27. Farrar FC, Harriman JE (1992) Density matrix theory and its application in spectroscopy. The Farragut Press Madison Vol 1
28. Fyfe CA, Gobbi GC, Kunowski J, Thomas JM, Ramdas S (1981) Nature 296: 350
29. Grimmer A-R, Cudinova AN (1972) Z. Chemie 12: 149
30. Jameson CI, de Dios AC (1993) Nuclear magnetics shieldings and molecular structure. ed. Tossell JA, NATO ASI series 386: 95
31. Slichter CP (1989) Principles of magnetic resonance. 3rd edn. Springer, Berlin, Heidelberg, New York
32. Spiess HW (1978) Rotation of molecules and nuclear spin relaxation in NMR-Basic principles and progress. 15: 55
33. Aleksandrov IV (1975) Teorija magnituoi relaksazija Moskva
34. Gobetto R, Harris RK, Apperley DC (1972) J. Magn. Reson. 96: 119
35. Duncan TM (1990) A compilation of chemical shift anisotropies. The Farragut Press, Chicago
36. Anet FAL, O'Leary DJ (1991) Concepts. Magn. Reson. 3: 193
37. Hansen AE, Bouman ThD (1993) Nuclear magnetic shieldings and molecul. structure. ed. Tossell JA , NATO ASI Series 386: 47
38. Jameson CI, de Dios AC (1993) private communication
39. Mason J, ed: Multinuclear NMR (1987) Plenum, New York
40. Herzfeld J, Berger AE (1980) J. Chem. Phys. 73: 6021
41. Robert JB, Wiesenfeld L (1982) Physics Report 86: 365
42. Grimmer A-R, unpublished results
43. Samoson A, Sun BQ (1992) Pines A in: Pulsed Magnetic Resonance: NMR, ESR, and Optics, p 80, Bagguley DMS, Ed., Clarendon, Oxford
44. Grimmer A-R, Lunk HJ, in preparation
45. Blümich B, Hagemeyer A, Schaefer D, Schmidt-Rohr K, Spiess HW (1990) Adv. Mat. 2: 72
46. Grimmer A-R (1978) Z. Chem. 18: 109
47. Power WP, Wasylishen RE (1991) Ann. Repts. NMR 23: 1
48. Heidemann D (1987) Ph.D. Thesis, Berlin
49. Andrew ER, Bersohn R (1950) J. Chem. Phys. 18: 159
50. Richards RE, Smith JAS (1952) Trans. Faraday 48: 675
51. Doremieux-Morin C (1976) J. Magn. Res. 21: 419
52. Doremieux-Morin C (1979) J. Magn. Res. 33: 505
53. Zeer EP, Zobov VE, Falaleev OV (1991) Novyi effekty v JaMR polikristallov, Navosibirsk
54. Grimmer A-R, Massiot D, unpublished results
55. Barfuss H (1984) XXII. Congress AMPERE, Zuerich 1984, abstracts 285
56. Müller D, unpublished results
57. Schadt RJ, Dong RY, Günther RY, Blümich B (1992) J. Magn. Reson. 96: 393
58. Llor A, Virlet J (1988) Chem Phys. Lett 152: 248
59. Wu Y, Sun BQ, Pines A, Samoson A, Lippmaa E (1990) J. Magn. Reson. 89: 297
60. Mueller KT, Sun BQ, Chingas GC, Zwanziger JW, Terao T, Pines A (1990) J. Magn. Reson. 86: 470
61. Stickney de Bourgas F, Waugh JS (1992) J. Magn. Res. 96: 280
62. Griffin RG, Aue WP, Haberkorn RA, Harbison GS, Herzfeld J, Menger EM, Munowitz MG, Olejniczak ET, Raleigh DP, Roberts JE, Ruben DJ, Schmidt A, Smith SO (1988) S. Vega in: Physics of NMR spectroscopy in biology and medicine, Proc. Int. School of Physics Enrico Fermi, Course C, Maraviglia B (Ed), North-Holland p 203
63. Hagemeyer A, Schmidt-Rohr K, Spiess HW (1989) Adv. Magn. Reson. 13: 85
64. Günther E, Blümich B, Spiess HW (1991) Chem. Phys. Lett. 154: 251
65. Abragam A, Goldman M (1982) Nuclear magnetism: Order and disorder, Clarendon, Oxford
66. Dixon WT (1982) J. Chem. Phys. 77: 1800; Titman JJ, Féaux de Lacroix S, Spiess HW (1993) J. Chem. Phys. 98: 3816
67. Caravatti P, Braunschweiler L, Ernst RR (1983) Chem. Phys. Lett. 100: 305; Caravatti P, Bodenhausen G, Ernst RR (1982) Chem. Phys. Lett. 89: 363
68. Schmidt-Rohr K, Clauss J, Blümich B, Spiess HW (1990) Magn. Reson. Chem 28: 3
69. Schmidt-Rohr K (1989) Hochauflösende NMR an Festkörpern und Untersuchung der Phasenstruktur fester Polymere, Diplomarbeit, Johannes-Gutenberg-Universität, Mainz
70. Chingas GC, Miller JB, Garroway AN (1986) J. Magn. Reson. 66: 530
71. Cory DG, Miller JB, Turner R, Garroway AN (1990) Molec. Phys. 70: 331
72. Mansfield P, Grannell PK (1973) J. Phys. C 6: L422
73. Mansfield P, Grannell PK (1975) Phys. Rev. B12: 3618

74. Cory DG (1990) Polymer Preprints 31: 149
75. Rhim WK, Ellman DD, Vaughan RW (1973) J. Chem. Phys. 58: 1772
76. Cory DG, Miller JB, Garroway AN (1991) J. Magn. Reson. 90: 205
77. Burum DP, Rhim WK (1979) J. Chem. Phys. 71: 944
78. Mansfield P (1971) J. Phys. C 4: 1444
79. Prigl R (1990): Hochauflösende Kernresonanz in Festkörpern: Prinzipielle und Praktische Grenzen der Multipuls-Technik, Diplomarbeit, Max-Planck-Institut für Medizinische Forschung, Heidelberg
80. Hartmann SR, Hahn EL (1962) Phys. Rev. 128: 2042
81. Burum DP (1990) Concepts Magn. Reson. 2: 213
82. Scheler G, Haubenreißer U, Rosenberger H (1981) J. Magn. Reson. 44: 134
83. Bronnimann CE, Hawkins BL, Zhang M, Maciel GE (1988) Anal. Chem. 60: 1743
84. Gerstein BC, Chou C, Pembleton RG, Wilson RC (1977) J. Phys. Chem. 81: 565
85. Harbison GS, Vogt V-D, Spiess HW (1987) J. Chem. Phys. 86: 1206
86. Grimmer A-R, Rosenberger H (1978) Z. Chem. 18: 378
87. Ratcliffe CI, Ripmeester JA, Tse JS (1985) Chem. Phys. Lett. 120: 427
88. Jelinski LW (1985) Adv. Mater. Science 15: 359
89. Müller K, Wassmer K-H, Kothe G (1990) Adv. Polym. Science 95: 1
90. Müller K, Meier P, Kothe G (1985) Progr. NMR Spectrosc. 17: 211
91. Spiess HW (1985) Adv. Polym. Science 66: 23
92. Spiess HW (1982) in: Developments in Oriented Polymers – 1, Ward IM (Ed.), Applied Science Publ., Barking
93. Grimmer A-R, Müller D, Neels J (1983) Z. Chem. 23: 140
94. Blümich B, Spiess HW (1988) Angew. Chem. Int. Ed. Engl. 27: 1655
95. Voelkel R (1988) Angew. Chem. Int. Ed. Eng. 27: 1468
96. Stejskal EO, Schaefer J, McKay RA (1977) J. Magn. Reson. 25: 569
97. Hentschel R, Schlitter J, Sillescu H, Spiess HW (1978) J. Chem. Phys. 68: 56
98. Ward IM (1975) Structure and properties of oriented polymers. Applied Science, London
99. Wefing S, Spiess HW (1988) J. Chem. Phys. 89: 1219
100. Wiesner U, Schmidt-Rohr K, Boeffel C, Pawelzik U, Spiess HW (1990) Adv. Materials 2: 484
101. Tang P, Santos RA, Harbison GS (1989) Adv. Magn. Reson. 13: 225
102. Callaghan PT (1984) Aust. J. Phys. 37: 359
103. Kärger J, Pfeifer H, Heink W (1988) Adv. Magn. Reson. 12: 1
104. Stilbs P (1987) Progr. NMR Spectroscopy 19: 1
105. Spiess HW (1980) J. Chem. Phys. 72: 6755
106. Schmidt C, Blümich B, Spiess HW (1988) J. Magn. Reson. 79: 269
107. Schaefer D, Spiess HW, Suter UW, Fleming WW (1990) Macromolecules 23: 3431
108. Blümich B, Hagemeyer A (1989) Chem. Phys. Lett. 161: 55
109. Anderson PW, Weiss PR (1954) J. Phys. Soc. Japan 9: 316
110. Blümich B, Kuhn W (1992) (eds.): Magnetic Resonance Microscopy. Verlag Chemie, Weinheim
111. Ackerman JL, Ellingson WA (1990) (eds.): Advanced Tomographic Imaging Methods for the Analysis of Materials. Materials Research Society, Symposium Proceedings Vol 217
112. Krestel E (1990): Imaging systems for medical diagnosis, Siemens AG, Berlin
113. Mansfield P, Morris PG (1982): NMR imaging in biomedicine. Adv. Magn. Reson. Suppl. 2, Academic, New York
114. Morris PG (1986): Nuclear magnetic resonance imaging in medicine and biology, Clarendon, Oxford
115. Wehrli FW, Schaw D, Kneeland JB (1988) ed., Biomedical Magnetic Resoanace Imaging. VCH Publishers, Weinheim
116. Hausser KH, Kalbitzer HR (1988) NMR in Medicine and Biology. Springer, Berlin, Heildeberg, New York
117. Damadian R (1971) Science 171: 1151
118. Damadian R (1981) ed., NMR–Basic Principles and Progress 19: NMR in Medicine, Springer, Berlin
119. Blümler P and Blümich B (1992) Mag. Reson. Imaging 10: 779
120. Samoilenko AA, Artemov DYu and Sibeldina LA (1988) JETP Lett 47: 348
121. Samoilenko AA, Zick K (1990) Bruker Report 1: 40
122. Carr HY, Purcell EM (1954) Phys. Rev. 94: 630
123. Meiboom S, Gill D (1958) Rev. Sci. Instrum. 29: 688
124. Blümich B, Blümler P, Jansen J (1992) Solid State Nuc. Magn. Reson. 1: 111

125. Strange JH (1990) Phil. Trans. R. Soc. Lond. A 333: 427
126. Cory DG, de Boer JC, Veeman WS (1989) Macromolecules, 22: 1618
127. Veeman WS, Cory DG (1989) Adv. Magn. Reson. 13: 43
128. Schauss G, Blümich B, Spiess HW (1991) J. Magn. Reson. 95: 437
129. Cory DG (1992) Solid State NMR Imaging. Annual reports of NMR 24: 87
130. Miller JB, Cory DG, Garroway AN (1989) Chem. Phys. Lett. 164: 1
131. Miller JB, Garroway AN (1989) J. Magn. Reson. 82: 529
132. McDonald PJ, Lonergan AR (1992) Physica B 176: 173
133. Cottrell SP, Halse MR, Strange JH (1990) Meas. Sci. Technol. 1: 624
134. McDonald PJ, Tokarcuk PF (1989) J. Phys. E: Sci. Instrum. 22: 948
135. de Luca F, de Simone BC, Lugeri N, Maraviglia B, Nuticelli C (1990) J. Magn. Reson. 90: 124
136. de Luca F, de Simone BC, Lugeri N, Maraviglia B (1992) Solid State Commun. 82: 151
137. Matsui S (1991) Chem. Phys. Lett. 179: 187
138. Takegoshi K, McDowell CA (1985) Chem. Phys. Lett. 116: 100
139. Hafner S, Demco DE, Kimmich R (1991) Meas. Sci. Technol. 2: 882
140. Weigand F, Blümler B, Blümich B, Spiess HW, to be published
141. Rommel E, Hafner S, Kimmich R (1990) J. Magn. Reson. 86: 264
142. Blümich B, Blümler P, Günther E, Schauss G, Spiess HW (1991) Makromol. Chem. Macromol. Symp. 44: 37
143. Blümich B, Blümler P, Günther E, Schauß G (1990) Bruker Report 2: 22
144. Demco DE, Hafner S, Kimmich R (1991) J. Magn. Reson. 94: 333
145. Kuhn W (1990) Angew. Chem. Int. Ed. Engl. 29: 1
146. Eccles CD, Callaghan PT (1986) J. Magn. Reson. 68: 393
147. Kumar A, Welti D, Ernst RR (1975) J. Magn. Reson. 18: 69
148. Chingas GC, Frydman L, Barrall, GA, Harwood JS in BLÜ1, p 374
149. Fleischer D, Fujara F (1993) NMR 30: xxx
150. Blümich B, Hagemeyer A, Schmidt-Rohr K, Spiess HW (1989) Ber. Bunsenges. Phys. Chem. 93: 1189

High-Resolution ^{13}C NMR Investigations of Local Dynamics in Bulk Polymers at Temperatures Below and Above the Glass Transition Temperature

Françoise Laupretre

Laboratoire de Physico-Chimie Structurale et Macromoléculaire associé au C.N.R.S., E.S.P.C.I., 10 rue Vauquelin, 75231 Paris cedex 05, France

Table of Contents

NMR Basic Principles and Progress, Vol. 30

In this chapter, we will first summarize the technical requirements for obtaining high-resolution solid-state ^{13}C NMR spectra of bulk polymers. We will then review the bases of the determination and analysis, in terms of local motions, of the spectrum line shape, proton–carbon dipolar interactions, chemical-shift anisotropies, relaxation times and line widths. A large part of the chapter will be devoted to the description of studies that demonstrate the capability of NMR to investigate the local dynamics of bulk polymers at temperatures above and below the glass transition temperature, respectively. At temperatures below the glass transition temperature, the example of poly(cyclohexyl methacrylate) will illustrate the importance of line width measurements, whereas the chemical-shift parameters and ^{13}C–^{1}H dipolar interactions will be the relevant parameters for the study of aromatic copolyesters and epoxy resins. In the case of bulk polymers at temperatures well above the glass transition temperature, we will first examine the different expressions that have been proposed for the orientation autocorrelation function associated with the polymer local dynamics. Tests of these expressions will be carried out through the determination of the ^{13}C spin–lattice relaxation times of a number of elastomers. Results thus obtained permit us to specify the main factors that control the local dynamics in bulk polymers at temperatures well above the glass-transition temperature. Then, NMR studies of compatible blends of polyvinylmethyl ether and polystyrene will show the possibility of investigating the relative influence of intramolecular interactions and intermolecular constraints on the local motions of a given blend component.

1 Introduction

In bulk polymer systems, the variation of the viscoelastic properties as a function of temperature points to the existence of several transitions. At high temperature, the glass transition phenomena arise from cooperative motions of the main chain, whereas, at lower temperatures, the secondary relaxations are known to be due to localized motions of small portions of the main chain or side-groups. Thus there appears to be a close relationship between some macroscopic mechanical data such as viscoelastic properties and molecular motions. The knowledge of the exact nature of the local motions that are involved in the various transitions of a polymer is therefore of major interest for a better understanding of its viscoelastic behavior.

As shown by the numerous studies described in the literature, high-resolution ^{13}C NMR has proven to be a most powerful tool for investigating local dynamics in polymers. Unlike fluorescence anisotropy or electron spin resonance techniques, it does not require any labeling of the molecule under study, and yields direct information on the compound under study. As a selective technique, it allows the observation of one signal per magnetically inequivalent carbon, and therefore the dynamic behavior of each part of a molecule can be followed independently. Moreover, many NMR parameters are sensitive to molecular motions. They include the different relaxation times as well as the spectrum line shape, the strength of the dipolar coupling and the chemical-shift anisotropy. The available spectral windows depend on the type of measurement which is performed. They range from about 10^{-1} Hz for slow processes to several hundreds of MHz for very fast modes. In the case of bulk polymers at temperatures well above the glass transition temperature, the fast processes of the local dynamics can be investigated by determining the spin–lattice relaxation time, $T_1(^{13}C)$, and the nuclear Overhauser enhancement, that probe modes in the LARMOR frequency region, whereas measurements probing slower motions are more appropriate for glassy state investigations.

In this chapter, we will first summarize the technical requirements for obtaining high-resolution solid-state ^{13}C NMR spectra of bulk polymers. We will then consider the sensitivity of the different NMR parameters to molecular motions. The last part of this chapter will be devoted to the description of several studies that clearly illustrate the capability of NMR to investigate the local dynamics of bulk polymers at temperatures above and below the glass transition temperature, T_g, respectively.

2 High-Resolution ^{13}C NMR Techniques

Whereas, modern NMR techniques permit high-resolution ^{13}C NMR spectra in bulk polymers to be obtained at temperatures either above or below their glass transition temperature, T_g, the strengths of the interactions that govern the spectral parameters are very different in both cases. At temperatures well above the glass transition temperature, the local motions of bulk polymers are fast cooperative processes of relatively large amplitude. As a consequence, the main tensorial interactions—chemical-shift anisotropy, homonuclear and heteronuclear dipolar couplings, and quadrupolar couplings in the case of spins higher than 1/2—are averaged to a large extent by fast local motions. For *cis*-1,4-polybutadiene at room temperature, for example, Cohen-Addad et al. [1] and English et al. [2, 3] have shown that most, i.e. 99% or more, of the dipolar interaction is averaged by rapid motions. Therefore, for bulk polymers at temperatures well above T_g, high-resolution ^{13}C NMR spectra can be obtained by using the conventional spectrometers for solution investigations. From an NMR point of view, bulk polymers in this temperature range and polymers in solution share a number of common features.

In contrast, motions that may occur in a glassy polymer are much slower than modes observed in the melt. Besides, they are very localized and involve only side groups or short sequences of the main chain. Therefore, the tensorial interactions listed above are only partly averaged, or even not averaged at all when no local modes exist. Therefore, in order to obtain high-resolution ^{13}C NMR spectra, one has to use the specific line-narrowing techniques that are described in Chapter 1. These techniques, that are based on proton dipolar decoupling (DD) and magic-angle sample spinning (MAS), perform an efficient averaging of the carbon–proton dipolar interactions and chemical-shift anisotropy and allow the recovery of high resolution. Moreover, in order to improve the sensitivity of the high-resolution solid-state ^{13}C NMR technique, most of the experiments discussed below will be performed by using the classical cross-polarization (CP) pulse sequence, discussed in Chapter 1. Basically, CP is achieved by spin-locking the protons with an rf field, H_{1H}, parallel to the proton rotating frame magnetization, while applying a second rf field, H_{1C}, to the carbons at the carbon resonance frequency under the Hartmann-Hahn condition: $\gamma_H H_{1H} = \gamma_C H_{1C}$, or $\omega_{1H} = \omega_{1C}$, in angular frequency units.

3 Dependence of Spectral Parameters on Local Dynamics

As already mentioned in the Introduction, the local dynamics can be studied by a number of NMR techniques. The spectrum line shape is determined by the number of inequivalent carbons in the molecule considered. When carbons are rendered equivalent by local motions, the line shape is strongly modified. Besides, the motional averaging effect on the internuclear distances and orientation of internuclear vectors with respect to the external magnetic field direction induces a decrease of the strength of the carbon–proton dipolar interactions. In the same way, the reorientation of the different carbons in the magnetic field is accompanied by a partial averaging of the chemical shift tensors. Therefore, selective measurement of these interactions which dominate the solid-state NMR behavior is of major interest for the investigation of local dynamics at temperatures below T_g. Finally, the effects of molecular motions on the carbon–proton dipolar interactions and chemical-shift anisotropies are also reflected by the values of the relaxation times. The main relaxation times which are of interest in high-resolution ^{13}C NMR are the spin–lattice relaxation time $T_1(^{13}C)$, the spin–lattice relaxation time in the rotating frame $T_{1\rho}(^{13}C)$ and the spin–spin relaxation time T_2. However, it must be noted that magnetic relaxation may be governed not only by dynamic phenomena but there may also be a contribution from static phenomena such as spin diffusion. Spin diffusion is the mutual exchange of spin state, or "flip-flops", between strongly dipolar-coupled nuclei which have the same precession frequency but anti-parallel spins. In the liquid state, the dipolar couplings are motionally averaged and the contribution of the spin-diffusion mechanism to the relaxation is negligible. In contrast, in the solid state at temperatures below T_g, nuclei are strongly coupled. Therefore, in this case, magnetic relaxation times cannot be interpreted in terms of local dynamics only, but the two contributions have to be separated.

In the following, we will review the bases of the determination and analysis of the above NMR parameters—spectrum line shape, proton–carbon dipolar interactions, chemical-shift anisotropies, relaxation times and line widths—in terms of local motions.

3.1 Spectrum Line Shape

In the same way as observed in solution NMR, chemical exchange can modify the line shape of high-resolution solid-state ^{13}C NMR spectra. Carbons which are magnetically inequivalent in the absence of motion and yield distinct peaks on the NMR spectrum can be rendered equivalent by specific motions, leading to a single NMR line. Between these two extreme situations of the slow exchange and of the rapid exchange, the spectrum line shape is strongly dependent on the rate of the motion in the range of 10^{-1}–10^6 Hz.

3.2 Chemical-Shift Anisotropy

The chemical shift of a nuclear spin is a tensorial quantity. Its value depends on the orientation of the electronic distribution about the nucleus with respect to the external magnetic field. In a liquid, due to the rapid molecular motions, this interaction is averaged to zero and the observed chemical shift is the trace of the tensor. In contrast, in a powder, in the absence of motions, all the orientations have the same probability and the signal obtained for each carbon is the sum of the elementary chemical shifts corresponding to the different orientations. When local motions occur in the bulk below T_g, they induce a partial averaging of the chemical-shift anisotropy.

As described in Chapter 1, when a solid sample is spun at an angular frequency, ω_r, about an axis making an angle, ψ, with the external field H_0, equal to the magic-angle, $\psi = 54.7°$, the resulting ^{13}C NMR spectrum depends on the extent of chemical-shift anisotropy modulation by magic-angle sample spinning. If the spinning speed $\omega_r/2\pi$ is large relative to the chemical-shift anisotropy $\Delta\sigma$ expressed in hertz, the spectrum consists of one peak per magnetically inequivalent carbon, the location of which is given by the trace of the shielding tensor. Conversely, if the spinning speed is smaller than the chemical-shift anisotropy, there appear, together with the main resonance, spinning sidebands at a distance $\omega_r/2\pi$ from each other. The envelope of the spinning sidebands is related to the shape of the chemical shift tensor [4].

The chemical shift tensor parameters can be obtained from the proton-decoupled ^{13}C NMR spectra recorded without magic-angle sample spinning. As the selectivity of this technique is poor, which may justify the use of ^{13}C-enriched samples, other methods have been proposed, based on either spinning out of the magic angle [5, 6] or slow magic-angle spinning [5, 7, 8]. In the latter case, by referring to the theoretical calculations performed by Herzfeld and Berger [4], the principal elements of the shielding tensor can be determined from the relative intensities of the spinning sidebands of different orders. Two-dimensional (2D) NMR techniques, that place the anisotropic information into a second-frequency dimension, while maintaining the isotropic spectrum in the first can also be used [9–19]. They can be classified into different groups: the ones that modulate interactions by reorienting the sample, and those that manipulate the interactions using transformations in spin as well as in ordinary space. In the first group of experiments, isotropic and anisotropic interactions become correlated by imposing a sudden change in the physical status of the sample, involving either a change in the spinning axis [9, 10], in the spinning speed of the sample [11] or in the orientation of a static sample with respect to the magnetic field [12]. In the second class of experiments, the chemical shift anisotropy is recovered by means of a multiple-pulse sequence synchronized with the spinning of the rotor [13–16]. Finally, a last group of experiments involves scaling of the shielding anisotropies by rf pulses together with restricting the sampling to the peaks of rotor echoes during the evolution period [17–19]. This approach yields high-resolution, isotropic spectra in the first dimension, while the anisotropies remain in the slow spinning sideband pattern in the second dimension.

A typical example of a partial motional averaging of the chemical shift anisotropy is that of an unprotonated aromatic carbon belonging to a *para*-substituted phenyl ring rotating about its local symmetry axis. For such a carbon, the principal elements of the tensor are parallel to the axes of the frame defined by the C_1–C_4 axis and the perpendicular to the plane of the phenyl ring. In the rotation about the *para* axis, the principal element parallel to this axis is not modified by the motion. On the opposite, the two components which are perpendicular to this axis are averaged by the motion. The resulting line shape corresponds to an axially symmetrical tensor and is markedly different from the rigid-lattice pattern. Such behavior has been observed in thermotropic polymers having a mesogen unit inside the main chain [20]. Below the crystal-smectic *C* transition temperature, the observed chemical-shift anisotropy (CSA) is characteristic of a rigid lattice. Above the crystal-smectic *C* transition temperature, it has the typical line shape of the symmetrical tensor described above, indicating the internal rotation of the aromatic rings about their local symmetry axis.

3.3 ^{13}C–^{1}H Dipolar Interaction

For a powder the expression of the ^{13}C–^{1}H dipolar interaction is given by:

$$\langle b^2 \rangle = \tfrac{4}{5}(\gamma_c \gamma_H \hbar^2 / r_{CH}^3)^2 \quad (1)$$

where r_{CH} is the carbon–proton distance.

The reduction in the strength of the dipolar interaction by molecular motions of frequencies comparable to or greater than the dipolar interaction itself is a measure of the amplitude of the motion. Essentially two techniques have been proposed for this purpose [21, 22]. The first method consists in measuring the intensity of the dipolar sideband patterns, obtained from dipolar rotational spin-echo ^{13}C NMR and arising from the heteronuclear dipolar interaction, while the homonuclear ^{1}H–^{1}H dipolar interactions are suppressed by multiple-pulse ^{1}H–^{1}H (WAHUHA) decoupling.

The strength of the ^{13}C–^{1}H dipolar coupling, $\langle b^2 \rangle$, can also be deduced from the rises of ^{13}C magnetization in cross-polarization experiments [21]. When ^{1}H–^{1}H homonuclear dipolar interactions are much stronger than ^{13}C–^{1}H heteronuclear dipolar interactions, the increase of magnetization as a function of the contact time is mainly exponential and its rate, $T_{CH}(SL)^{-1}$, can be calculated exactly [23]. In contrast, when carbons are strongly coupled to protons, the rises of polarization can no longer be described by an exponential law. At the very beginning of the contact, there occurs a coherent energy transfer between the carbon of interest and the strongly coupled protons. This oscillatory transfer is damped by the coupling with the more remote protons. For very short contact times, the carbon and its *n* strongly coupled protons can be considered as an isolated CH_n system. At the end of a very short contact, *t*, the magnetization $M(t)$ can be described by an oscillatory function whose expression is deduced from results obtained for a C–H group [24] and extended to CH_2 groups by using the calculation for the magnetization in liquid AX_n systems [25]:

$$M(t)/M(\infty) = \sin^2 \sqrt{n\langle b^2 \rangle}\,\frac{t}{4} \quad (2)$$

Therefore $\langle b^2 \rangle$ can be estimated in a very simple way by measuring the contact time $t_{1/2}$ necessary to obtain half of the maximum polarization $M(\infty)$:

$$\sqrt{\langle b^2 \rangle} = \frac{\pi}{\sqrt{n}t_{1/2}} \tag{3}$$

For $r_{CH} = 1.09$ Å, $t_{1/2} = 28\,\mu s$ for a rigid CH group and $t_{1/2} = 20\,\mu s$ for a rigid CH_2 group. Experimental $t_{1/2}$ values longer than these rigid-lattice values are evidence for a reduction of the ^{13}C–^{1}H dipolar coupling by motional processes whose frequencies are higher than 10^5 Hz [21].

3.4 Relaxation Times and Line Widths

3.4.1 Spin–Lattice Relaxation Time in the Rotating Frame, $T_{1\rho}(^{13}C)$

Due to the low gyromagnetic ratio and natural abundance of the ^{13}C spins, the spin diffusion between carbon-13 nuclei is slow and usually negligible in polymers as compared with the dynamic contribution to spin–lattice relaxation. Therefore $T_{1\rho}(^{13}C)$ is usually not determined by ^{13}C–^{13}C spin diffusion. However, as a consequence of the pulse sequence used to measure $T_{1\rho}(^{13}C)$, there may exist among the mechanisms involved in the relaxation a contribution from the spin diffusion between carbon-13 and proton spins [26–30].

One of the cross-polarization pulse sequences used to measure $T_{1\rho}(^{13}C)$ is shown in Fig. 1. During the evolution time, Δt, of this pulse sequence, the ^{13}C magnetization is spin-locked along H_{1C}. Any reorientation of the ^{13}C–^{1}H internuclear vectors induces fluctuations of the dipolar local fields, and the ω_{1C} component of these fluctuating fields participates in the relaxation of the ^{13}C spins. This is a "spin–lattice" relaxation mechanism which is related to molecular motions and is characterized by the relaxation time [26–30]:

$$(T_{1\rho}(^{13}C))^{-1}_{\text{spin-lattice}} = \langle \Delta M^2_{CH} \rangle_m J(\omega_{1C}) \tag{4}$$

where $\langle \Delta M^2_{CH} \rangle_m$ is the part of the ^{13}C–^{1}H second moment which is averaged by the motions and $J(\omega_{1C})$ is the spectral density associated with the fluctuating fields at the frequency ω_{1C}: $J(\omega) = \frac{1}{2} \int_{-\infty}^{+\infty} G(t) e^{i\omega t} dt$, where $G(t)$ is the normalized second-order spherical harmonic autocorrelation function.

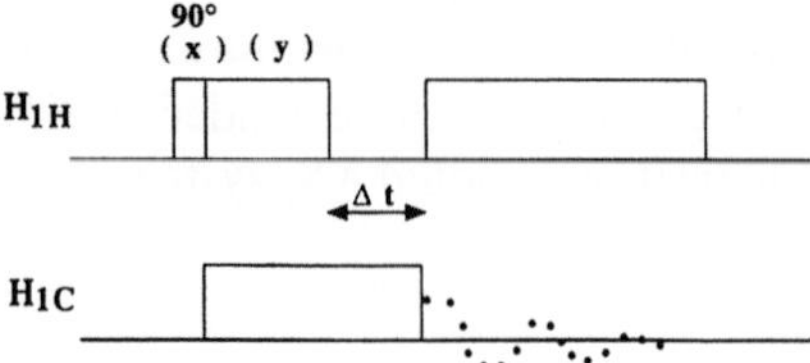

Fig. 1. Pulse sequence used for $T_{1\rho}(^{13}C)$ determination. After the contact, the ^{13}C magnetization evolves in the presence of the rf field, H_{1C}, applied to the carbon spins during a time Δt. The decrease of the magnetization as a function of Δt is characterized by the $T_{1\rho}(^{13}C)$ relaxation time (after [29])

During the same Δt period, there is no rf field applied to the protons. Therefore the protons are in their dipolar local field. Their magnetization decreases with a relaxation time T_{1D} which is strongly dependent on the speed of the rotor. For rapid magic-angle spinning, the relaxation time T_{1D} is very short ($<100\,\mu$s) and the proton magnetization decreases very quickly. Under these conditions, a magnetization transfer occurs characterized by the time T_{CH}(ADRF) [26–30], between the ^{13}C spins, which have a noticeable magnetization, and the proton nuclei which have lost their magnetization. The complete expression for $T_{1\rho}(^{13}\mathrm{C})$, as measured by the pulse sequence represented in Fig. 1 is:

$$(T_{1\rho}(^{13}\mathrm{C}))^{-1} = (T_{1\rho}(^{13}\mathrm{C}))^{-1}_{\text{spin-lattice}} + T_{CH}(\mathrm{ADRF})^{-1} \qquad (5)$$

where T_{CH}(ADRF) essentially represents the dipolar fluctuations and not the local dynamics. It can be determined by a nine-step procedure [31] including a measurement of the proton local dipolar fields [32].

In rigid materials such as crystalline polymers where there are nearly no motions:

$$T_{1\rho}(^{13}\mathrm{C}) = T_{CH}(\mathrm{ADRF}) \qquad (6)$$

In contrast, when the spectral density at ω_{1C} is important, the $T_{1p}(^{13}\mathrm{C})$ relaxation time essentially reflects the local dynamics in the ω_{1C} frequency range.

3.4.2 Line Width and T_{2m} Spin–Spin Relaxation Time

Line-broadening mechanisms in glassy polymers have already been reviewed [33]. Some of them, the static ones, i.e. bulk susceptibility of the sample, chemical-shift dispersions due to packing effects, bond distortions, and conformational inequivalence, induce only a relatively small effect, $(\pi T_{2\text{res}})^{-1}$, of the order of 2 to 6 ppm. More important is the line broadening arising from relaxation mechanisms such as motional modulation of the chemical-shift anisotropy [34] and motional modulation of the dipolar carbon–proton coupling [35].

For carbons having a high chemical-shift anisotropy, the motional modulation of the chemical-shift anisotropy induces a line broadening, $(\pi T_{2\sigma})^{-1}$, that is maximum when the rate of molecular motion is equal to the sample spinning speed.

In the case of carbons having a strong dipolar carbon–proton coupling under suitable conditions of magic-angle setting and proton decoupling irradiation, the main cause of motional line broadening is the modulation of the dipolar ^{13}C–^{1}H coupling. This mechanism gives maximum line broadening when the rate of molecular motion is equal to the proton decoupling rf field strength ω_{1H} expressed in angular frequency units. This motional line broadening $(\pi T_{2m})^{-1}$ may be much larger than that due to the previous static effects. Under the conditions that the proton irradiation is applied exactly on resonance and that the sample spinning rate is much smaller than the proton decoupling field strength, the transverse relaxation time resulting from this mechanism and contributing an amount of $(\pi T_{2m})^{-1}$ to the line width is [35]:

$$(T_{2m})^{-1} = \langle \Delta M^2_{CH} \rangle_m J(\omega_{1H}) \tag{7}$$

Under the Hartmann-Hahn matching condition for cross-polarized spectra: $\omega_{1H} = \omega_{1C}$, so that T_{2m} is equal to $T_{1\rho}(^{13}C)_{spin-lattice}$:

$$T_{2m} = T_{1\rho}(^{13}C)_{spin-lattice} \tag{8}$$

Therefore the dynamic contributions to the relaxation times $T_{1\rho}(^{13}C)$ and T_{2m}, which are due to the motional modulation of the dipolar coupling, are equal. For a carbon strongly coupled with a proton and placed in a not too intense magnetic field so that the motional broadening may be not negligible as compared to the static line width, the range of motional frequencies which are accessible from the two relaxation parameters are quite similar (10^4–10^7 Hz). A motion in this frequency range, which is able to dominate the spin-diffusion magnetization transfer, will also induce a broadening of the lines corresponding to the carbons that are involved in this process.

Finally, in the general case where both mechanisms are effective, the observed full line width at half height $(\pi T_2)^{-1}$ may be written as:

$$\frac{1}{\pi T_2} = \frac{1}{\pi T_{2m}} + \frac{1}{\pi T_{2\sigma}} + \frac{1}{\pi T_{2res}} \tag{9}$$

3.4.3 *Spin–Lattice Relaxation Time*

As for $T_{1\rho}(^{13}C)$, $T_1(^{13}C)$ is usually not determined by ^{13}C–^{13}C spin-diffusion, even in the solid state.

With the assumption of a purely ^{13}C–^{1}H dipolar relaxation mechanism, the spin–lattice relaxation time $T_1(^{13}C)$ under ^1H decoupling conditions, and nuclear Overhauser enhancement (NOE) obtained from a ^{13}C experiment are given by the well-known expression [36]:

$$(nT_1(^{13}C))^{-1} = (\hbar^2\gamma_C^2\gamma_H^2/10{r_{CH}}^6)(J(\omega_H - \omega_C) + 3J(\omega_C) + 6J(\omega_H + \omega_C)) \tag{10}$$

$$\mathrm{NOE} = \frac{\gamma_H}{\gamma_C} \frac{6J(\omega_H + \omega_C) - J(\omega_H - \omega_C)}{J(\omega_H - \omega_C) + 3J(\omega_C) + 6J(\omega_H + \omega_C)} \tag{11}$$

where n is the number of protons directly bound to the considered carbon, ω_H and ω_C are the ^1H and ^{13}C resonance frequencies and r_{CH} is the internuclear distance.

3.5 *Conclusions*

As can be seen from the expressions given in the above section, $T_1(^{13}C)$ and NOE probe motions in the Larmor frequency range. $T_1(^{13}C)$ and NOE determinations will therefore be of major interest for the investigation of local dynamics in bulk polymers in their very mobile state at temperatures well above the glass transition temperature. In contrast, to study the lower frequency motions that may occur below T_g, line shape analysis and measurements of the tensorial interactions, line widths and $T_{1\rho}(^{13}C)$ relaxation times will be more appropriate.

4. Experimental Studies of Local Dynamics in Bulk Polymers at Temperatures Below the Glass Transition Temperature

The experimental results reported in this section will show how the different NMR parameters can be used to investigate the local dynamics of bulk polymers at temperatures below the glass-transition temperature. The example of poly(cyclohexyl methacrylate) will illustrate the attractiveness of T_{2m} measurements, whereas the chemical shift parameters and $^{13}C-^{1}H$ dipolar interactions will be the relevant parameters for the study of aromatic copolyesters and epoxy resins.

4.1 Local Motions Associated with the γ-Relaxation of Poly(cyclohexyl methacrylate)

The secondary relaxations of polymethacrylates have been extensively studied by Heijboer by using mechanical techniques [37]. Figure 2 shows the shear modulus, G', and the damping, tan δ, of poly(cyclohexyl methacrylate) as a function of temperature at different frequencies. The sharp decrease of G' at high temperature is the manifestation of the glass transition phenomena. The smaller decrease that occurs in the region of $-80\,°C$ at 1Hz, together with the corresponding peak in the tan δ variation vs temperature, are assigned to a secondary relaxation usually called γ-relaxation in the poly(cyclohexyl methacrylate) series. They indicate the existence of a local motion below T_g. Although there is no direct evidence of the nature of this motion, comparison of the mechanical behaviors of poly(cyclohexyl methacrylate) and poly(phenyl methacrylate) has led to the conclusion that the γ-peak is due to the cyclohexyl ring [37].

To support this conclusion, MAS/CP/DD ^{13}C NMR spectra of solid poly(cyclohexyl methacrylate) have been obtained at 12 MHz as a function of temperature [38]. They are shown in Fig. 3a, together with the numerotation of the ring carbons of poly(cyclohexyl methacrylate). They consist of several lines which have been identified from left to right, in order of increasing magnetic field, as a line due to the carboxyl carbon, a line due to the methine group of the side ring, a broad line arising from the main-chain methylene group, a line due to the main-chain quaternary carbon, and finally a complex pattern, whose shape depends on temperature, corresponding to the cyclohexyl methylene

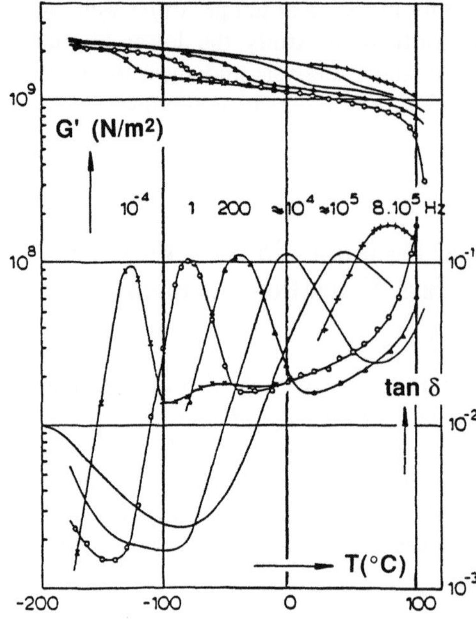

Fig. 2. Shear modulus, G', and damping, $\tan\delta$, of poly(cyclohexyl methacrylate) as a function of temperature at different frequencies (after [37])

carbon and methyl carbon resonances. At $-50\,°C$, the three cyclohexyl methylene carbon peaks are narrow and well-resolved. They are in sharp contrast to the room-temperature spectrum, which is so broadened that the resolution has completely disappeared. Between these two extremes, intermediate features are observed at the other temperatures. This broadening of aliphatic protonated carbon lines, which exceeds a few ppm, together with its characteristic variation as a function of temperature, is a clear indication of modulation of the ^{13}C–^{1}H dipolar coupling by motions of the side ring in the ω_{1H} frequency region.

In order to determine the transverse relaxation times T_{2m}, the broadened cyclohexyl methylene and methyl carbon lines were decomposed, using Lorentzian line shapes. The chemical shifts of the three methylene lines were determined from the low-temperature ($-50\,°C$) spectrum. The methyl resonance location was taken from the MAS/CP/DD ^{13}C NMR spectrum of poly(cycloheptyl methacrylate) [21]. As the experimental MAS/CP/DD ^{13}C NMR spectra were recorded under the condition of optimum carbon–proton contact time, the intensity of the individual lines was assumed to be proportional to the number of resonant carbons.

Comparison of experimental and simulated spectra is shown in Fig. 3. T_{2res} was estimated as 6×10^{-3}s from the methylene ring carbon line width of poly(cycloheptyl methacrylate) recorded under similar experimental conditions. In this polymer at room temperature, the ring motions have proved to be very rapid processes, having too high a frequency to broaden the corresponding lines

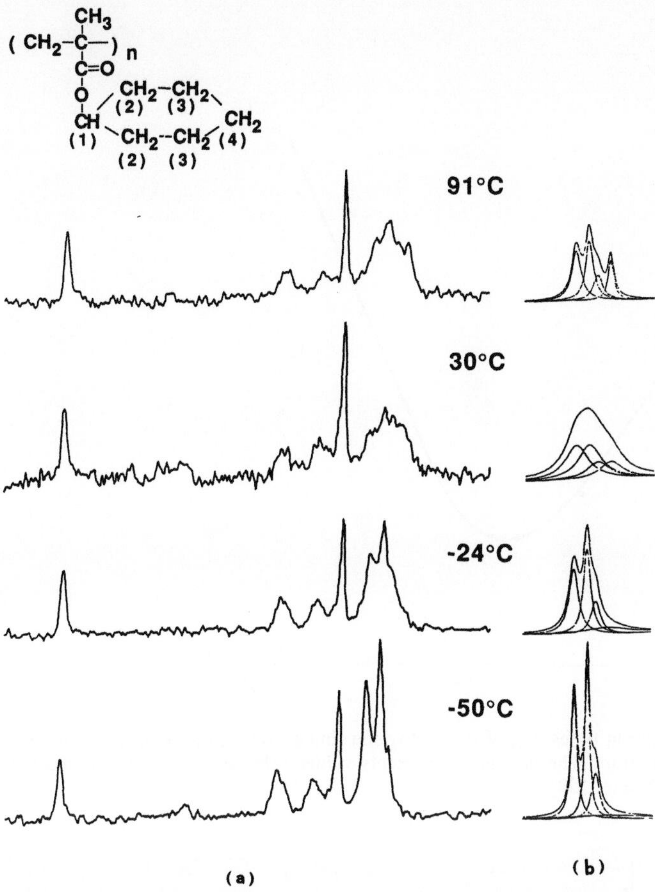

Fig. 3a, b. Experimental (**a**) and simulated (**b**) 12-MHz MAS/CP/DD ^{13}C NMR spectra of solid poly(cyclohexyl methacrylate) at different temperatures (after [38])

of MAS/CP/DD ^{13}C NMR spectra, and therefore the line widths of the methylene ring carbon peaks are purely static in origin.

As an example, the transverse relaxation time T_{2m} of the methylene ring carbon C2 thus obtained is reported in Fig. 4. Its dependence on temperature is similar to that of the theoretical curve predicted by expression [7]. The T_{2m} minimum is observed around 20 °C, which indicates that, at this temperature, the frequencies of the ring motions are of the order of ω_{1H}, i.e. $32 \times 2\pi$ kHz. A similar behavior has been observed for the methylene ring carbon C3. Although results obtained on the C4 ring carbon are not very accurate, the line of the C4 ring carbon seems to be less broadened than those of carbons C2 and C3 as pointed out by the simulated spectra plotted in Fig. 3. Moreover, the widths of the main-chain methylene carbon and of the C1 ring carbon lines do not seem to be temperature dependent. All these data indicate that the observed

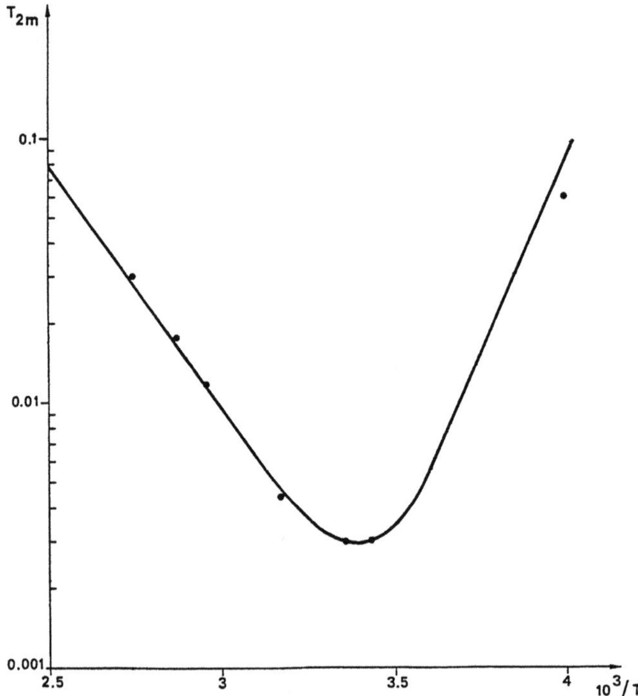

Fig. 4. Transverse relaxation times T_{2m} of the methylene ring carbon C2 as a function of $10^3/T$. T is the absolute temperature. The *full curve* represents values calculated by using the motional model depicted in Fig. 5 (after [38])

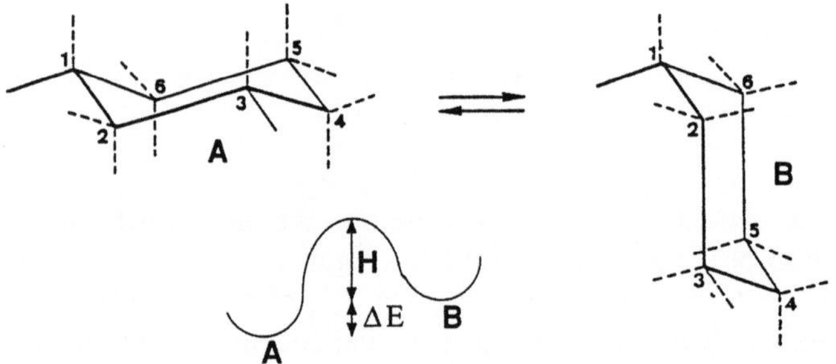

Fig. 5. Chair–chair inversion of a cyclohexyl ring with carbon C1 in a fixed position and two-potential well model for the cyclohexyl inversion (after [38])

dynamic process is a ring motion that mainly affects the C2-H and C3-H internuclear vectors.

The above results have been interpreted in terms of a simple model in which the polymer main chain is fixed whereas the cyclohexyl ring, which is rigidly bound to the carboxyl moiety, experiences chair–chair inversions (Fig. 5). Such chair–chair inversions have already been proposed by Heijboer, on the basis of conformational energy calculations, to account for the mechanical behavior of poly(cyclohexyl methacrylate) [37]. They can be represented mathematically by a two-state system with equilibrium positions A and B whose energies are E_A and E_B. H and $H + \Delta E$ are the energy barriers separating states B and A, and A and B respectively. By assuming Boltzmann statistics and supposing that the flips from A to B and B to A constitute a stationary process with correlation time τ_c, the autocorrelation function associated to this model can be written as [38]:

$$G(t) = \frac{(a^2 + 2ax + 1) + 2a(1 - x)\exp(-t/\tau_c)}{10(1 + a)^2} \tag{12}$$

a is the ratio of the equilibrium populations: $a = \exp(\Delta E/RT)$ and $x = P_2(\cos\theta)$, where θ is the angle between the C-H vectors in the A and B positions.

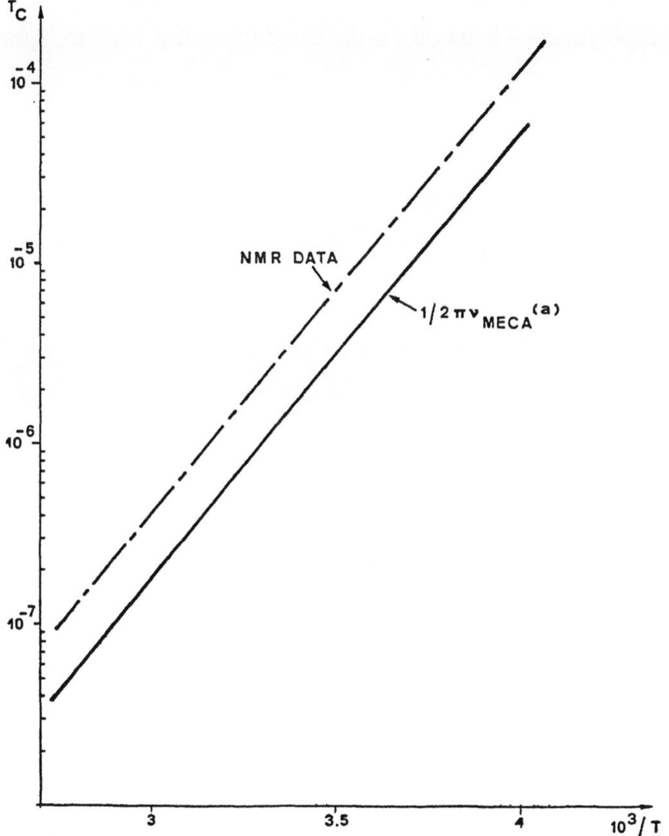

Fig. 6. Correlation times τ_C calculated from data reported in Fig. 4. NMR results are compared with mechanical data reported in [37] (after [38])

The best fit between experimental T_{2m} and calculated ones has been obtained for $\Delta E \sim 12.7$ kJ/mol and $H = 47.4$ kJ/mol. It is shown in Fig. 4. Variation of τ_c deduced from the T_{2m} values is plotted in Fig. 6 as a function of temperature. It is compared with the dependence of the location of the γ-relaxation maximum on frequency, ν, and absolute temperature, T, exhibited in Fig. 2. This location is well described by the following equation:

$$\nu = \nu_0 \exp(-E_a/RT) \tag{13}$$

where the apparent activation energy, E_a, is 47.2 kJ/mol and log ν_0 is 13.0 [37]. Agreement of results obtained from the NMR and mechanical experiments is very good. It confirms the conclusions reached by Heijboer [37] identifying the molecular process responsible for the γ-transition of poly(cyclohexyl methacrylate) as a motional mode of the cyclohexyl ring.

4.2 *Local Motions of Aromatic Copolyesters*

The thermotropic copolyester *A* is based on the three following units in equal proportions:

unit I

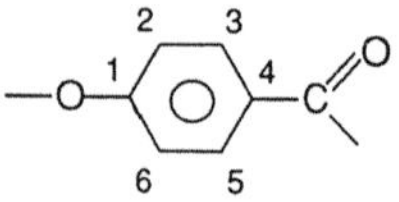

unit II

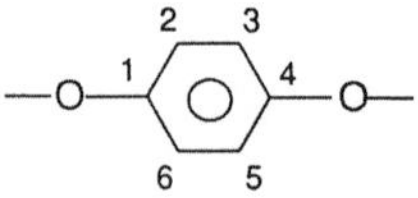

unit III

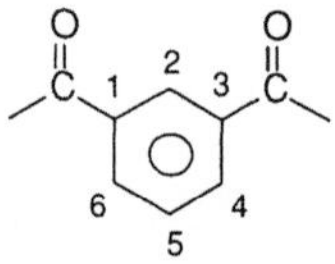

The mesogenic groups are the *para*-substituted type I and II units. Type III units act as chain disruptors. From a mechanical point of view, this copolyester is known to exhibit two secondary β and γ relaxations which induce noticeable decreases of its shear modulus with increasing temperature [39]. For an NMR analysis, it must be noted that it consists of a succession of carboxyl groups and aromatic rings, whose carbons all have a large chemical shift anisotropy that should be very sensitive to local motions.

The 75 MHz MAS/CP/DD ^{13}C NMR spectrum of a non-oriented sample of copolyester A, obtained at room temperature with a spinning speed of around 4000 Hz, is plotted in Fig. 7 [40]. It shows both main lines and first- and

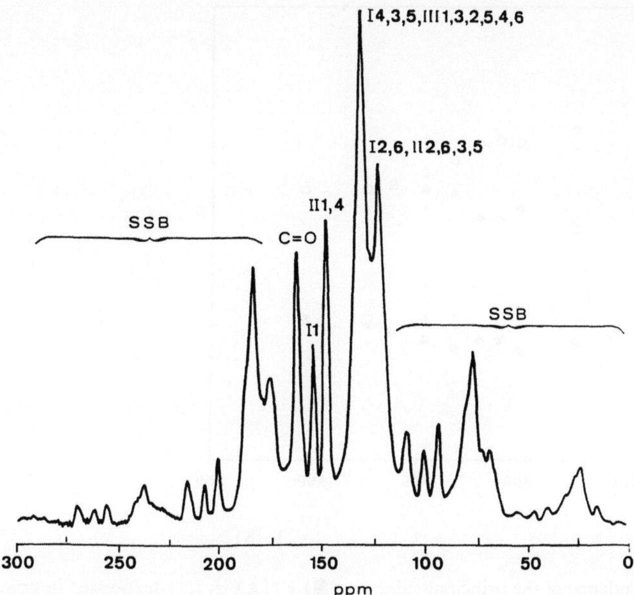

Fig. 7. MAS/CP/DD ^{13}C NMR spectrum of copolyester A at room temperature (spinning speed: 4000 Hz, contact time: 1 ms) (after [40])

second-order spinning sidebands. The 164 ppm peak represents the resonances of all the carboxyl carbons of the copolyester units. The 154 ppm line is characteristic of carbon 1 of the type I unit (I1). The 148 ppm line corresponds to carbons 1 and 4 of the type II unit (II1,4). One can therefore obtain specific information on the motional behavior of rings I and II, by following the behavior of the 154 and 148 ppm lines, respectively. On the other hand, all the resonances of the carbons of the type III unit belong to composite lines, so that no specific information on the motion of the *meta*-substituted ring can be derived from the spectrum of the protonated copolyester A. Therefore, to study the type III ring motion, the copolyester A has been selectively deuterated on the type I and II units, leaving the type III rings protonated.

Analysis of the MAS/CP/DD ^{13}C NMR spectra of a non-oriented sample of copolyester A shows that the peak positions are independent of temperature in the range from 248 to 520 K, whereas the relative intensities of the spinning sidebands decrease with increasing temperature. The decrease in the spinning sideband intensities is a clear indication of a progressive averaging of the chemical shift tensors and, as a consequence, of the occurrence of local motions.

The principal values σ_1, σ_2 and σ_3 of the chemical shift tensors for the different carbons were determined from the relative intensities of the spinning sidebands by using the method of Herzfeld and Berger [4]. Results obtained for the 154 ppm line of carbon I1 are plotted in Fig. 8(a). Between 248 K and 425 K, σ_1 remains constant whereas σ_3 decreases and σ_2 increases until they

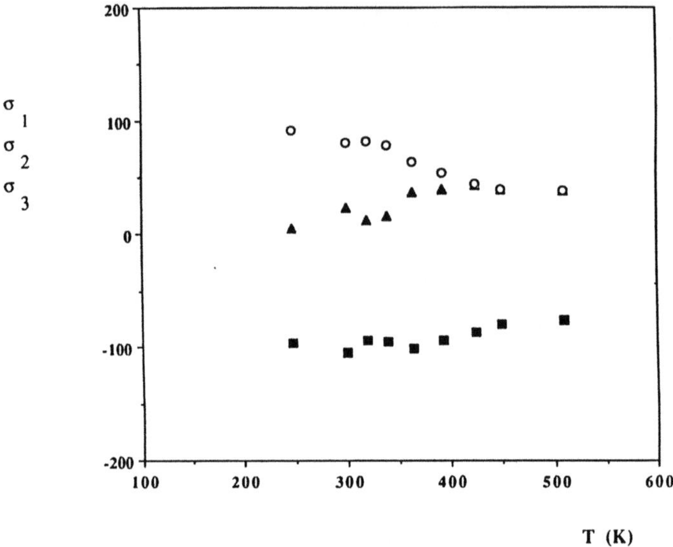

Fig. 8a. Temperature dependence of the principal values σ_1 (■), σ_2 (▲), σ_3 (○), (expressed in ppm) of the chemical shift tensor for the carbon I1 of polymer A

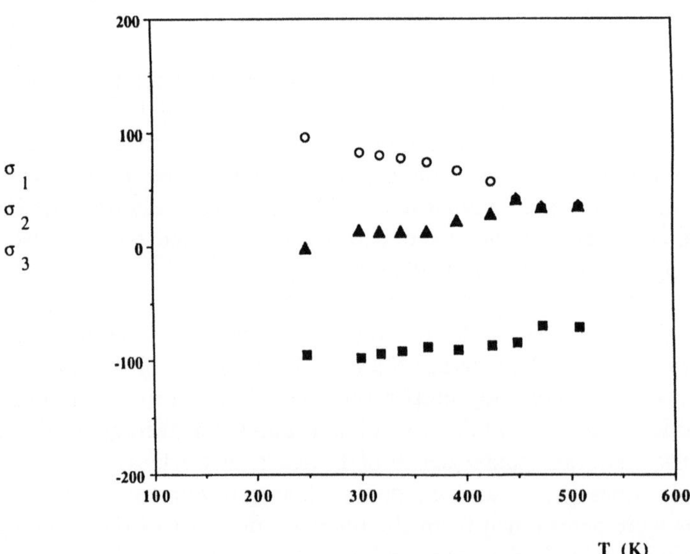

Fig. 8b. Temperature dependence of the principal values σ_1 (■), σ_2 (▲), σ_3 (○), (expressed in ppm) of the chemical shift tensor for the carbon II1,4 of polymer A (after [40])

reach a common value at about 425 K. At higher temperatures, there occurs a slight increase in the σ_1 value and a decrease in the common value of σ_2 and σ_3. Assuming that the directions of the principal axes of the chemical shift tensor of carbon I1 are identical to those of other aromatic carbons previously studied [41, 42], i.e. the least shielded element, σ_1, is in the aromatic plane pointing radially out from the ring whereas the most shielded one, σ_3, is perpendicular to the ring plane, data plotted in Fig. 8(a) at temperatures up to 425 K are consistent with the existence of motions about the *para* axis of the ring. Indeed, processes about the *para* axis of the ring lead to a partial averaging of the σ_2 and σ_3 components, which are perpendicular to the axis of rotation, while leaving the σ_1 principal value, that is parallel to the axis of rotation, unchanged. Above 425 K, results reported in Fig. 8(a) show a partial averaging of σ_1, which corresponds to a reorientation of the *para* axis of the ring.

The results obtained between 248 K and 425 K can be interpreted in terms of rapid oscillations about the *para* axis of the ring between increasing angles α and $-\alpha$, corresponding to a 2α amplitude of motion. The rigid-lattice values ($\alpha = 0$) were taken to be equal to those measured at 248 K for carbon I1, an assumption corroborated by the fact that the 248 K chemical-shift parameters are quite similar to those determined for the corresponding carbon of the small model molecule methyl 1–4 acetoxybenzoate in its crystalline state [40]. Results of the calculations are plotted in Fig. 9. They show a regular increase of α from 248 K up to a complete ring flip ($\alpha = 90°$) at 425 K.

Above 425 K, the dynamics of the local symmetry axis of the ring can be described in terms of an oscillation about an axis perpendicular to the σ_1 direction. Taking into account the fact that σ_1 has already slightly increased

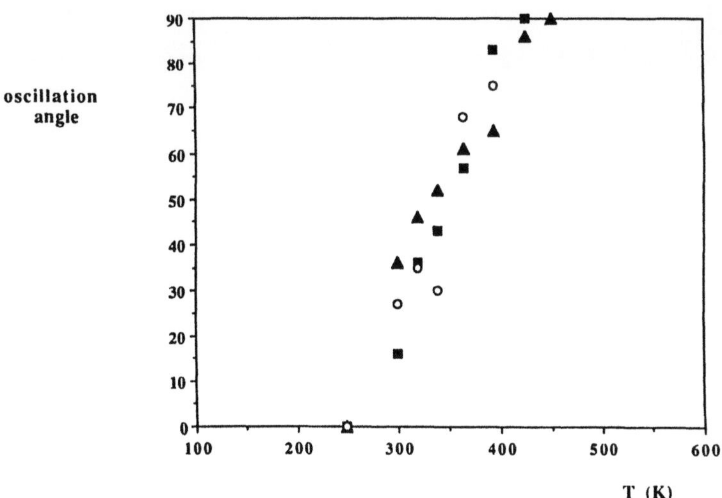

Fig. 9. Variation of the amplitude of oscillation of the type I aromatic units (■), type II aromatic units (▲) and C=O groups (○) as a function of temperature (after [40])

below 425 K, the analysis shows that the amplitude of this oscillation varies from 20° at 425 K to 35° at 510 K.

The dependence on temperature of σ_1, σ_2 and σ_3 for the carbons 1 and 4 of unit II, determined from the method of Herzfeld and Berger, is plotted in Fig. 8(b). It is essentially the same as that observed for the principal values of the shielding tensor of carbon I1. Therefore, the results can be interpreted in terms of the same type of local motions, i.e. an oscillation about the local symmetry axis of the phenyl ring that is completed about 425 K and a reorientation of this axis at high temperature.

The evolution of the $\sigma_3 - \sigma_1$ difference for the 4,6 carbon pair of the selectively deuterated copolyester is shown in Fig. 10 as a function of temperature. It is a decreasing function of temperature that can be interpreted in terms of a ring oscillation whose amplitude varies between the angles γ and $-\gamma$ and whose axis joins the carbons 1 and 3 as represented below:

Assuming such an oscillation, γ values have been calculated by using as rigid-lattice parameters, chemical-shift parameters measured for the 4,6 carbon pair either for the selectively deuterated copolyester at 123 K, or for crystalline

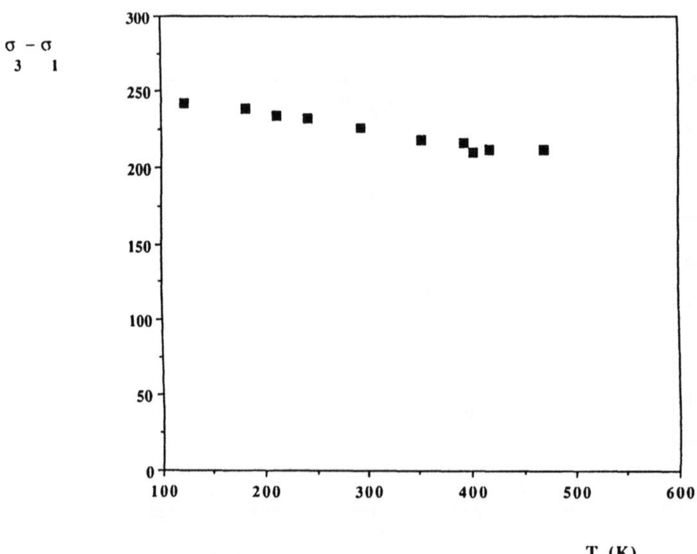

Fig. 10. Temperature dependence of the $\sigma_3 - \sigma_1$ difference (expressed in ppm) for the carbons III4,6 of the selectively deuterated copolyester A (after [40])

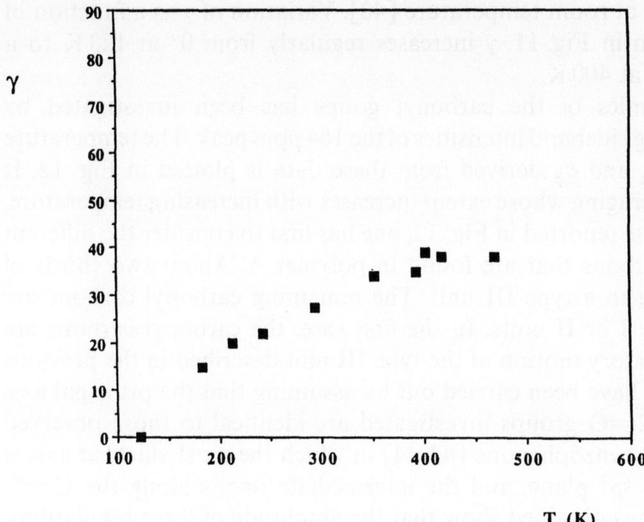

Fig. 11. Variation of the amplitude of oscillation, γ(°), of the type III aromatic units as a function of temperature (after [40])

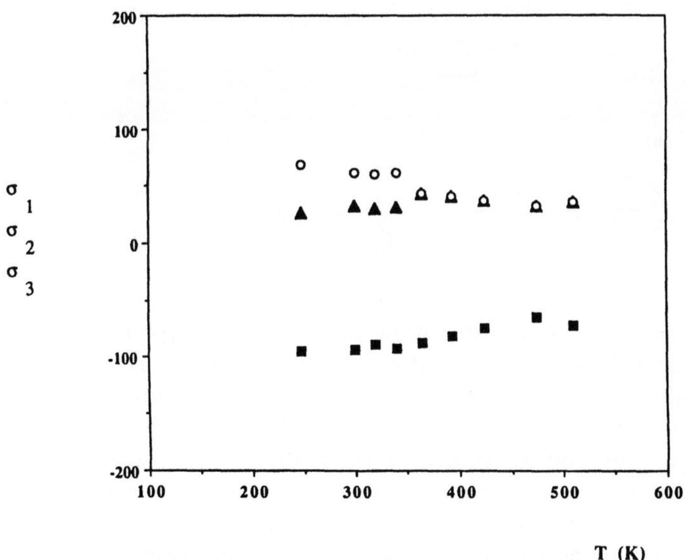

Fig. 12. Temperature dependence of the principal values σ_1 (■), σ_2 (▲), σ_3 (○), (expressed in ppm) of the chemical shift tensor for the C=O carbons of polymer A (after [40])

dimethylisophthalate at room temperature [40]. Variation of γ as a function of temperature is shown in Fig. 11. γ increases regularly from 0° at 123 K to a limiting value of 40° at 400 K.

The local dynamics of the carbonyl goups has been investigated by analyzing the spinning sideband intensities of the 164 ppm peak. The temperature dependence of σ_1, σ_2 and σ_3 derived from these data is plotted in Fig. 12. It shows a motional averaging whose extent increases with increasing temperature. To interpret the results reported in Fig. 12, one has first to consider the different types of carbonyl carbons that are found in polymer A. About two thirds of these carbons belong to a type III unit. The remaining carbonyl carbons are located between type I or II units. In the first case, the carbonyl carbons are involved in the oscillatory motion of the type III unit described in the previous section. Calculations have been carried out by assuming that the principal axis orientations in the C=O groups investigated are identical to those observed in acetophenone and benzophenone [43, 44] in which the most shielded axis is perpendicular to the sp^2 plane, and the intermediate one is along the C=O bond [42]. Results thus obtained show that the amplitude of these oscillations, given in Fig. 11, is too small to be responsible for the motional averaging evidenced in Fig. 12 above 320 K. Therefore, a second type of motion must exist which involves at least part of the carbonyl carbons.

Consideration of molecular models and local geometries points to the likely existence of motions about the axes as schematized below:

(a)

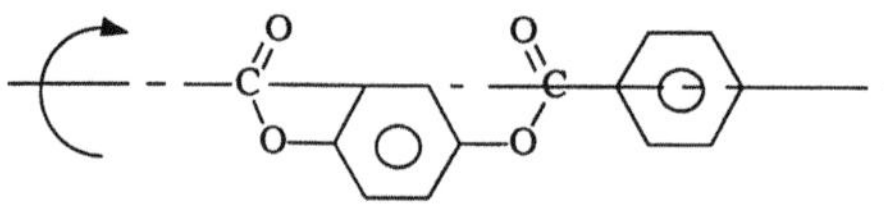

(b)

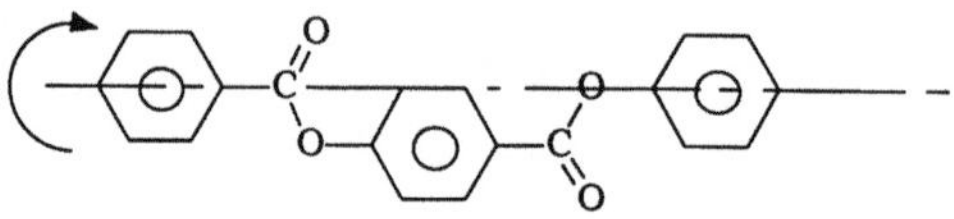

Such motions induce a translation of the *para* axis of the type I or II rings. They can only occur if the carbonyl carbons are in a given *cis* (a) or *trans* (b) configuration with respect to the phenyl ring depending on the unit considered. However, it is of interest to notice that if one unit is in the correct configuration allowing this type of motion to occur, then the motion of the considered unit will induce a *cis-trans* conformational change of the next unit which might in turn be placed into the right configuration. The configuration that favors these processes corresponds to the maximum extension of the chain, which is consistent with the nematic character of the polymer. Due to the lack of information about the actual conformations of polymer A and in order to simplify further calculations, it has been assumed that all the C=O carbons are involved in such a reorientation.

To estimate the amplitude of this process, it has been assumed that the 248 K NMR spectrum corresponds to a quasi rigid-lattice behavior, which is likely in terms of the activation energy such a process would require, and that the measured chemical shift anisotropy at higher temperatures is the average of the contributions of the two types of carbonyl carbons. Motional amplitudes obtained under these assumptions are plotted in Fig. 9 together with the temperature dependence of the amplitudes of the oscillations of the type I and II rings about their *para* axis. It has to be noted that the three motional amplitudes have similar values and temperature dependences until 425 K. Such a result indicates that the motions of the rings and C=O groups are correlated, which is consistent with the motional model adopted. Above 425 K, there occurs an additional chain reorientation.

The above study of the motional averaging of the ^{13}C chemical shift tensors points out the existence of different types of motions in the aromatic copolyester A. At low temperature, only small-amplitude oscillations of the *meta*-substituted rings occur. Above 300 K, the two *para*-substituted rings exhibit identical reorientations about their local symmetry axis, which are coupled with the motions of the neighboring carbonyl carbons. The amplitude of these processes increases regularly until 425 K where the *para*-substituted rings are involved in full ring flips. Finally, above 425 K, one observes, in addition to the previous modes, a reorientation of the main-chain axis.

It is of interest to compare these conclusions to the manifestations of the glass transition phenomenon and β and γ secondary relaxations observed by dynamic mechanical experiments [39]. There is a close similarity between the temperatures at which the glass transition manifests itself through dynamic mechanical measurements in the 10^3–10^5 Hz range and the temperature at which the reorientation of the main-chain axis is first observed by NMR. The β and γ secondary relaxations are observed by viscoelastic techniques at about 340 K and 265 K, respectively, for an experimental frequency of 10^3 Hz. Comparing these results with the NMR results described above, it seems that the β process can be identified with the coupled motions of the *para*-substituted rings and carbonyl carbons, and the γ-process with the oscillations of the *meta*-substituted rings. To confirm this assignment, further NMR and dynamic mechanical experiments on copolyesters made of the same three reccurring units with different proportions are now in progress.

4.3 *Local Motions of Model Epoxy Resins*

In the last decade, numerous studies have been devoted to the determination of the relationships existing between the chemical structure and viscoelastic properties of epoxy-amine networks [45–53]. In addition to the glass transition temperature, T_g, that markedly depends on both the crosslink density and chain flexibility, epoxy-amine networks exhibit secondary relaxation processes. The molecular origin of the β-relaxation in epoxy-amine networks is still an open question. Mechanical measurements performed on series of resins built from different types of amine and epoxy units have lead to the conclusion that this secondary transition is due to the motion of the hydroxypropylether group [45]. On the other hand, high-resolution solid-state ^{13}C NMR experiments [54] have pointed out the existence of phenyl ring motions in a frequency range comparable with that of the β-process. In Fig. 13 are shown the MAS/CP/DD ^{13}C NMR spectra of an epoxy resin studied by Garroway et al. [54]. At 151 K the resonances of the carbons *ortho* to the oxygen are split into two resolved peaks. The resonance of the *meta* carbons is just barely split at 151 K. At higher temperatures each set of peaks merges into a single line and that line continues to narrow even at the highest temperature of 352 K. The observed coalescence of these spectral lines is an indication of 180° rotations of the phenylene rings. The full temperature dependence has been described in terms of a distribution of correlation times or, equivalently, a nonexponential autocorrelation function. Such an approach has led to a precise characterization of the ring motion in terms of correlation time distribution, whose central values depend on the exact model considered, and an activation energy of about 63 kJ/mol [54].

To obtain direct evidence of the specific motions of the aliphatic units, $t_{1/2}$ determinations have been performed as a function of temperature for the

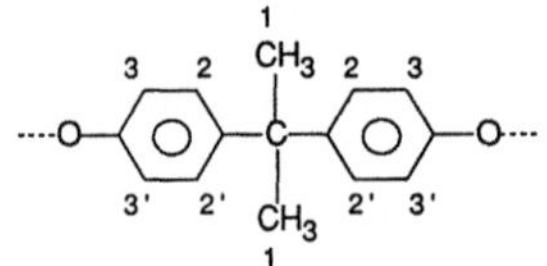

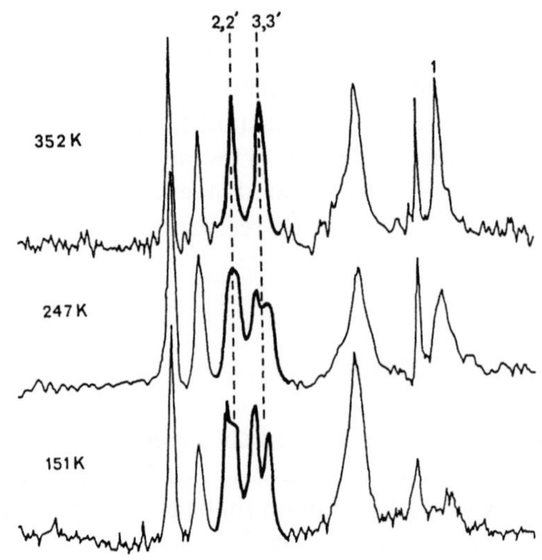

Fig. 13. Variable-temperature ^{13}C MAS/CP/DD NMR spectra of an epoxy resin based on diglycidyl ether of bisphenol A (DGEBA) cured with piperidine (after [54])

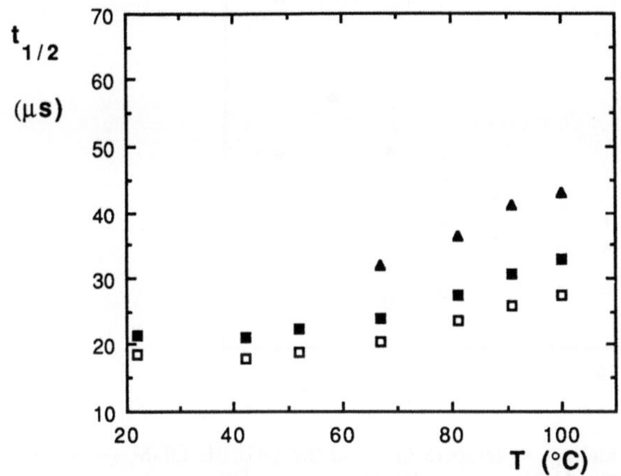

Fig. 14. Schematic formula and $t_{1/2}$ vs temperature T in the DGEBA-HMDA system (after [56]). ■: CH_2O—CHOH units; □: CH_2N units; ▲: HMDA central CH_2 groups

—CHOH—CH_2O— and —CH_2N— units of epoxy-amine resins based on diglycidylethers of either bisphenol A (DGEBA) or butanediol (DGEBU) and cured in stoichiometric proportions with hexamethylene diamine (HMDA) or diaminodiphenylmethane (DDM) [55, 56]. A post-curing step was applied to yield an epoxy-amine reaction as complete as possible. Under the above conditions, the diglycidylethers of bisphenol A and butanediol are known to produce no side reactions with amines and the number of unreacted epoxy groups is quite low [55]. Therefore, as a first approximation, the materials under study can be considered as "model" epoxy networks. Examples of the chemical structure of such systems are given in Figs. 14 and 15.

Examples of $t_{1/2}$ variations as a function of temperature are shown in Figs. 14 and 15. In all the considered structures, $t_{1/2}$ is almost constant in the range 30 °C–60 °C for both —CHOH—CH_2O— and —CH_2N— groups. At a higher

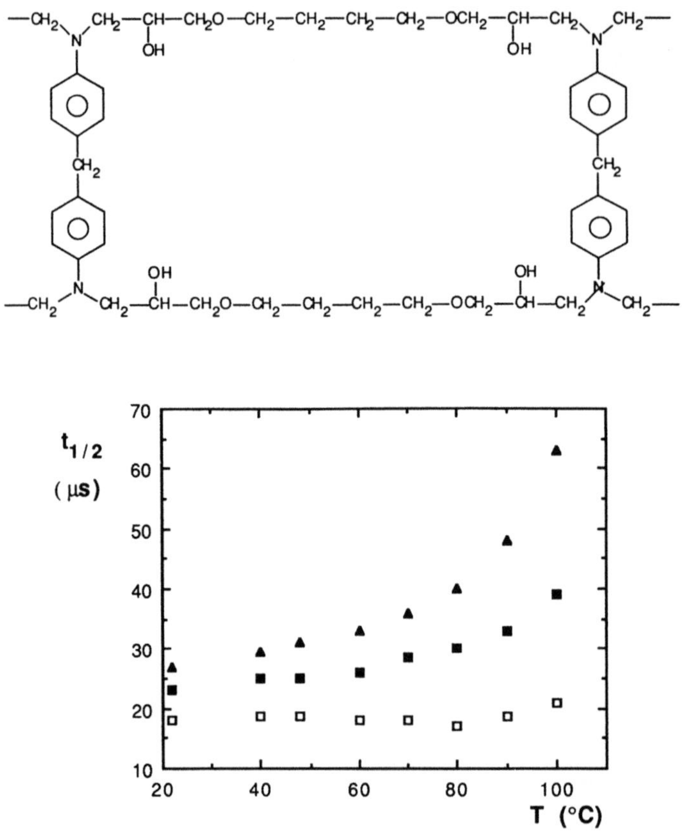

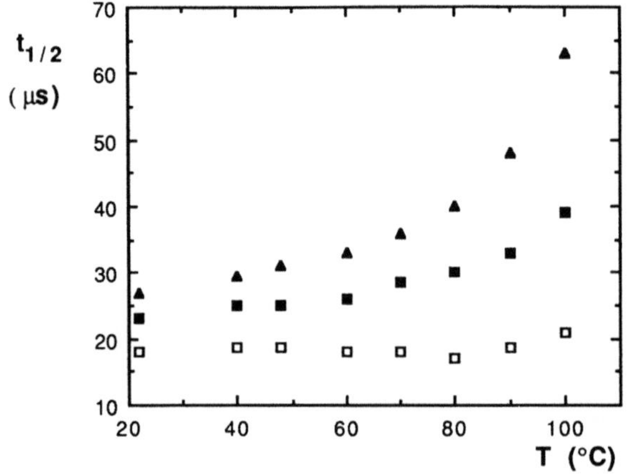

Fig. 15. Schematic formula and $t_{1/2}$ vs temperature T in the DGEBU-DDM system (after [56]). ■: CH_2O—CHOH units; □: CH_2N units; ▲: DGEBU central CH_2 groups

temperature, whose exact value depends on the considered unit, $t_{1/2}$ increases, indicating an increase in the mobility of the corresponding group. The increase in the —CH_2N— mobility is observed at a higher temperature for an aromatic diamine than for an aliphatic one. Similarly, the motion of the hydroxypropylether group is slowed down by the presence of the bulky bisphenol A unit, as compared with the more flexible butanediol unit. As expected, the mobility of a $(CH_2)_4$ sequence is higher in an epoxy unit than in an amine one. It must be noted that all the motions observed in these $t_{1/2}$ experiments are related to sub-T_g processes.

Values of the ^{13}C–^{1}H dipolar interaction have been calculated from $t_{1/2}$ data and interpreted in terms of elementary motions. In Table I are shown the percentages of —CHOH—CH_2O— and —CH_2N— groups involved in isolated conformational jumps at 100 °C for the different networks. The most mobile —CHOH—CH_2O— units are those of the DGEBU-DDM resin: they benefit from the neighborhood of the flexible $(CH_2)_4$ groups. The most mobile

Table 1. Percentages of $—CH_2N—$ and $—CHOH—CH_2O—$ units involved in isolated conformational jumps at 100 °C. (after [55, 56])

sample	$—CH_2N—$	$—CHOH—CH_2O—$
DGEBU-DDM	10	100
DGEBA-DDM	10	10
DGEBA-HMDA	70	70

$—CH_2N—$ units are those of the DGEBA-HMDA resin which are next to the amine $(CH_2)_6$ sequence. The $—CH_2N—$ groups of the DGEBU-DDM network have a reduced mobility due to the fact that they belong to a DDM unit. The least mobile system is the DGEBA-DDM resin in which both the amine and epoxy units are rigid.

Comparison with the mechanical analysis performed on these systems [53] indicates that the motions of the hydroxylpropylether groups and crosslinking points likely participate to the β-transition. Moreover, the motions of the central CH_2 groups of the DGEBU units, which occur at temperatures lower than those assigned to the β-transition, can be related to the mechanical γ-transition observed in these systems [51].

5. Experimental Studies of Local Dynamics in Bulk Polymers at Temperatures Well Above the Glass Transition Temperature

5.1 *Expressions for the Polymer Autocorrelation Functions*

As can be seen from Eqs. (10) and (11), for a protonated carbon whose internuclear distance, r_{CH}, is equal to the bond length and therefore well-known, the determinations of the spin–lattice relaxation time $T_1(^{13}C)$ and nuclear Overhauser enhancement, NOE, can be used as a test for the autocorrelation function $G(t)$ and its Fourier transform, the spectral density, $J(\omega)$, in the high-frequency domain. In this regard, it is of interest to review the different expressions that have been proposed to represent the local dynamics of a bulk polymer at temperatures well above the glass transition temperature.

For rapid isotropic rotation of the internuclear vector, $J(\omega)$ can be written as:

$$J(\omega) = \frac{\tau}{1 + \omega^2\tau^2} \tag{14}$$

where τ is the correlation time of the motion.

For amorphous polymers at temperatures well above the glass transition temperature, the high-frequency local dynamics consist of segmental modes that involve a small number of monomer units and are much faster than the overall tumbling of a high molecular weight compound: in the example of *cis*-1, 4-polybutadiene at room temperature mentioned in the Introduction [1–3], the local motions are fast enough to participate to the spin–lattice relaxation, whereas the complete reorientation is due to modes with correlation times of the order of 10^{-3} s, much too slow to take part in the $T_1(^{13}C)$ relaxation. Due to the constraint resulting from the connectivity of the chain, these local motions are usually too complicated to be described by a simple isotropic rotation. Indeed, fluorescence anisotropy decay experiments, which directly yield the orientation autocorrelation function, have shown that the experimental data obtained on anthracene-labeled polybutadiene and polyisoprene in solution or in the melt cannot be represented by simple motional models. To account for the connectivity of the polymer backbone, specific autocorrelation functions, based on conformational changes that propagate along the chain according to a damped diffusional process, have been derived for local chain motions in solution [57–60]. As shown by fluorescence anisotropy decay studies carried out on polybutadiene or polyisoprene, these specific autocorrelation functions can also represent the experimental data recorded in the bulk state [61, 62].

Among the various expressions that are based on a conformational jump model and have been proposed for the orientation autocorrelation function of a polymer chain, the formula derived by Hall and Helfand (HH) [57] leads to a very good agreement with fluorescence anisotropy decay data. It is written as:

$$G(t) = \exp(-t/\tau_2)\exp(-t/\tau_1)I_0(t/\tau_1) \qquad (15)$$

the Fourier transform of which is:

$$J(\omega) = \mathrm{Re}[1/(\alpha + i\beta)^{1/2}] \qquad (16)$$

$$\text{with: } \alpha = \tau_2^{-2} + 2\tau_1^{-1}\tau_2^{-1} - \omega^2 \quad \text{and} \quad \beta = -2\omega(\tau_1^{-1} + \tau_2^{-1})$$

I_0 is the modified Bessel function of order 0, τ_1 is the correlation time associated with correlated jumps responsible for orientation diffusion along the chain, and τ_2 corresponds to damping which consists either of nonpropagative specific motions or of distortions of the chain with respect to its most stable conformations. However, it must be pointed out that the molecular origin of the damping is still not well identified. Molecular processes such as specific isolated jumps, fluctuations of internal rotation angles, and chemical defects (head-to-head linking, stereochemical sequences, side groups$\cdots$) are expected to affect the propagation of conformational changes along the chain sequence leading to a damping effect.

A more complete expression for the orientation autocorrelation function has been established by Dejean, Lauprêtre and Monnerie [63]. The Dejean-Lauprêtre-Monnerie (DLM) orientation autocorrelation function describes the segmental motions in terms of damped conformational jumps, as proposed by Hall and Helfand, and takes also into account independent librations of the internuclear vectors represented, as suggested by Howarth [64] by a random anisotropic fast reorientation of the CH vector inside a cone of half-angle Θ

and axis the rest position of the internuclear vector. The resulting orientation autocorrelation function can be written as:

$$G(t) = (1-a)\exp(-t/\tau_2)\exp(-t/\tau_1)I_0(t/\tau_1) + a\exp(-t/\tau_0)\exp(-t/\tau_2)\exp(-t/\tau_1)I_0(t/\tau_1) \quad (17)$$

where the correlation time τ_1 characterizes the conformational jumps, the correlation time τ_2 is associated with the damping and τ_0 is the characteristic time of the libration.

a is related to the half-angle Θ of the libration cone through the relation:

$$(1-a) = ((\cos\Theta - \cos^3\Theta)/2(1-\cos\Theta))^2 \quad (18)$$

Assuming that τ_0 is much shorter than τ_1 and τ_2 (a situation which is often encountered), the second term in Eq. (17) can be simplified and $G(t)$ written as:

$$G(t) = (1-a)\exp(-t/\tau_2)\exp(-t/\tau_1)\,I_0(t/\tau_1) + a\exp(-t/\tau_0) \quad (19)$$

Under this condition the spin–lattice relaxation time $T_1(^{13}\mathrm{C})$ is written as [63]:

$$(nT_1(^{13}\mathrm{C}))^{-1} = ((1-a)\hbar^2\gamma_C^2\gamma_H^2(J_{HH}(\omega_H-\omega_C) + 3J_{HH}(\omega_C) + 6J_{HH}(\omega_H+\omega_C)) + a\hbar^2\gamma_C^2\gamma_H^2\,(J_0(\omega_H-\omega_C) + 3J_0(\omega_C) + 6J_0(\omega_H+\omega_C)))/10r_{CH}^6 \quad (20)$$

$J_{HH}(\omega)$ and $J_0(\omega)$ are defined as:

$$J_{HH}(\omega) = (\alpha + i\beta)^{-1/2}$$

$$J_0(\omega) = \tau_0/(1+\omega^2\tau_0^2)$$

with: $\alpha = \tau_2^{-2} + 2\tau_1^{-1}\tau_2^{-1} - \omega^2$ and $\beta = -2\omega(\tau_1^{-1} + \tau_2^{-1})$

The first term, which is proportional to $1-a$, contains only factors originating from the HH function and thus depends only on τ_1 and τ_2. The second term, which is proportional to a, corresponds to the libration of limited angular extent and depends on τ_0. Under the assumption of fast librations and segmental motions, $\tau_0 \ll \tau_1$, τ_2, $(\omega_H + \omega_C)\tau_1 < 1$, the second term in expression 20 can be neglected, giving:

$$(nT_1(^{13}\mathrm{C}))^{-1} = (1-a)\hbar^2\gamma_C^2\gamma_H^2(J_{HH}(\omega_H-\omega_C) + 3J_{HH}(\omega_C) + 6J_{HH}(\omega_H+\omega_C))/10r_{CH}^6 \quad (21)$$

5.2 Spin–Lattice Relaxation Time Studies of Local Dynamics

Figure 16 shows the behavior of $nT_1(^{13}\mathrm{C})$ as a function of the reciprocal of temperature T for the CH and CH_2 carbons of bulk poly(vinylmethyl ether) at temperatures well above the glass transition temperature [63]. The $nT_1(^{13}\mathrm{C})$

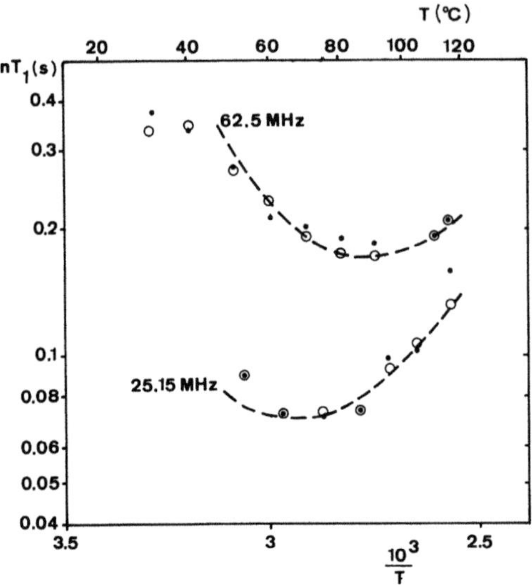

Fig. 16. 62.5- and 25.15-MHz ^{13}C spin–lattice relaxation times $nT_1(^{13}C)$ in bulk poly(vinylmethyl ether): (○) CH carbon; (●) CH_2 carbon; (---) best fit calculated from DLM autocorrelation function: $a = 0.40$, $\tau_1/\tau_0 = 200$, $\tau_2/\tau_1 = 2$ (after [63])

Table 2. Comparison of experimental and calculated $nT_1(^{13}C)$ (second) values at the minimum in bulk poly(vinylmethyl ether) (after [63])

ω_C	62.5 MHz	25.15 MHz
experimental	0.177	0.070
isotropic model	0.100	0.040
HH model	0.128	0.050

minimum is observed at 90 °C at 62.5 MHz and 70 °C at 25.15 MHz. $nT_1(^{13}C)$ values at the minimum are 0.177 and 0.070 s at 62.5 and 25.15 MHz respectively. These experimental values of the $nT_1(^{13}C)$ minima are listed in Table 2 together with those calculated from the isotropic and HH models with r_{CH} taken as 1.08 Å for the unsaturated CH group and 1.09 Å for the saturated CH and CH_2 groups [65]. It can be seen that the experimental values are always higher than the calculated ones, in proportions ranging from about 35% in the case of the HH model to 75% in the case of isotropic reorientation. It is important to note that this large discrepancy cannot be accounted for by internuclear distance imprecisions.

The nT_1 (^{13}C) values at the $nT_1(^{13}C)$ minimum are listed in Table 3 for a number of amorphous polymers in bulk at temperatures well above the glass transition temperature. For all the polymers considered, the same result is observed: at the minimum, $nT_1(^{13}C)$ is always higher than the values that are derived from the isotropic and HH models [63, 66–68].

To account for the large deviations between calculated and experimental $nT_1(^{13}C)$ values at the minimum, two different assumptions can be made. The first assumption is the existence, inside the polymer chain, of an additional

Table 3. Experimental $nT_1(^{13}C)$ (second) values at the minimum in bulk polymers. (after [63, 66–68])

	saturated CH		CH_2	
	25.15 MHz	62.5 MHz	25.15 MHz	62.5 MHz
poly(vinylmethyl ether)	0.070	0.177	0.070	0.177
poly(propylene oxide)	0.068	0.168	0.096	0.250
polyisobutylene			0.060	0.18

	unsaturated CH		CH_2	
	25.15 MHz	62.5 MHz	25.15 MHz	62.5 MHz
cis-1,4-polyisoprene	0.064	0.154	0.080	0.21
			0.095	0.22
cis-1,4-polybutadiene	0.072	0.166	0.095	0.24

motion that is not considered in the previous models and that contributes to a partial reorientation of the CH vectors with a characteristic time that differs from the correlation time for orientational diffusion along the chain. The alternative explanation is that the chain segmental motions have to be described by a distribution of correlation times. As shown in Table 3, for all polymers except poly(vinylmethyl ether), the $nT_1(^{13}C)$ values at a given experimental frequency are different for adjacent CH and CH_2 carbons of the polymer main chain. Therefore, a different distribution of correlation times would be required for each of these adjacent sites, which is not consistent with the concept of cooperative motions involving a few monomer bonds. If one now considers the existence of an additional motion, the assignment of this additional motion to a process slower than the orientation diffusion along the chain associated with τ_1 must be precluded since results derived from fluorescence anisotropy decay [61] and NMR [1–3] on polybutadiene have shown that such slow modes contribute only to a very small extent to the decay of the orientation autocorrelation function in this temperature range. Therefore, this additional motion has to be a faster mode and thus more local than the orientation diffusion process along the chain. A similar fast process has recently been observed by neutron scattering experiments [69, 70]. It can be assigned to molecular librations of limited extent of the CH vectors about their equilibrium conformation and corresponds to oscillations inside a potential well.

For bulk poly(vinylmethyl ether), fit of $nT_1(^{13}C)$ values at the $nT_1(^{13}C)$ minimum by the DLM orientation autocorrelation function, which is precisely based on segmental motions and independent librations, yields $a = 0.40$ and $\Theta = 33°$. With this parameter a thus determined, the best fit obtained from Eq. 20 is shown in Fig. 16 for the same polymer. As can be concluded from this figure, the overall agreement between theoretical prediction and the experimental data is very good. Only, in the temperature range of 50–60 °C, where the measurements are less precise due to the broadening of the lines, do deviations occur.

Table 4. $T_{ref/NMR}$ reference temperatures (K) and half-angle Θ_{CH} and Θ_{CH_2}(°) of the libration cone for the CH and CH_2 carbons (after [74])

	Θ_{CH}	Θ_{CH_2}	$T_{ref/NMR}$
poly(vinylmethyl ether)	33°	33°	344
poly(propylene oxide)	25°	37°	270
polyisobutylene		23°	333
cis-1,4-polyisoprene	20°	33°, 37°	297
cis-1,4-polybutadiene	26°	36°	234

It is of interest to note that, as shown by results reported in [63], the relaxation data obtained from poly(vinylmethyl ether) in $CDCl_3$ solution are described by exactly the same values of the ratios of the correlation times τ_1/τ_0 and τ_2/τ_1 and the angle of the libration cone Θ as those listed in Table 4 for the bulk local dynamics. Only the temperature dependence of the correlation times is different. Such a similarity in behavior shows that the very nature of the motions involved in the magnetic relaxation at a given frequency is identical in bulk at temperatures well above T_g and in solution. For example, in poly(vinylmethyl ether) the libration which is revealed by the high value of the $T_1(^{13}C)$ minimum has the same amplitude in both cases.

For all the polymers listed in Table 3 that have been investigated in bulk at temperatures well above the glass transition temperature, the DLM autocorrelation function gave good fits of the experimental data [63, 66–68]. The values of the half-angle of the libration cone Θ determined from NMR experiments are listed in Table 4. They are smaller for a methine carbon than for a methylene carbon except for poly(vinylmethyl ether). Θ is large for polypropylene oxide whereas it takes the smallest value for polyisobutylene. These results indicate that Θ is strongly related to the steric hindrance at the considered site; the larger the steric hindrance, the smaller the half-angle of the cone, as can be expected.

Although the τ_1/τ_0 ratio is not determined to a good accuracy when it reaches large values, interpretation of $T_1(^{13}C)$ data shows that τ_0 is at least 150 or 200 times shorter than τ_1 for each polymer under investigation. This implies that the anisotropic mode which is observed in addition to the segmental reorientation in the NMR relaxation experiments is indeed a very fast process. The high frequency of this motion is one of the main reasons for its assignment to a libration of the internuclear vector about its rest position.

5.3 *Temperature Dependence of Local Dynamics at Temperatures Well Above the Glass Transition Temperature*

In the above description of local motions, τ_1 characterizes the segmental modes. In order to know whether these segmental motions observed by NMR in bulk at temperatures well above the glass transition temperature belong to the glass transition processes, it is of interest to compare the variations of τ_1 as a function of temperature with the predictions of the Williams-Landel-Ferry (WLF) equation [71]. The WLF equation describes the frequency dependence of the motional processes associated with the glass transition phenomena. It can be written as [72]:

$$\log[\tau_C(T)/\tau_C(T_g)] = -C_1^g(T - T_g)/(C_2^g + T - T_g) \tag{22}$$

where $\tau_C(T)$ is the viscoelastic relaxation time at temperature T, and $\tau_C(T_g)$ is the viscoelastic relaxation time at T_g, which serves here as a reference. C_1^g and C_2^g can be expressed as:

$$C_1^g = \frac{B}{2.303 f_g} \quad C_2^g = \frac{f_g}{\alpha} \quad C_1^g C_2^g = \frac{B}{2.303\alpha} \tag{23}$$

where f_g is the fractional free volume at T_g, α the thermal expansion coefficient of the free volume and $B \sim 1$. Equation (22) can be generalized to any arbitrary reference temperature T_0:

$$\log \tau_C(T) = -\frac{C_1^0(T - T_0)}{C_2^0 + T - T_0} + \log \tau_C(T_0) \tag{24}$$

$$\text{with: } C_1^0 C_2^0 = C_1^g C_2^g \tag{25}$$

$$\text{and: } T_0 - C_2^0 = T_g - C_2^g \tag{26}$$

Using the temperature $T_\infty = T_g - C_2^g$ for which $\tau_C(T)$ tends to infinity, the WLF equation can be written:

$$\log[\tau_C(T)/\tau_C(T_g)] = -C_1^g + C_1^g C_2^g/(T - T_\infty) \tag{27}$$

T_∞ and the product $C_1^g C_2^g$ are constants characteristic of a given polymer. They do not depend on the reference temperature. Variation of $\log[\tau_C(T)/\tau_C(T_g)]$ as a function of $1/(T - T_\infty)$ is linear, and a change in the reference temperature only induces a translation of the line without any modification of its slope.

WLF coefficients for bulk poly(vinylmethyl ether) are $T_\infty = 188$ K, $C_1^g = 11.46$, and $C_1^g C_2^g = 671.74$ K [73]. In Fig. 17 are plotted the variations of $\log[\tau_C(T)] + A$, where A is an arbitrary constant, and of $\log(2\pi\tau_1)$ as a function of $(T - T_\infty)$ for bulk poly(vinylmethyl ether). The slopes of the two lines are quite similar, which shows that the segmental motions associated with the τ_1 process have the same temperature dependence as the viscoelastic relaxation times. Therefore, the segmental motions described by τ_1 belong to the processes that are involved in the glass transition phenomena. Identical results have been obtained for poly(propylene oxide) [66], *cis*-1,4-polybutadiene and *cis*-1,4-polyisoprene [67]. In the case of polyisobutylene, the agreement between the variations of $\log[\tau_C(T)] + A$ and of $\log(2\pi\tau_1)$ as function of $(T - T_\infty)$ is not so good. In this case, the temperature dependence of the segmental motions as observed by NMR can be understood by considering both glass transition and secondary relaxation processes [68].

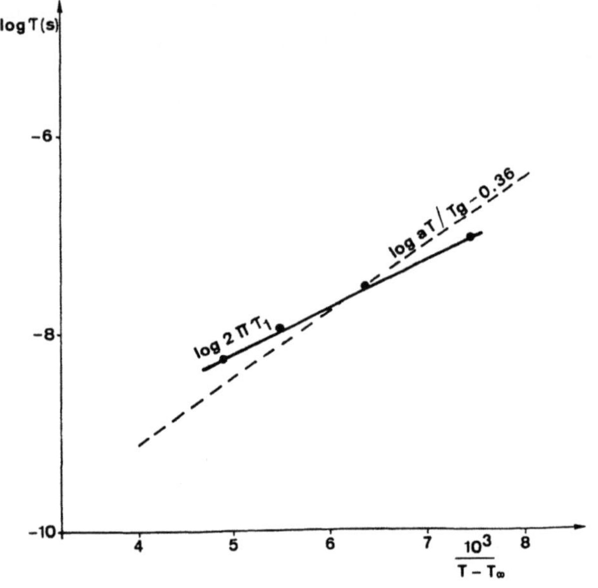

Fig. 17. Comparison of $\log(2\pi\tau_1)$(——) and ($\log a_{T/Tg} - 0.36$) (---) dependences on $10^3/(T - T_\infty)$ (after [63]).

5.4 *Factors Controlling the Local Dynamics at Temperatures Well Above the Glass Transition Temperature*

Data plotted in Fig. 18 show that each polymer has its own dependence of log τ_1 as a function of $(T - T_\infty)$. Moreover, at a constant $(T - T_\infty)$ difference, the segmental mobility depends on the polymer considered. These results indicate that the differences in segmental mobility at a given temperature observed for

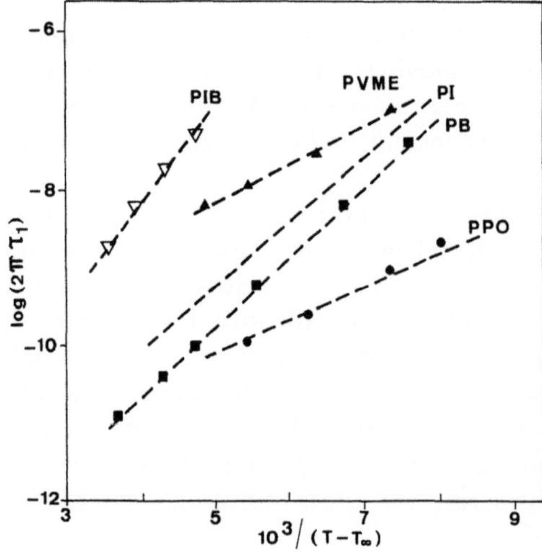

Fig. 18. Dependences of log $(2\pi\tau_1)$ (---) derived from $T_1(^{13}C)$ NMR experiments, as a function of $10^3/(T - T_\infty)$ (after [74])

the above polymers cannot be interpreted in terms of the differences in the T_∞ temperatures. Similar results are observed when comparing the $\log \tau_1$ variations as a function of $(T - T_g)$ for the different polymers: at a given $(T - T_g)$ difference, each polymer has its own dependence of $\log \tau_1$ as a function of $(T - T_g)$. Although the τ_1 processes are controlled by the segmental motions of the polymer chains involved in the glass transition phenomena, the polymer matrices do not share the same local dynamics at a given $(T - T_\infty)$ or $(T - T_g)$ difference. Therefore, neither T_∞ nor the glass transition temperature T_g can be considered as good descriptors for rescaling the segmental motions in bulk polymers at temperatures well above T_g.

It is of interest to define a reference state in which all the polymers are in equivalent states from the point of view of their local mobility. In order to define such equivalent states, one has to look for a property in relation with the frequency window of each experiment. For the ^{13}C NMR relaxation experiments, the temperature, $T_{ref/NMR}$, at which the spin–lattice relaxation time $T_1(^{13}C)$ reaches its minimum may constitute such an appropriate reference state. The $T_1(^{13}C)$ minimum is directly related to the local mobility in the frequency range defined by $(\omega_H - \omega_C)$, (ω_C) and $(\omega_H + \omega_C)$ and can be easily determined experimentally. It is independent of the model used to describe the local dynamics. However, due to the flat character of the $T_1(^{13}C)$ minimum, the uncertainty in determining the experimental values of $T_{ref/NMR}$ is quite large. Moreover the precise value of τ_1 at the $T_1(^{13}C)$ minimum slightly depends on the τ_2/τ_1 ratio. An alternative definition of the NMR reference temperature is the temperature $T_{ref/NMR10^{-9}}$ at which τ_1 is equal to 10^{-9} s, in the center of the $T_1(^{13}C)$ frequency window. In the latter case, $T_{ref/NMR10^{-9}}$ is model dependent. $T_{ref/NMR}$ reference temperatures have been obtained at 25.15 MHz for several polymers [74]. They are listed in Table 4 and show that $(T_{ref/NMR} - T_g)$ strongly varies from one polymer to an other. For example $T_{ref/NMR} - T_g$ is 36 °C higher in PIB than in PI, which implies that the same mobility in terms of correlation time τ_1 is obtained at 36 °C higher in PIB than in PI.

The next step of this approach is to relate the reference temperatures to data obtained from viscoelastic experiments. Evaluation of fractional free volumes f_{ref} at the $T_{ref/NMR}$ reference temperature clearly shows that the same fractional free volume is not required for all polymers to perform conformational jumps characterized from NMR by the same value of the correlation time τ_1.

The free volumes $v_f(T_{ref})$ at the reference temperatures can be derived from f_{ref} and Van Krevelen's data on amorphous polymers [75]. They represent the free volume per mole of repeat unit of the polymer divided by the number of main-chain atoms per repeat unit. Once more, the values differ from one polymer to another, which implies that other molecular parameters have to be considered.

Another quantity of interest for the local dynamics is the monomeric friction coefficient ζ_0 which characterizes the resistance encountered by a monomer unit moving through its surroundings [76]. It has been shown to follow the WLF law. The variations of the monomeric friction coefficient ζ_0 as a function of temperature are plotted in Fig. 19 for several polymers. PI and PIB data have

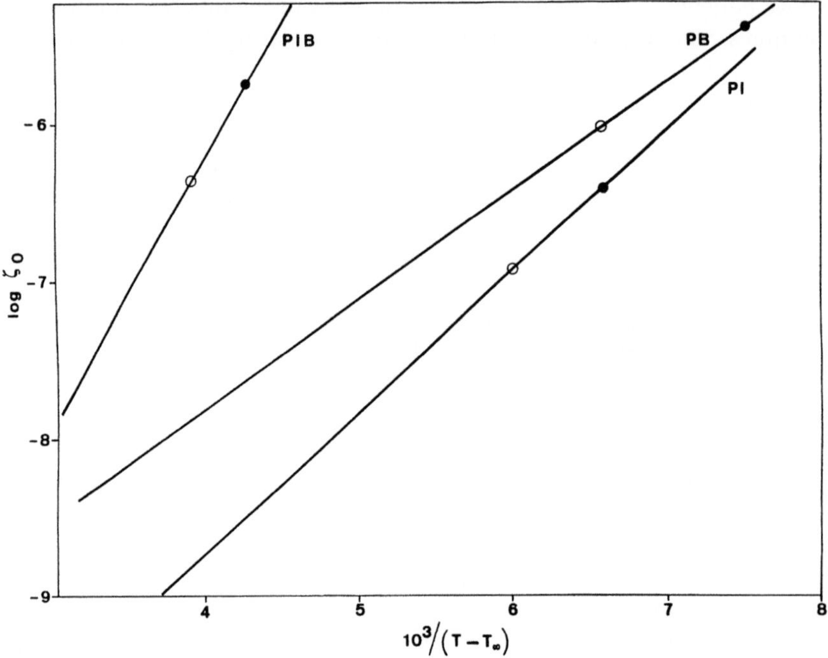

Fig. 19. Plots of log ζ_0 as a function of $10^3/(T - T_\infty)$ and $\log \zeta_0$ values at the $T_{ref/NMR}$ (●) and $T_{ref/NMR\,10^{-9}}$ (○) reference temperatures of the polymer matrices (after [74])

been taken from Ferry's book [72]. For PB, the approximation $\zeta_0 = \zeta_1$, where ζ_1 is the friction coefficient of a foreign molecule of like size, has been used. For a given $(T - T_\infty)$ difference, the ζ_0 value strongly depends on the considered polymer. However, $\log \zeta_0$ at $T_{ref/NMR}$ and $T_{ref/NMR\,10^{-9}}$ have the same order of magnitude for the polymers considered. Such a result implies that at these reference temperatures the polymers share a nearly common value of ζ_0. Therefore, the monomeric friction coefficient of the bulk polymer appears as one of the main factors that control the local dynamics in bulk polymers at temperatures well above the glass transition temperature. Extensive discussion concerning the role of ζ_0 can be found in [74] where comparison between NMR and excimer fluorescence studies of local dynamics in bulk polymers at temperatures well above T_g shows that the amplitude of the local jumps observed by NMR depends on the precise chemical nature of the polymer chain, i.e. on both intramolecular constraints and intermolecular interactions.

5.5 *Local Dynamics in Compatible Polymer Blends: Relative Influence of Intramolecular Constraints and Intermolecular Interactions*

In a blend of two given compatible homopolymers, the intramolecular constraints are constant, fixed by the chemical structure of the components independently of the blend composition, whereas the intermolecular interactions depend on the blend composition. Therefore, the study of compatible blends as a function of their composition is of special interest for investigating the relative influence of these two contributions on the local dynamics not only in the case of blends, but also in the larger field of bulk polymers.

The polystyrene (PS)/poly(vinylmethyl ether) (PVME) blends are compatible at temperatures below their critical solution temperature (LCST). They show only one glass transition temperature, $T_{g\,blend}$, whose values, as observed by DSC, are listed in Table 5 for a PS chain of Mn = 19160 and a PVME chain of Mn = 46500 and various compositions. The local dynamics of the PS and PVME chains in these PS/PVME blends have been studied by MAS/DD ^{13}C NMR at temperatures well below the LCST in order to avoid density fluctuations [77].

As an example, the 75MHz CP/MAS/DD ^{13}C NMR spectra of the 52% wt PS sample are shown in Fig. 20 as a function of temperature. Of particular interest are the peaks due to the aromatic carbons of polystyrene and the PVME methine carbons: on increasing temperature, large line broadening first occurs and a further decrease of the line width at high temperatures. The temperatures at which a maximum line broadening occurs are listed in Table 5. They depend on the carbon line under study and the composition of the blend.

The line widths at mid-height, Δ_{O-CH}, measured at 75 MHz for the PVME methine carbon from the MAS/CP/DD ^{13}C NMR spectra are reported in Fig. 21 for the 38%, 52% and 67% wt PS blends. At high temperature, the measured line widths are independent of the nature (cross-polarization or direct polarization (DP) with long recycle times) of the pulse sequence used. Since the CP pulse sequence accentuates the contribution of the most rigid carbons, whereas the direct polarization experiment with long recycle delays leads to the observation of the whole sample with the same contribution from each carbon-13

Table 5. Glass transition temperatures of the blends, $T_{g\,blend}$, (as measured by DSC) and temperatures (°C) at which the maximum Δ_{O-CH}, Δ_{arom} and Δ_{chain} line broadenings are observed for the PVME methine carbon and PS protonated aromatic and chain carbons, respectively. (after [77])

% wt PS	38	52	67
$T_{g\,blend}$	−19	−6	19
Temperature of the maximum Δ_{O-CH} line broadening	20	29	40
Temperature of the maximum Δ_{arom} line broadening	55	75	98
Temperature of the maximum Δ_{chain} line broadening			105

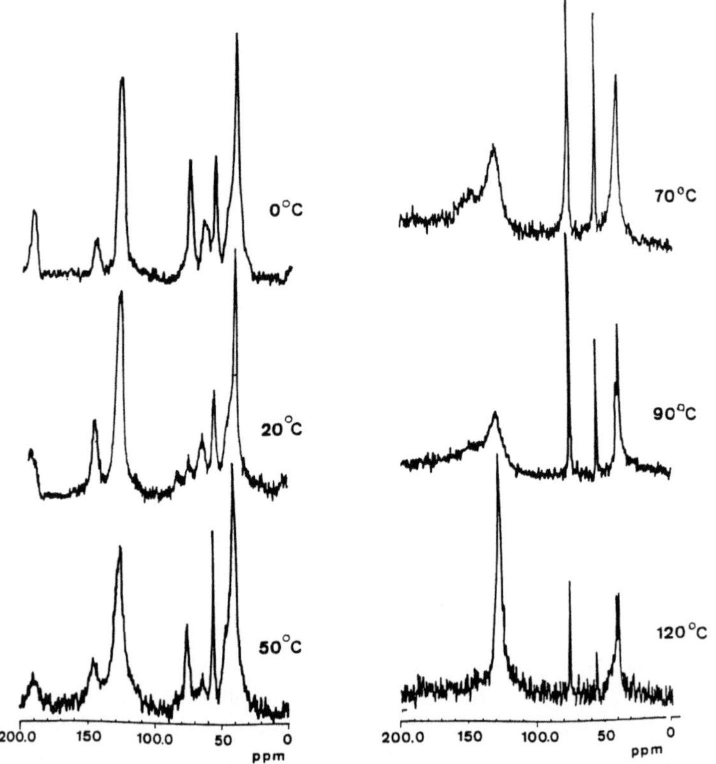

Fig. 20. High-resolution solid-state CP/MAS/DD ^{13}C NMR spectra of the 52% wt PS sample recorded at 75 MHz as a function of temperature, T. (contact time: 1 ms) (after [77])

nucleus, the close similarity of the line widths measured by the two NMR techniques for the PVME methine carbon at high temperatures is significant. It implies that, at these temperatures, the blend samples are not highly heterogeneous. This point is in agreement with conclusions derived from fluorescence emission and small-angle neutron scattering studies of PS/PVME blends [78, 79] that have shown that, at temperatures sufficiently lower than the phase coexistence temperature, the system fluctuations are low.

The order of magnitude of the maximum Δ_{O-CH} line broadening is greater than the static line broadenings [33]. For the PVME methine carbon, which is coupled to a directly bonded proton and whose chemical-shift anisotropy is weak, the observed line broadenings can be interpreted in terms of a static contribution and motional modulation of the dipolar carbon–proton coupling. They show the classical $(\pi T_{2\,m})^{-1} + (\pi T_{2\,res})^{-1}$ dependence on temperature and frequency of molecular motions and are a clear indication of molecular processes involving the PVME methine carbon with a frequency at the temperature of

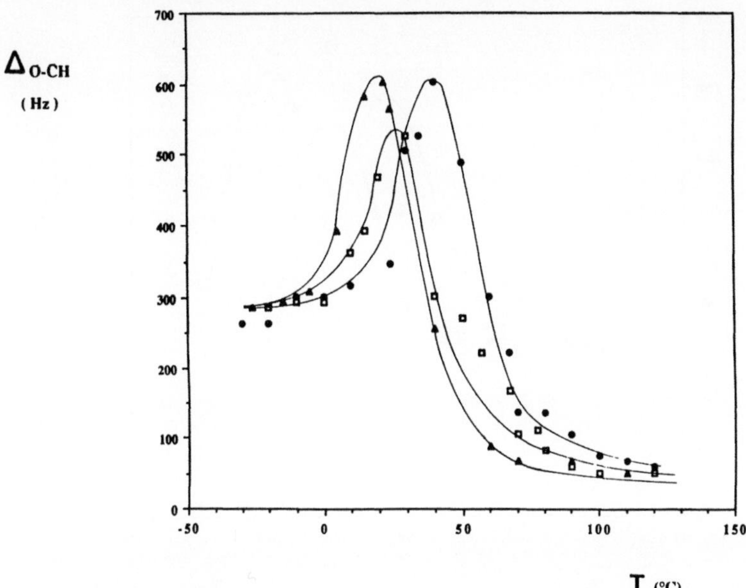

Fig. 21. Variation of the 75 MHz Δ_{O-CH} line width as a function of temperature (after [77]). ▲: 38% wt PS sample; □: 52% wt PS sample; ●: 67% wt PS sample

the maximum line broadening of the order of the strength of the proton decoupling field, i.e. 62.5 kHz.

The line widths at mid height, Δ_{arom}, determined at 75 MHz from the MAS/CP/DD ^{13}C NMR spectra for the PS protonated aromatic carbon lines are shown in Fig. 22. As for the PVME methine carbon line widths, experiments performed at high temperature have shown that they are independent of the nature (CP or DP) of the pulse sequence used. The absolute values and temperature dependence of Δ_{arom} cannot be accounted for by static effects only and point out the existence of motions of the PS phenyl rings. However, in this case the precise interpretation of the phenomena is not straightforward, since a protonated aromatic carbon has both quite strong chemical-shift anisotropy and dipolar carbon–proton coupling. Therefore, both the motional modulation of the chemical-shift anisotropy and motional modulation of the dipolar coupling may be active relaxation mechanisms for this carbon and maximum line broadenings are expected for motions either of the order of the spinning speed, i.e. 4 kHz, or of the strength of the proton decoupling field, i.e. 62.5 kHz. In this regard, it is of interest to compare the behavior of the lines assigned to the PS protonated and non-protonated aromatic carbons since the relative contribution of the modulation of the chemical-shift anisotropy to the maximum line broadening is larger for non-protonated aromatic carbons than for protonated aromatic carbons. Although the accuracy of the measurements of the substituted carbon line width is quite limited, results displayed in Fig. 23 clearly show that

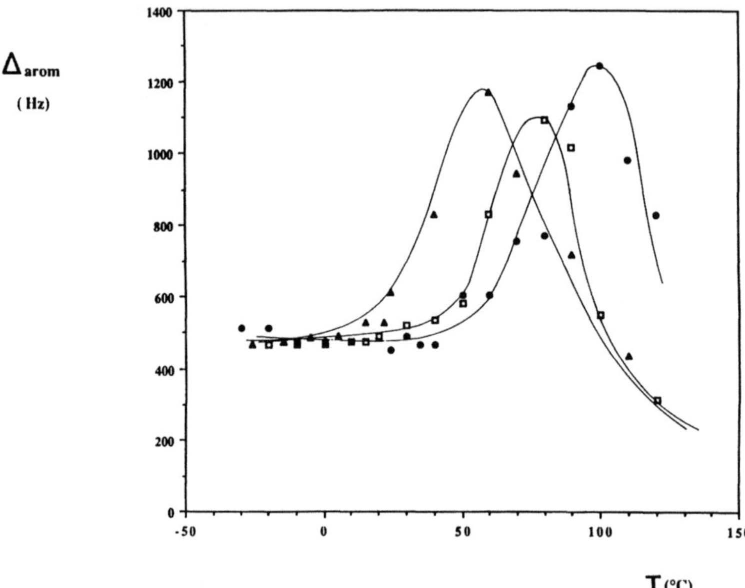

Fig. 22. Variation of the 75 MHz Δ_{arom} line width as a function of temperature (after [77]). ▲: 38% wt PS sample; □: 52% wt PS sample; ●: 67% wt PS sample

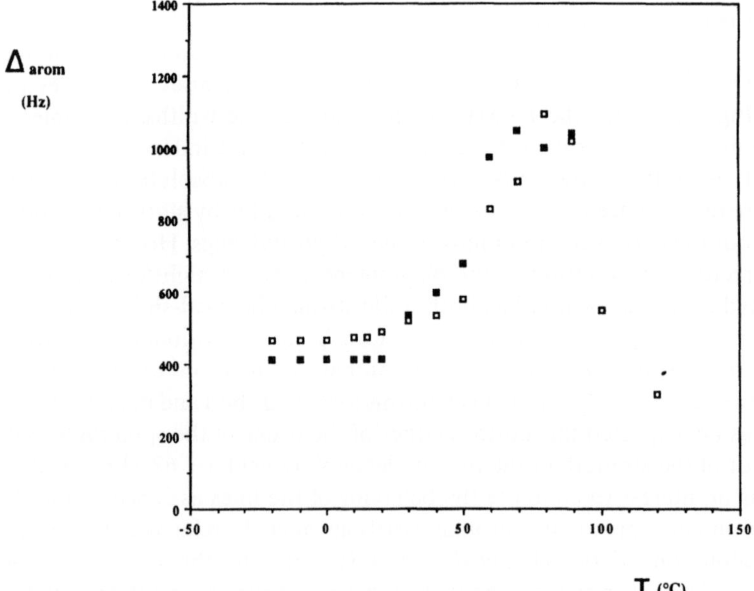

Fig. 23. Variation of the 75 MHz line width of the protonated (□) and non-protonated (■) aromatic carbons as a function of temperature in the 52% wt PS sample (after [77])

PS protonated and non-protonated aromatic carbons have a parallel temperature dependence of their line broadenings with quite similar temperatures at the maximum line broadening. In addition, the line widths of both peaks are quite similar. A likely explanation for such a similarity in the behavior of the protonated and non-protonated carbon lines is that the motional modulation of the chemical-shift anisotropy plays a major role in the line broadening of all the PS aromatic carbon lines. If this is the case, the motions involving the phenyl rings have frequencies in the neighborhood of 4 kHz at the maximum line broadening. However, it must be noted that, although the two mechanisms probe motions in somewhat different frequency ranges, this frequency difference corresponds to a temperature difference of only a few degrees, of the order of the accuracy of the determination of the temperature of the maximum line broadening.

Information on the behavior of the PS main-chain carbons has been derived from 20 μs-contact time CP experiments. During such a short contact time only carbons with a strong $^{13}C-^{1}H$ dipolar coupling, i.e. PS protonated carbons that have a relatively rigid behavior, can acquire some magnetization. Temperature dependence of the line widths of the PS main-chain carbons is parallel to the temperature dependence observed for the PS aromatic carbons, with a maximum at a few degrees above the temperature of the Δ_{arom} maximum as expected from the higher (62.5 kHz) observation frequency of motions.

The results reported above lead to the identification of different types of local modes in the PS/PVME blends. For PVME, the behavior of the methine carbon indicates the existence of local main-chain processes. In the case of PS, the aromatic and main-chain carbon lines have parallel temperature dependence with extrema at very similar temperatures. These observations imply that the PS main-chain and side-rings are involved in correlated motions.

It is of interest to note that the maxima of the line broadenings occur at much lower temperatures in the PVME chain than in the PS one. Therefore, although the blends are compatible as proved by the existence of a single glass transition temperature observed by DSC, $T_{g\,blend}$, the PVME and PS chains in the PS/PVME blends do not share the same local chain dynamics at a given $(T - T_{g\,blend})$ difference [77, 80]. In the present experiments, the frequencies of the motions observed for the PS aromatic carbons and PVME methine carbon are centered around 4 and 62.5 kHz, respectively. For the same observation frequency of 62.5 kHz, the difference in behavior of these two carbons would be still larger.

Another important point is the dependence of the temperature of the maximum line broadenings on the composition of the blend, which is observed for the PVME component as well as for the PS component (Table 5). This result indicates that the local dynamics of each homopolymer in the blend is not only determined by intramolecular parameters, but also depends on intermolecular interactions.

The local motions observed in the range of a few kHz or tens of kHz in the above experiments occur at temperatures somewhat higher than the glass transition temperature of the compatible blend. Therefore, they are likely related to the motional processes associated with the glass transition phenomena. For homopolymers, the temperature dependence of the correlation times of the motional processes associated with the glass transition phenomenon is well-represented by the WLF equation [71, 72]. For polymer blends, the availability of viscoelastic data is limited. In the specific case of PS/PVME blends, the viscoelastic behavior of the blend as a whole can be represented to some extent by a WLF-type equation [81]. One can also expect the viscoelastic responses of the individual components, which are free-volume dependent, to obey a WLF-type equation. Such a WLF equation for the PVME chains inside the blend can be written as:

$$\log \tau_C(T)_{\mathrm{PVME}} = -\frac{C^g_{1(\mathrm{PVME})}(T - T_{\mathrm{g\,blend}})}{C^g_{2(\mathrm{PVME})} + T - T_{\mathrm{g\,blend}}} + \log \tau_C(T_{\mathrm{g\,blend}})_{\mathrm{PVME}} \tag{28}$$

and for the PS chains inside the blend:

$$\log \tau_C(T)_{\mathrm{PS}} = -\frac{C^g_{1(\mathrm{PS})}(T - T_{\mathrm{g\,blend}})}{C^g_{2(\mathrm{PS})} + T - T_{\mathrm{g\,blend}}} + \log \tau_C(T_{\mathrm{g\,blend}})_{\mathrm{PS}} \tag{29}$$

where $C^g_{1(\mathrm{PVME})}$, $C^g_{2(\mathrm{PVME})}$, $C^g_{1(\mathrm{PS})}$, $C^g_{2(\mathrm{PS})}$, $\tau_C(T_{\mathrm{g\,blend}})_{\mathrm{PVME}}$ and $\tau_C(T_{\mathrm{g\,blend}})_{\mathrm{PS}}$ are unknown functions of the blend composition.

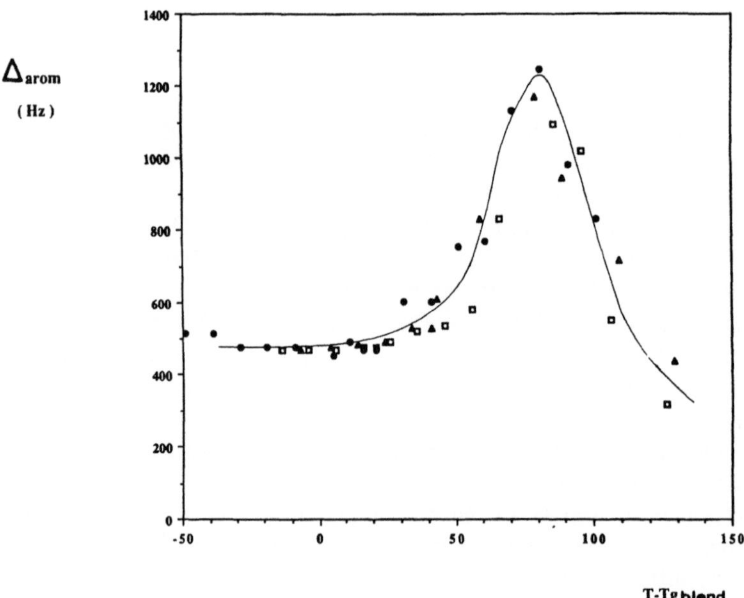

Fig. 24. Variation of the 75 MHz Δ_{arom} line width as a function of $(T - T_{\mathrm{g\,blend}})$ (after [77]). ▲: 38% wt PS sample; □: 52% wt PS sample; ●: 67% wt PS sample

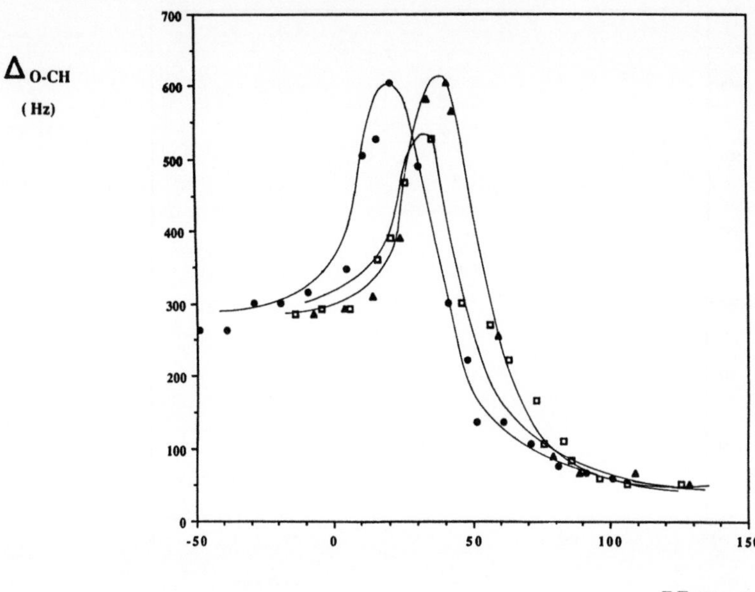

Fig. 25. Variation of the 75 MHz Δ_{O-CH} line width as a function of $(T - T_{g\ blend})$ (after [77]). ▲: 38% wt PS sample; □: 52% wt PS sample; ●: 67% wt PS sample

Variations of Δ_{arom} and Δ_{O-CH} as a function of $(T - T_{g\ blend})$ are shown in Figs. 24 and 25. The Δ_{arom} dependence on $(T - T_{g\ blend})$ is identical for the three blends under study, which implies that, within the sensitivity of the ^{13}C NMR experiments, the motions of the PS aromatic carbons can be described by the same values of the $C^g_{1\ (PS)}$, $C^g_{2\ (PS)}$, and $\tau_C(T_{g\ blend})_{PS}$ coefficients, independently of the blend composition. Since the broadenings of the PS lines show similar temperature dependences whatever the nature (aromatic or aliphatic, side-chain or main-chain) of the carbon under consideration, the local motions of the PS chains in the blends can also be represented by the same values of the $C^g_{1\ (PS)}$, $C^g_{2\ (PS)}$ and $\tau_C(T_{g\ blend})_{PS}$ coefficients independently of the blend composition. Therefore, as determined by the local dynamic behavior of the PS chains, the fractional free volume at the glass transition of the blend and the expansion coefficient of the free volume do not depend on the blend composition.

In contrast, the variation with $(T - T_{g\ blend})$ of the PVME Δ_{O-CH} line widths depends on the blend considered. At the temperature of the maximum Δ_{O-CH} line broadening, the local motions undergone by the PVME methine carbons have the same correlation time whatever the blend composition. Variations of Δ_{O-CH} as a function of $T - T_{max}$, where T_{max} is the temperature of the maximum line broadening, are plotted in Fig. 26. Results thus obtained are almost identical for the three blends under consideration. Such behavior implies that the local motions of the PVME chains in the three blends, with $T_0 = T_{max}$ as the reference

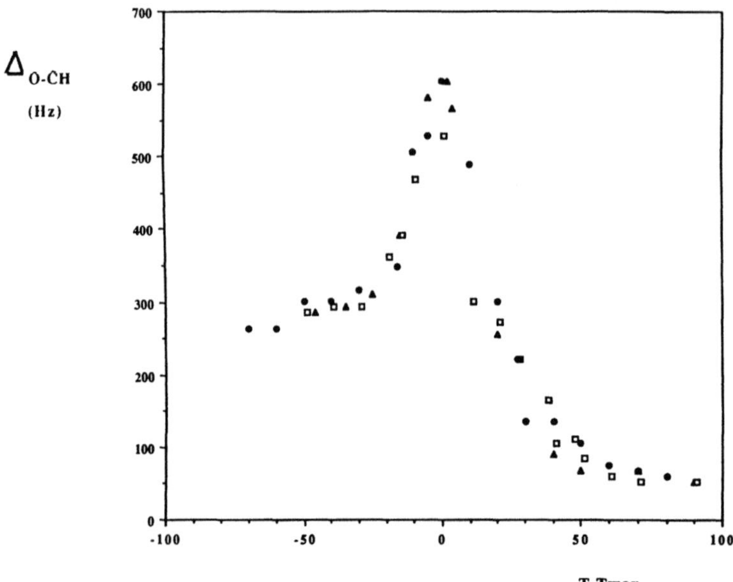

Fig. 26. Variation of the 75 MHz Δ_{O-CH} line width as a function of $(T - T_{max})$ (after [77]). ▲: 38% wt PS sample; □: 52% wt PS sample; ●: 67% wt PS sample

temperature, can be described by quite the same values of the WLF coefficients $- C^0_{1\,(PVME)}$, $C^0_{2\,(PVME)}$ and $\tau_C(T_0)_{PVME}$—where the reference temperature $T_0 = T_{max}$ depends on the blend composition. These results, combined with Eq. (25), lead to the important conclusion that the $C^g_{1(PVME)}C^g_{2(PVME)}$ product is independent of the blend composition.

In spite of the identical values of the $C^g_{1(PVME)}C^g_{2(PVME)}$ product, it is important to notice that the value of the differences $(T_{max} - T_{g\,blend})$ varies according to the blend composition (Table 5). Rewriting Eq. (26) as: $T_0 - T_g = C^0_2 - C^g_2$, and considering that, at $T_0 = T_{max}$, the values of $C^0_{2(PVME)}$ are independent of the blend composition, yields the conclusion that $C^g_{2\,(PVME)}$ has to depend on the composition. Use of Eq. (25) leads to the same conclusion for $C^g_{1(PVME)}$. The latter results imply that the fractional free volume at $T_{g\,blend}$, as detected by the dynamics of the PVME chains, changes with blend composition. According to data reported in Table 5, the fractional free volume for the PVME units at the glass transition of the blend would increase with increasing PS content in the blend.

Information regarding the effect of the blend composition on $\tau_C(T_{g\,blend})_{PVME}$ can be obtained by applying Eq. (28) at $T = T_0 = T_{max}$:

$$\log \tau_C(T_{g\,blend})_{PVME} = \frac{C^g_{1(PVME)}(T_0 - T_{g\,blend})}{C^g_{2(PVME)} + T_0 - T_{g\,blend}} + \log \tau_C(T_0)_{PVME}$$

$$= \frac{C^g_{1(PVME)}(T_0 - T_{g\,blend})}{C^0_{2(PVME)}} + \log \tau_C(T_0)_{PVME} \qquad (30)$$

Since $\log \tau_C(T_0)_{PVME}$ and $C^0_{2\,(PVME)}$ are independent of blend composition at $T_0 = T_{max}$, results obtained show that $\tau_C(T_{g\,blend})_{PVME}$ is a decreasing function of PS content, in agreement with data displayed in Fig. 25. In other words, when the proportion of PS in the blends increases, the local motions of the PVME chains are faster and faster when the temperature reaches the glass transition of the blend.

The difference in the motional behavior of the PS and PVME chains can be analyzed in terms of the relative sensitivity of the local motions to intramolecular constraints and intermolecular interactions. The fact that the $C^g_{1\,(PS)}$, $C^g_{2\,(PS)}$ and $\tau_C(T_{g\,blend})_{PS}$ coefficients are independent of the blend composition would imply that, for the local motions of the PS units, which is the more rigid species in the blend, the intramolecular constraints are the dominant factors. On the contrary, for the local motions of the more flexible PVME chains, the intramolecular contribution is reflected through the faster dynamics of the PVME chains as compared to the PS one. Besides, the composition dependence of the $C^g_{1(PVME)}$, $C^g_{2(PVME)}$, and $\tau_C(T_{g\,blend})_{PVME}$ coefficients would indicate a significant effect of the intermolecular interactions.

6 Conclusions

The examples quoted in this chapter give a clear indication that high-resolution ^{13}C NMR is a very powerful tool for studying local dynamics in bulk polymers. Of course, other experimental techniques can provide information in these fields. However, what is specific to ^{13}C NMR is its high selectivity allied with the natural abundance of carbon-13 nuclei. A typical example, demonstrating the particular interest of this method, is the study of secondary relaxations in polymer systems: dynamic mechanical measurements lead to a description of these phenomena in terms of temperatures and frequencies, which can be conveniently summarized on a relaxation map. However, the precise identification of the motional processes which are involved in the secondary relaxations can only be deduced from a comparison of the mechanical behavior of different systems. High-resolution solid-state ^{13}C NMR used in the temperature and frequency ranges of interest, which can tell how the different NMR parameters of a given carbon are modified by the motion, is thus able to provide an unambiguous identification of the chemical groups involved in the motions responsible for these transitions. Similar examples showing the power of the technique can also be taken among the studies dealing with bulk polymers at temperatures well above the glass transition temperature. Results obtained from spin–lattice relaxation time determinations on a large number of polymer melts have shown that the local motions can be described in terms of a damped diffusion of orientations along the chain sequence and independent librations. The relative influence of intramolecular constraints and intermolecular interactions has been detected either by comparing NMR data with results obtained from excimer fluorescence studies or by following the behavior of the individual components in compatible blends as a function of the blend composition. However, at the present time, whatever the temperature below or above the glass-transition temperature, and except for very simple processes such as the chair–chair inversion of the cyclohexyl ring of poly(cyclohexyl methacrylate), one cannot derive the exact geometry of the motion from ^{13}C NMR experiments. In this domain, molecular dynamics simulations should be able to bring important complementary information.

References

1. Cohen-Addad JP, Faure JP (1974) J. Chem. Phys. 61: 2440
2. English AD, Dybowski CR (1984) Macromolecules 17: 446
3. English AD (1985) Macromolecules 18: 178
4. Herzfeld J, Berger A (1980) J. Chem. Phys. 73: 6021
5. Stejskal EO, Schaefer J , McKay RA (1977) J. Magn. Reson. 25: 569
6. Frydman L, Chingas GC, Lee YK, Grandinetti PJ, Eastman MA, Barrall GA, Pines A (1992) J. Chem. Phys. 97: 4800
7. Maricq M, Waugh JS (1977) Chem. Phys. Lett. 47: 327
8. Waugh JS, Maricq M, Cantor R (1978) J..Magn. Reson. 29: 183
9. Bax A, Szeverenyi NM, Maciel GE (1983) J. Magn. Reson. 55: 494
10. Terao T, Fujii T, Onodera T, Saika A (1984) Chem. Phys. Lett. 107: 145
11. Zeigler RC, Wind RA, Maciel GE (1988) J. Magn. Reson. 79: 299
12. Bax A, Szeverenyi NM, Maciel GE (1983) J. Magn. Reson. 52: 147
13. Lippmaa E, Alla M, Tuherm T (1976) Proceedings of the 19th Congress Ampere, Heidelberg
14. Yarim-Agaev Y, Tutunjian PM, Waugh JS (1982) J. Magn. Reson. 47: 51
15. Bax A, Szeverenyi NM, Maciel GE (1983) J. Magn. Reson. 51: 400
16. Tycko R, Dabbagh G, Mirau PA (1989) J. Magn. Reson. 85: 265
17. Aue WP, Ruben DJ, Griffin RG (1981) J. Magn. Reson. 43: 472
18. Aue WP, Ruben DJ, Griffin RG (1984) J. Magn. Reson. 80: 1729
19. Kolbert AC, Raleigh DP, Levitt MH, Griffin RG (1989) J. Chem. Phys. 90: 679
20. Sergot P, Laupretre F, Louis C, Virlet J (1981) Polymer 22: 1150
21. Laupretre F, Monnerie L, Virlet J (1984) Macromolecules 17: 1397
22. Schaefer J, McKay RA, Stejskal EO, Dixon WT (1983) J. Magn. Reson. 52, 123
23. Mehring M (1976) High resolution NMR spectroscopy in solids. Springer Berlin Heidelberg New York (NMR Basic Principles and Progress. vol 11)
24. Muller L, Kumar A, Baumann T, Ernst RR (1974) Phys. Rev. Lett. 32: 1402
25. Chingas GC, Garroway AN, Bertrand RD, Moniz WB (1981) J. Chem. Phys. 74: 127
26. Garroway AN, Moniz WB, Resing HA (1979) Faraday Symp. Chem. Soc. 13: 63
27. Garroway AN, Vanderhart DL, Earl WL (1981) Philos. Trans. R. Soc. London, Ser. A 299: 609
28. Stejskal EO, Schaefer J, Steger TR (1979) Faraday Discuss. Chem. Soc., Faraday Symposium 13: 56
29. Schaefer J, Stejskal EO, Steger TR, Sefcik MD, McKay RA (1980) Macromolecules 13: 1121
30. Vanderhart DL, Garroway AN (1979) J. Chem. Phys. 71: 2773
31. Schaefer J, Sefcik MD, Stejskal EO, McKay RA (1984) Macromolecules 17: 1118
32. Schaefer J, Sefcik MD, Stejskal EO, McKay RA (1981) Macromolecules 14: 280
33. Vanderhart DL, Earl WL, Garroway AN (1981) J. Magn. Reson. 44: 361
34. Suwelack D, Rothwell WP, Waugh JS (1980) J. Chem. Phys. 73: 2559
35. Rothwell WP, Waugh JS (1981) J. Chem. Phys. 74: 2721
36. Allerhand A, Doddrell D, Komoroski RJ (1971) J. Chem. Phys. 55: 189
37. Heijboer J (1972) Ph.D. Thesis, University of Leyden
38. Laupretre F, Virlet J, Bayle JP (1985) Macromolecules 18: 1846

39. Monnerie L, Halary JL, private communication
40. Gerard A, Laupretre F, Monnerie L (1993) Macromolecules 26: 3313
41. Pausak S, Pines A, Waugh JS (1973) J. Chem. Phys. 59: 591
42. Veeman WS (1984) Progress in NMR Spectroscopy 16: 193
43. Kempf J, Spiess HW, Haeberlen U, Zimmermann H (1972) Chem. Phys. Lett. 17: 39
44. Dongen Torman J, Veeman WS, De Boer E (1978) J. Magn. Res. 32: 49
45. Dammont FR, Kwei TK (1967) J. Polym. Sci., Part A2, 5: 761
46. Pogany GA (1970) Europ. Polym. J. 6: 343
47. Murayama T, Bell JP (1970) J. Polym. Sci., Part A2, 8: 437
48. Ochi M, Iesako H, Shimbo M (1985) Polymer 26: 457
49. Ochi M, Iesako H, Shimbo M (1986) J. Polym. Sci. B, 24: 1271
50. Ochi M, Yoshizumi M, Shimbo M (1987) J. Polym. Sci. B 25: 1817
51. Charlesworth JM (1988) Polym. Eng. Sci. 28: 221
52. Charlesworth JM (1988) Polym. Eng. Sci. 28: 230
53. Cukierman S, Halary JL, Monnerie L (1991) J. of non-cryst. Solids 131–133: 898
54. Garroway AN, Ritchey WM, Moniz WB (1982) Macromolecules 15: 1051
55. Eustache RP (1990) thesis, University of Paris 6
56. Laupretre F, Eustache RP, Monnerie L (1992) Polymer Preprints 33: 136
57. Hall CK, Helfand E (1982) J. Chem. Phys. 77: 3275
58. Viovy JL, Monnerie L, Brochon JC (1983) Macromolecules 16: 1845
59. Bendler JT, Yaris R (1978) Macromolecules 11: 650
60. Jones AA, Stockmayer WH (1975) J. Polym. Sci., Polym. Phys. Ed. 15: 847
61. Viovy JL, Monnerie L, Merola F (1985) Macromolecules 18: 1130
62. Veissier V (1987) Thesis, Université Pierre et Marie Curie, Paris
63. Dejean de la Batie R, Laupretre F, Monnerie L (1988) Macromolecules 21: 2045
64. Howarth OW (1980) J. Chem. Soc., Faraday Trans. 2, 76: 1219
65. Pople JA, Gordon M (1967) J. Am. Chem. Soc. 89: 4253
66. Dejean de la Batie R, Laupretre F, Monnerie L (1988) Macromolecules 21: 2052
67. Dejean de la Batie R, Laupretre F, Monnerie L (1989) Macromolecules 22: 122
68. Dejean de la Batie R, Laupretre F, Monnerie L (1989) Macromolecules 22: 2617
69. Buchenau U, Monkenbusch M, Stamm M, Majkrzak CF, Nucker N (1987) Workshop on polymer in dense systems, Grenoble, September 23–25
70. Kanaya T, Kaji K, Inoue K (1991) Macromolecules 24: 1826
71. Williams ML, Landel RF, Ferry JD (1955) J. Am. Chem. Soc. 77: 3701
72. Ferry JD (1980) In: Viscoelastic properties of polymers, 3rd edn Wiley, New York
73. Faivre JP (1985) Thesis, Université Pierre et Marie Curie, Paris
74. Laupretre F, Bokobza L, Monnerie L (1993) Polymer 34: 468
75. Van Krevelen DW (1972) Properties of polymers. Correlations with chemical structure. Elsevier, Amsterdam
76. Bueche F (1952) J. Chem. Phys. 20: 1959
77. Le Menestrel C, Kenwright A, Sergot P, Laupretre F, Monnerie L (1992) Macromolecules 25: 3020
78. Halary JL, Ubrich JM, Monnerie L, Yang H, Stein RS (1985) Polymer Commun. 26: 73
79. Halary JL, Ben Cheikh Larbi F, Oudin P, Monnerie L (1988) Makromol. Chem. 189: 2117
80. Takegoshi K, Hikichi K (1991) J. Chem. Phys. 94: 3200
81. Schneider HA, Wirbser J (1989) Polymer Preprints 30, 1: 54 and references therein.

Xenon NMR Spectroscopy

Dan Raftery[1] and Bradley F. Chmelka[2]

[1]Department of Chemistry, University of Pennsylvania, Philadelphia, Pennsylvania 19104–6323, USA
[2]Department of Chemical and Nuclear Engineering, University of California, Santa Barbara, California 93106–5080, USA

Table of Contents

Xenon NMR spectroscopy has become an important and widely used tool for characterizing complex chemical systems and host phases. Issues central to new and existing uses of xenon NMR in materials research are emphasized, including xenon mass transport considerations, the sensitivity of xenon's NMR parameters to local and macroscopic material structure and dynamics, and novel optical polarization techniques. After reviewing the foundations of ^{129}Xe and ^{131}Xe NMR spectroscopy methods, a number of applications are presented, particularly those focusing on new developments in the field.

NMR Basic Principles and Progress, Vol. 30

1 Introduction

Xenon has many interesting properties that have made it valuable to an impressive variety of physics, chemistry and materials science investigations. Important experimental measurements of many of these properties, both atomic and molecular, have come from xenon nuclear magnetic resonance (NMR) spectroscopy, which is a sensitive probe of local environment. Together with xenon's relative inertness, the sensitivity of xenon's NMR parameters to local bonding, symmetry, and motion allows a variety of different interactions to be probed, often non-invasively [1]. The large polarizability of xenon's electron cloud accounts for its facile adsorption properties and its large resonance (chemical) shift range. Two xenon isotopes exist that are accessible to NMR investigation: spin 1/2 ^{129}Xe, which comprises the vast majority of the xenon NMR literature (26.44% natural abundance, 0.021 sensitivity relative to ^{1}H); and spin 3/2 ^{131}Xe, which has a quadrupolar moment (21.18% natural abundance, 0.0028 sensitivity relative to ^{1}H).

This review will focus on the physics, chemistry, and diagnostic utility of xenon's interactions with its environment, as established by NMR measurements of its behavior in the presence of covalently bound ligands, host lattices, or bulk phases. Xenon NMR data yield insight into the types of interactions that xenon probe atoms experience, including the symmetry or geometry of xenon's local environment. For this reason, much of the xenon NMR literature deals with the elucidation of the structure of different materials that interact with xenon. Such studies span a diverse range, and include such important and complex classes of materials as proteins, catalysts, inclusion compounds, polymers, and so forth. Research applications of xenon NMR have become more prevalent after the discovery [2–7] that xenon could be used to determine structural aspects of materials with large surface areas. This has been driven mainly by interest in determining the structures, adsorption/catalytic properties, and guest/host interactions of molecular sieves, clathrates, and other high surface area materials. The application of xenon NMR to microporous materials has recently been reviewed [8–11], so our discussion here will be brief. Special emphasis will be placed on newly developed xenon optical-pumping techniques, which permit dramatic enhancement in sensitivity with corresponding applications to low-surface-area materials [12]. We will outline current and future directions for xenon NMR research and contrast experiments from an array of different fields of science.

1.1 Xenon Resonance Shifts

Probably the most important NMR parameter of ^{129}Xe is its resonance (chemical) shift, which has a range of over 7500 ppm (parts per million) that is enormous compared to the chemical shift range of ^{1}H (10 ppm) or ^{13}C (150 ppm).

Xenon's large and extremely polarizable electron cloud makes its NMR chemical shift an especially sensitive measure of local atomic interactions. As discussed by Ramsey [13], the chemical shift is composed of two parts, a diamagnetic shift originally derived by Lamb [14] and a paramagnetic contribution, which produce opposing effects on the nuclear resonance frequency. The chemical shift σ may be written as a sum of the two respective terms [13, 15]:

$$\sigma = \frac{e^2}{3mc^2}\int \frac{\rho}{r}\,dr - \frac{4}{3\Delta E}\left\langle 0 \left| \sum_{\mathrm{j,k}} L_{\mathrm{j}} \cdot \frac{L_{\mathrm{k}}}{r_{\mathrm{k}}^3} \right| 0 \right\rangle, \tag{1}$$

where ρ is the electron density, ΔE is the average excited-state energy, L_{j} and L_{k} are the angular momentum operators, and r_{k} is the electron-nuclear distance summed over all excited states. Diamagnetic shifts occur from the interaction of the electron orbitals with applied magnetic fields: the electrons precess in a way that opposes the field (Lenz' law). This effect is said to "shield" the nucleus from the applied field, as it reduces the resonance frequency of the nucleus. The diamagnetic term has been calculated for isolated atoms; for xenon, it is about 5600 ppm to lower frequency relative to the bare nucleus [17].

More important to chemistry are the paramagnetic shifts that arise from excited electronic state mixing and which generally increase the resonance frequency of the xenon nucleus. The paramagnetic term gives rise to the observed resonance shift in interactions of xenon with itself (a pressure-dependent shift

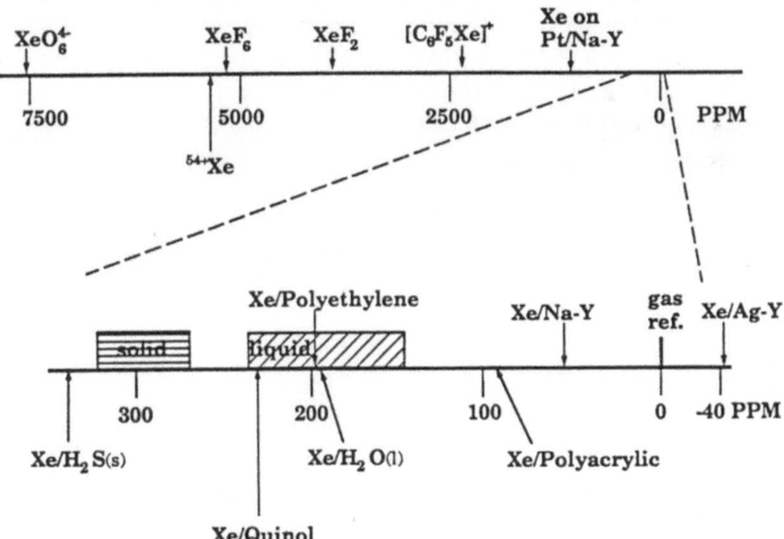

Fig. 1. ^{129}Xe NMR chemical shift scale for several representative xenon compounds, phases, and environments. The scale has been expanded at the low frequency end to show phase- and adsorption-dependent chemical shifts. Xenon at infinite dilution, a situation approximated by gas-phase atomic xenon at low pressure, has been defined as having a chemical shift of zero as a reference.

[18, 19]) and other species. Paramagnetic terms generally produce shifts more deshielded than the bare nucleus, as in xenon compounds such as XeO_6^{4-} [1]. Calculations of such shifts can be challenging [16], especially for those involving molecular bonds where two-centered (or more) interactions must be considered. Ramsey showed that the calculation of the paramagnetic term could be made more tractable by approximating the sum over the excited states by an average energy ΔE [13]. The choice of the reference frame for the calculations (which is equivalent to the choice of gauge) is also an important consideration, because it establishes the way in which the paramagnetic and diamagnetic terms comprise the chemical shift [15]. In Fig. 1, ^{129}Xe chemical shifts are shown for a variety of different xenon phases and chemical species. Note the unusual behavior of xenon adsorbed on Ag-Y zeolite for which a negative ^{129}Xe chemical shift (greater shielding) has been recorded relative to atomic xenon [20]!

1.2 Relaxation Effects

The dependence of nuclear relaxation times on the temperature and nature of the sample provide information on local xenon structure and dynamics. As originally discussed by Bloembergen [21], spin-lattice relaxation times T_1 are governed by the interactions of the nuclear spin with the random energy fluctuations of the environment (collectively called the "lattice") at the resonance frequency of the nuclear spin. The lattice is responsible for causing random spin-flips, which bring the nuclear spin system to an equilibrium that can be described by a Boltzmann distribution at the lattice temperature during a characteristic time T_1. Random fluctuations of some perturbation to the spin system, such as a magnetic field H, are described with a correlation function that has an exponentially decaying form:

$$\overline{H(T)H(T+t)} = \overline{H^2} e^{-t/\tau_c} \quad , \tag{2}$$

where the overbar refers to an ensemble average during the time period T to $T+t$. Real Fourier transformation leads to an expression for the frequency response of the relaxation rate [15, 22]:

$$\frac{1}{T_1} = \frac{2\gamma^2 \overline{H^2}}{3} \frac{\tau_c}{1+\omega^2\tau_c^2} \quad , \tag{3}$$

where τ_c is the correlation time and $\overline{H^2}$ is the time average value of a random, fluctuating magnetic field in the transverse (usually the x or y) direction. Spin–spin relaxation times, T_2, are similar in form to that for T_1, except that fluctuations at zero frequency (Hz) additionally contribute to T_2 relaxation. Energy conserving flip-flops of dipole–dipole-coupled spins, such as found in solids, cause rapid T_2 relaxation, for instance.

Figure 2 shows the dependence of ^{129}Xe relaxation times on the correlation time of fluctuating magnetic fields experienced by xenon in different bulk or adsorbed phases. The well known features of the plot arise from either slow

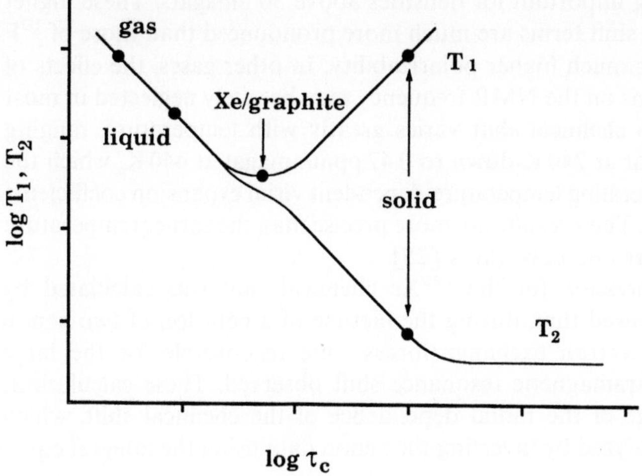

Fig. 2. T_1 and T_2 relaxation times plotted schematically as a function of correlation time τ_c, as discussed in Ref. [21]. Different xenon phases are shown in their appropriate region of the plot.

motions, such as those found in solids or at low temperatures, which lead to long T_1 relaxation times and broad lines (short T_2s), or fast motions, such as found in gas or liquid phases, which lead to long T_1s and narrow lines (relatively long T_2s). The T_1 minimum occurs where the correlation times (the mean collision time for gas-phase relaxation) is equal to the Larmor period, $\tau_c\omega = 1$. As examples, solid xenon is located in the slow-motion regime, gaseous and liquid xenon are in the fast-motion (also termed the "motionally-narrowed") regime, and xenon adsorbed on graphite is near the T_1 minimum for a typical high-field Larmor frequency.

2 NMR Studies of Bulk Xenon

2.1 Gas Phase Experiments

Following the first observations of ^{129}Xe by NMR [23], gas phase studies concentrated on relaxation and chemical shift measurements as functions of pressure and temperature. Jameson et al. [19] expressed the ^{129}Xe chemical shift as a virial expansion of xenon density ρ in the gas phase to describe the nonlinear behavior:

$$\sigma = \sigma_0 + \sigma_1\rho + \sigma_2\rho^2 + \cdots. \tag{4}$$

These results yield a value of 0.55 ppm chemical shift per amagat (the density of xenon under standard conditions at 298 K and 1 atm, $\sim 2.5 \times 10^{19}$ atoms/cm^3) for the linear term, with higher order contributions from three or more body interactions becoming important for densities above 50 amagats. These higher order ^{129}Xe chemical shift terms are much more pronounced than those of ^{19}F or ^{1}H, due to xenon's much higher polarizability. In other gases, the effects of three-body interactions on the NMR frequency may be safely neglected in most situations. The ^{129}Xe chemical shift varies greatly with temperature, ranging from 0.68 ppm/amagat at 240 K down to 0.47 ppm/amagat at 440 K, which the Jamesons fit by incorporating temperature-dependent virial expansion coefficients $\sigma_i(T)$ into Eq. (4) [24]. Their results are more precise than the earlier temperature measurements by Carr and coworkers [25].

A theoretical expression for the ^{129}Xe chemical shift was calculated by Adrian [26], who showed that, during the lifetime of a collision of two xenon atoms, short-range electron exchange forces were responsible for the large density-dependent paramagnetic resonance shift observed. These calculations depend on knowledge of the radial dependence of the chemical shift, which Jameson [27] has analyzed by inverting the xenon data using the integral equation:

$$\sigma_1(T) = \int_0^\infty \sigma(r) e^{-V(r)/kT} r^2 \, dr \quad , \tag{5}$$

where $V(r)$ is the Xe–Xe pair potential [28]. From this treatment, Jameson [27] showed that σ_1 is not a monotonically decreasing function of the radius, as had been assumed earlier by Adrian [26]. In fact, the chemical shift function has a form similar to $V^2(r)$, which may be modelled using a Lennard-Jones potential, for instance. Adrian also calculated the temperature dependence of the xenon chemical shift using a pair-distribution function and a chemical shift expression for two interacting xenon atoms [29]. His results display the large temperature dependence evidenced by experiments, plus the interesting effect of a maximum in the $\sigma_1(T)$ function at a temperature around 540 K, which to date has not been verified experimentally.

A number of measurements have been made of ^{129}Xe relaxation times in pure xenon gas. Hunt and Carr showed that the relaxation time of gas-phase ^{129}Xe is inversely proportional to the pressure, with $T_1 = 2 \times 10^5$ s/ρ (in amagats) [30], indicating that the correlation times are short (in the "motionally narrowed" regime). Torrey [31] showed that the spin–rotation interaction, where the nuclear spin couples to the angular momentum of a short-lived diatomic species, accounted for Hunt and Carr's relaxation data. As a result, xenon relaxes much faster than would be expected if dipole–dipole interactions were the dominant relaxation mechanism. Later work by Shizgal [32], using a Lennard-Jones 6–12 potential, built on the work of Adrian by calculating a ^{129}Xe relaxation time that is within 40% of the experimental result.

The relaxation of ^{131}Xe is dominated by collision-induced nuclear quadrupole interactions that couple the nuclear spin to the rotational angular momentum of the colliding atomic pair [33]. Adrian showed that exchange forces are the dominant contribution to spin relaxation in quadrupolar ^{131}Xe, as well as the key to the large xenon chemical shifts observed in both ^{129}Xe and ^{131}Xe. His calculations for ^{131}Xe relaxation are in agreement with measurements by Brinkmann and coworkers [34], who recorded a pressure-dependent relaxation time of $T_1 = 26\,s/\rho$ (in amagats).

2.2 *Xenon with Other Gases*

To elucidate the interatomic potential surface of xenon interacting with other atoms or molecules, Jameson, Jameson and coworkers have made numerous NMR investigations of mixtures of xenon with other gases. These studies have included xenon mixtures with other rare gases (Kr and Ar) [35]; molecules with tetrahedral symmetry (so-called spherical tops: CH_4, CF_4 and SiF_4) [36]; nonspherical molecules (NO_2, C_2H_4 and BF_3) [37]; linear molecules (CO, N_2, HBr, HCl, CO_2 and C_2H_2) [37, 38]; and paramagnetic species (O_2 and NO) [39-41]. From their earlier work on pressure- and temperature-dependent virial expansions of the xenon chemical shift in pure xenon gas, the Jamesons were able to examine xenon in other gas mixtures by accounting for and removing the xenon–xenon contribution. For the spherical top molecules, the results are as expected: the temperature dependence has the same form as the xenon–xenon virial coefficients, though smaller, probably as a result of the smaller polarizabilities of these molecules. Normalizing the $\sigma(r)$ functions in terms of the equilibrium collision distance and the well depth appears to be satisfactory for xenon in mixtures with similar partners, such as the other rare gases, or the spherical top molecules. However, between these groups, the differences are substantial, so that it has not been possible to generalize the treatment [36].

The properties of ^{129}Xe in mixtures with linear molecules reveal much different behavior. As shown in Fig. 3, the second virial coefficients for CO and N_2 are *increasing* functions of the temperature, unlike all of the other (diamagnetic) buffer gases studied for which the second virial coefficient decreases. According to the authors, this does not indicate that the functional form of $\sigma_1(r)$ is radically different in the presence of linear molecules, but simply that the average collision radius for xenon with CO and N_2 is much different (smaller) than for the other collision partners studied [38].

The xenon chemical shifts are also quite different in the presence of paramagnetic buffer gases, such as NO and O_2. Calculations of xenon in paramagnetic gas mixtures were made in 1972 by Buckingham and Kollman [42], who used the Xe–NO and Xe–O_2 collision pairs to establish that the Fermi-contact interaction is responsible in both cases for the unusually large resonance shifts observed in mixtures of these gases. The valence electrons of the paramagnetic species have a large spin density at the xenon nucleus due to exchange enhancement, as dictated by the Pauli exclusion principle. (This will be discussed more fully in the optical-pumping discussion which follows.)

The Jamesons and coworkers have determined the density-dependent chemical shifts of ^{129}Xe due to the presence of O_2 and NO, finding that

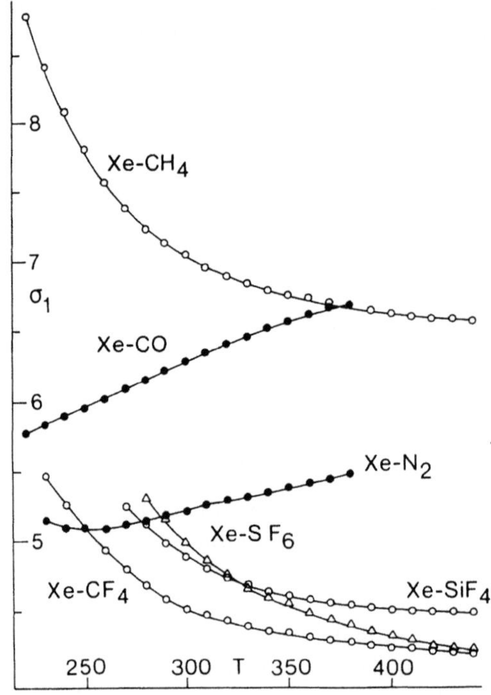

Fig. 3. Temperature dependences of ^{129}Xe second virial coefficients in binary mixtures of xenon and other gases. Used with permission from Ref. [38].

$\sigma_1 = -1.25$ ppm/amagat for oxygen and -0.92 ppm/amagat for NO buffer gases [39]. These are much larger (by factors of 6–10) than for a xenon–argon mixture, even though the argon atoms have very nearly the same size and polarizability as the two paramagnetic species. Data on temperature-dependent resonance shifts [43] support the proposition that the paramagnetic oxygen species cause a Fermi-contact-induced shift. Related work by these authors established that NO had a similar effect on the ^{129}Xe resonance frequency, although not as large in magnitude [40]. The contact term, given by:

$$\sigma_1(T) = \frac{-4\pi\overline{\mu^2_{\text{eff}}}}{9kT} \int \rho_{\text{spin}} \mathrm{e}^{-V(r)/kT} \mathrm{d}x^3 \qquad (6)$$

is proportional to the electron spin density of the paramagnetic species found at the xenon nucleus, averaged over the collision pair configurations, and is larger for NO than O_2. The two unpaired electrons in oxygen, however, make σ_1 larger overall for O_2. In addition, gas-phase ^{129}Xe relaxation in the presence of 0.005 to 0.87 mole fraction O_2 is dominated by intermolecular nuclear spin-electron spin dipole–dipole coupling, which can shorten ^{129}Xe T_1 times by several orders of magnitude [41].

2.3 Solid and Liquid Xenon

^{129}Xe chemical shifts, relaxation times, and self-diffusion coefficients for bulk xenon in both the liquid and the solid state have been investigated by Norberg and coworkers [44-47]. For example, in Fig. 4, large variations are observed in the ^{129}Xe spin–spin relaxation time T_2 of liquid and solid xenon as a function of temperature [44]. Abrupt changes in the slope of the plot in Fig. 4 correspond to Xe phase transitions, which produce structural changes that affect the local dynamical behavior of xenon atoms. In the liquid state, ^{129}Xe T_2's are relatively long, and the line widths are motionally narrowed. Diminished mobility of the xenon atoms in the solid phase, however, results in lower T_2 values and broader lines. Below the normal melting point ($T = 163$ K), the xenon line width increases monotonically until about 118 K, where the rigid lattice value of 300 Hz is reached. The observed line width is in agreement with the Van Vleck formulation for a dipole–dipole coupled solid. For solid xenon, the density dependence has been measured for the chemical shift by Brinkmann and Carr, who reported a linear variation of 6.0 ppm/amagat [48]. This is in agreement with theoretical work [49] and was later reproduced by Cowgill and Norberg [47].

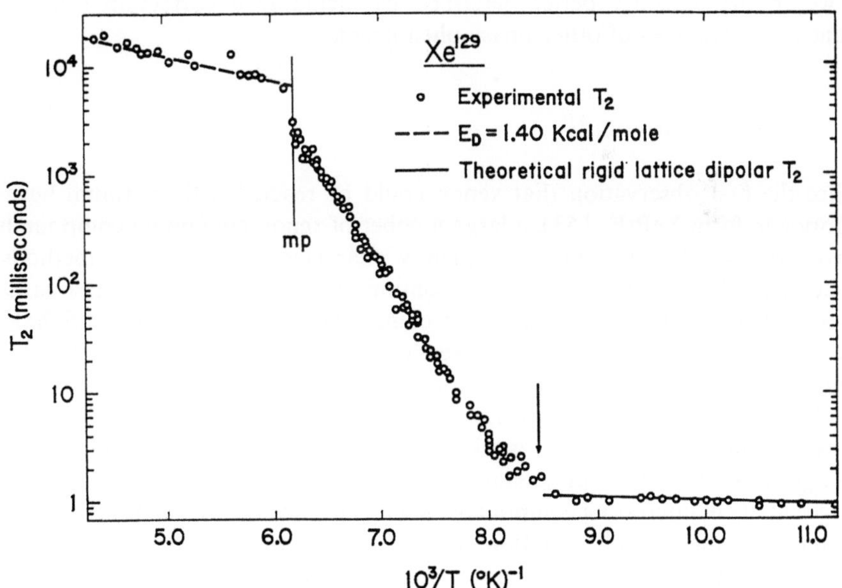

Fig. 4. ^{129}Xe spin–spin relaxation times (T_2) measured for liquid and solid xenon. The discontinuities observed for the slope in the different regions of the plot reflect changes in local xenon dynamics across phase transitions. Points corresponding to the melting point (163 K) and a secondary solid phase transition (118 K, see *arrow*) are clearly evident. Used with permission from Ref. [44].

Relaxation times for solid ^{131}Xe measured by Warren and Norberg [45] are in qualitative agreement with quadrupolar relaxation caused by a two-phonon mediated Raman process [50] for solids with a Debye distribution of phonon energies. The Debye temperature for solid xenon is estimated to be 55 K. Single-phonon relaxation processes are limited by the numbers of phonons at ω or 2ω, whereas a two-phonon Raman process couples the nuclear quadrupole spin to pairs of phonons with energies $\gamma_1 - \gamma_2 = \omega$ or 2ω and thus can involve all of the phonon frequencies. Van der Waals and exchange forces are responsible for creating the fluctuating electric field gradients that produce relaxation. At temperatures near the melting point, the T_1 data deviate from the theory, a fact which the authors ascribe to quadrupolar relaxation caused by diffusion to paramagnetic impurity centers. Warren and Norberg [46] also studied both ^{129}Xe and ^{131}Xe dipolar and quadrupolar echos at 4.2 K. They found that the spin–spin relaxation time T_2 of solid ^{131}Xe is as short as 8 ms at such low temperatures.

Carr and coworkers have used ^{131}Xe NMR to examine the xenon liquid-vapor coexistence curve near the critical temprature T_c [51]. In a limited range near the critical point, a single power law $B\epsilon^{\beta}$ describes the shape of the liquid-vapor coexistence curve $[(\rho_\ell - \rho_v)/\rho_c]$ over the reduced temperature range $[(T_c - T)/T_c]\epsilon$. A value of $\beta = 0.317$ compares well with the value predicted by separate model calculations [52]. Over a broader temperature range, however, the data are better described by including a correction term, as standard in analyses of other phase phenomena.

2.4 Xenon Compounds

Since the first observation that xenon could be reacted with platinum hexafluoride to form $XePtF_6$ [53], a large number of xenon-containing compounds have been synthesized and subsequently characterized by several methods, including ^{129}Xe NMR [1]. Early on, continuous wave (CW) double-resonance "spin-tickling" experiments on xenon compounds such as $XeOF_4$ and XeF_2 were performed by irradiating the xenon resonance while observing the ^{19}F signal [54]. The first direct observations of ^{129}Xe in chemical compounds were made by Seppelt and Rupp for XeF_4, XeF_6, XeO_3, $Xe(OSeF_5)_2$, $Xe(OTeF_5)_2$, plus others [55]. More recent work by Schrobilgen and coworkers [1, 56–58] has gone far to help us understand the structure and bonding of xenon in a large variety of interesting compounds. Xenon chemical shifts can be understood in terms of the formal oxidation state of xenon, ionic bond character, and solvent effects. Synthesis of xenon compounds with metal ligands, such as $XeF_2 \cdot nWOF_4$ and $XeF_2 \cdot nMoOF_4$ [57], as well as xenon-nitrogen bonds in $FXeN(SO_2F)_2$ [58] exemplify the interesting chemistry inherent in this work.

Two major trends describe the ^{129}Xe chemical shifts observed in these substances. Except for Xe^{IV} (xenon with formal charge of +4) in compounds such as XeF_4 and XeF_3^+, the chemical shift increases with the oxidation state [1] according to $Xe^{VIII} > Xe^{IV} > Xe^{VI} > Xe^{II}$, as suggested by Jameson and Gutowsky in their calculations [16]. Oxygen ligands tend to increase the chemical shift, probably due to the ionic character of the Xe^{+}—O^{-} bond [1]. Also evident in the spectra are highly resolved J couplings to ^{19}F, which can be as large as 8 kHz, and to other ligands, although fluorine couplings are the strongest. Axial ligands have much larger couplings than equatorial ligands, which can have zero or even negative J couplings [1]. Xenon chemical shifts also show a sensitivity to the polarity of the solvents for these compounds. Schrobilgen suggests, for example, that XeF_2 may be solvated by weak bonds through the fluorine ligand, such as:

$$F - \mathrm{Xe} - F \cdots \mathrm{H} - F.$$

Recent synthetic efforts have produced a number of new xenon compounds with several novel features. Naumann and Tyrra [59] have reported the first stable xenon-carbon bond in C_6F_5Xe, which they characterized using ^{129}Xe NMR. Turowsky and Seppelt [60] have synthesized the first xenon-containing polymer (—Xe—O—TeF_4—O—$)_n$, which is insoluble in all common solvents and stable to 353 K. In addition, synthesis of the first stable xenon carboxylate, pentafluorophenylxenon (II) pentafluorobenzoate, has produced a relatively shielded xenon environment with a ^{129}Xe frequency shifted −2029.7 ppm with respect to XeF_2 [61]. This Xe-carboxylate compound has an unusual covalent (rather than ionic) Xe—O bond, which also is the longest Xe—O bond known (0.2367 nm). ^{129}Xe NMR is among the routine tools now used for the structural characterization of such new xenon compounds.

3 Xenon as a Probe of Host Structures

Much of the current xenon NMR literature deals with the use of xenon as a probe of material structure. As mentioned in the Introduction, this body of work has as its thematic origin in the pioneering experiments of Ito and Fraissard [2, 6] and of Ripmeester [3-5, 7], who applied xenon NMR to the study of microporous zeolites and clathrates, respectively. Since then, numerous applications of xenon NMR to the study of liquids, polymers and other less ordered materials have appeared.

3.1 Liquid Solutions

Several studies of xenon in solution have pursued an understanding of the interactions between xenon atoms and the solvent, primarily through measurement of the solvent's influence on ^{129}Xe or ^{131}Xe shift and/or relaxation rates. Stengle et al., for example, have investigated the relaxation of ^{131}Xe in a variety of solvents [62] and have compared several theories and additional Monte-Carlo calculations to explain their relaxation data. The theory that best fits their data for xenon in polar liquids is one due to Hertz [63], who considered electric-field-gradient-induced relaxation with the solvent modeled as a collection of fluctuating point dipoles. Determination of the correlation time τ_c using such theories is generally difficult, because the relaxation rate depends on both the electric field gradient (efg) and τ_c. The results of the study [62] show that a correlation time model based solely on solvent molecule tumbling does not adequately describe the ^{131}Xe relaxation data; other motions are, thus, expected to contribute. For the liquids examined, the xenon chemical shift was not affected by the electric dipole moment of the solvent molecules.

Rummens has used a model to correlate the chemical shift with a solvent function $f(n)$ based on the index of refraction n [64]:

$$f(n) = \left(\frac{n^2 - 1}{2n^2 + 1}\right)^2. \tag{7}$$

This so-called "reaction field" treatment assumes that a solvent field, induced by spontaneous electric moments in the solute, alters the magnetic shielding of the solute [65]. The reaction field model can predict the ^{129}Xe chemical shift behavior quite well for xenon dissolved in chemically similar solvents [66], as shown for example in Fig. 5 for a series of straight chain alcohols studied by Stengle et al. [67]. Good agreement has similarly been observed for xenon

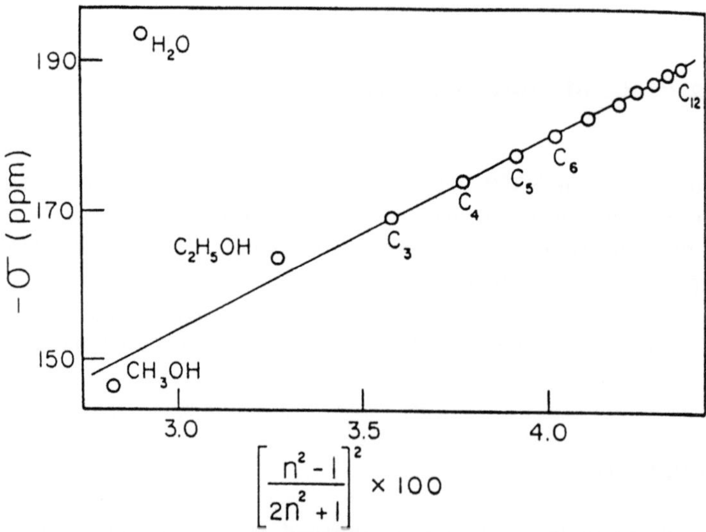

Fig. 5. Chemical shifts of ^{129}Xe in water and twelve straight chain alcohols from methanol to dodecanol. The shifts are plotted together with the reaction field function (*solid-line*) suggested by Rummens [64]. All shifts are downfield from the xenon gas reference. Used with permission from Ref. [67].

dissolved in an homologous series of *n*-alkane solutions [68] and in mixtures of organic liquids [69]. In water, the ^{129}Xe chemical shift is anomalously high, and does not fit the reaction field model at all for reasons thought to be due to the unusual nature of the hydration shell around each solvated xenon atom. Recently, Reisse and coworkers [70] have developed a new theoretical model using xenon-solvent dispersion energies to account for ^{129}Xe chemical shifts. This new theory has been applied to a variety of xenon-solvent systems and appears to correlate solvent character with the ^{129}Xe chemical shift somewhat better than the reaction field model.

Several related xenon-solvent studies have recently appeared which use ^{129}Xe as a structural probe of biological systems. For example, Stengle, Williamson, and coworkers [71] have extended their investigations of xenon in various solvents to include myoglobin, $(Myr)_2$lec lipid bilayers, and *Torpedo californica* electric fish membranes in aqueous solutions. Solvated xenon atoms in these systems undergo rapid exchange between solvent and protein environments on the ca. 10^{-3} s time scale of the NMR experiment. In the case of myoglobin, the authors estimate that approximately 10 xenon atoms are associated with each protein molecule, in addition to the xenon binding site known from X-ray data. In $(Myr)_2$lec vesicles, xenon at 308 K shows a distinct peak from the aqueous solution, in addition to the peak associated with the protein. These peaks coalesce at 323 K, as the xenon atoms undergo more rapid exchange at the higher temperature.

Additionally, Tilton and Kuntz have studied several xenon-protein systems and determined the rates of association by measuring the ^{129}Xe line width as a function of temperature and protein concentration [72]. They determined that the association rate is approximately $6 \times 10^7\,M^{-1}s^{-1}$ for xenon with methemoglobin and $1 \times 10^7\,M^{-1}s^{-1}$ for xenon with metmyoglobin. The xenon chemical shift was found to be sensitive to the iron oxidation and spin states, with the so-called "met" state (Fe^{3+}) more deshielded by approximately 5–10 ppm from the other iron states of myoglobin.

3.2 Amorphous Polymers

As the viscosity of a solution increases, the mobility of solvated xenon atoms diminishes. In amorphous polymers, this can be exploited to investigate structural heterogeneities that may exist in systems that are not "uniform". (As discussed below, the relevant length scale is the distance diffused by a xenon atom during a NMR experiment.) Investigation of polymer microstructure and dynamics by ^{129}Xe NMR is currently an area of increasing activity, as difficulties imposed by the viscous and disordered nature of polymeric systems are being overcome, especially through the application of powerful two-dimensional (2D) NMR methods.

The solubility of xenon in many polymers allows measurement of local structural and dynamical features associated with free volume sites occupied by solvated xenon guests. Stengle and Williamson [68], for instance, have introduced xenon into low density polyethylene (LDPE) at a pressure of 8 atm and observed a ^{129}Xe resonance at approximately 200 ppm. They attribute this signal to xenon in amorphous regions of the polyethylene sample, consistent with their measurements of ^{129}Xe shifts in *n*-alkane solutions, which could be described by a reaction field model [64], as discussed above. At temperatures above the glass transition T_g, mobile polymer segments permit high xenon mobility between different free volume environments, producing a motionally narrowed ^{129}Xe signal [68, 73, 74]. Suppression of polymer motions below T_g, however, traps the xenon atoms in various sites, producing inhomogeneously broadened ^{129}Xe peaks [75]. Variable temperature ^{129}Xe NMR measurements of xenon mobility, based primarily on observations concerning ^{129}Xe line widths, can provide a sensitive means of monitoring coupled changes in chain or sidegroup motions and gas transport properties in amorphous polymers. These have been investigated, for example, for xenon in poly(ethyl methacrylate) near its glass transition [68]. Along similar lines, Kennedy [76] has examined rubber curing using ^{129}Xe NMR and found that sulphur cross-linking produces greater material homogeneity, as evidenced by a reduction in the number of features in the motionally narrowed ^{129}Xe NMR spectrum.

The utility of ^{129}Xe NMR for resolving structural heterogeneities relies in general on slow xenon exchange between distinct sample regions. Brownstein et al. [77] found that xenon line widths were sensitive to the local environment of amorphous polymers by monitoring peaks in the ^{129}Xe spectrum that were assigned to distinct regions of a polystyrene-polyisoprene block copolymer sample. They presented evidence that the observed ^{129}Xe line width (up to 6 kHz) was due primarily to xenon diffusion between the two polymer domains and calculated an average xenon diffusion coefficient of $D = 3 \times 10^{-11}\ m^2/s$, slightly less than values measured for nitrogen and argon. More recently, Walton and coworkers have resolved distinct peaks in two-component polymer blends, allowing miscibility issues, such as estimation of domain sizes, to be investigated [73].

Such structural and mobility studies have recently been given important impetus through application of two-dimensional exchange NMR methods [78, 79] to xenon in heterogeneous polymers. Kentgens et al. [80] have performed the first 2D exchange experiments on ^{129}Xe in linear low density polyethylene (LLDPE) and in polycarbonate, systems which display distinctly different xenon mobilities. As shown in Fig. 6(a), a diagonal ^{129}Xe exchange spectrum results for xenon in LLDPE using a mixing time of 2 s, indicating the absence of xenon exchange between different regions within this time period. Conversely, the circular 2D ^{129}Xe contour pattern in Figure 6(b) for xenon in amorphous polycarbonate reveals that xenon atoms sample essentially all available local environments during a 1-s mixing time.

Tomaselli et al. [74] have produced a good example of the information available from the ^{129}Xe 2D exchange experiment, measuring xenon diffusivities and exchange between heterogeneous domains of polystyrene and poly(vinyl-

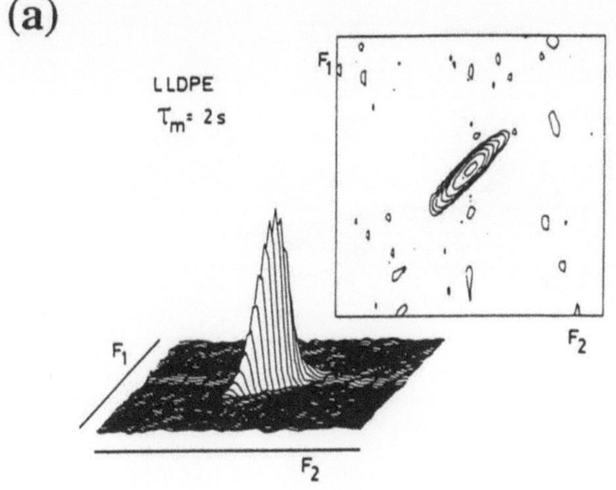

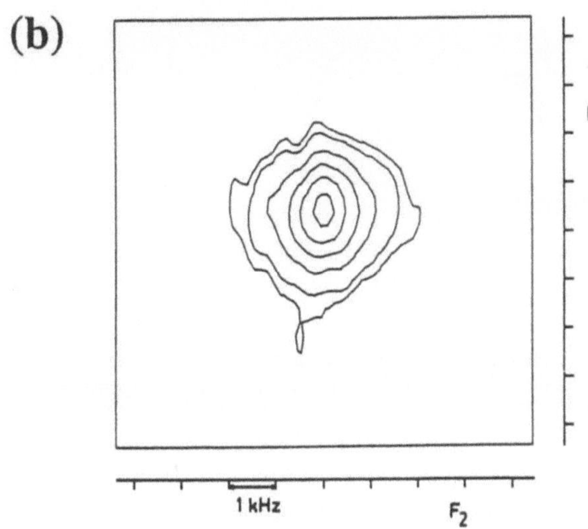

Fig. 6a. ^{129}Xe 2D exchange spectrum of xenon in linear low density polyethylene. The diagonal spectrum indicates that there is apparently no exchange of xenon between regions with different local environments during a mixing time of 2 s. **b)** ^{129}Xe 2D exchange spectrum of xenon in polycarbonate. The broad circular pattern indicates that xenon atoms sample essentially all local environments in the amorphous polymer during a mixing time of 1 s. Used with permission from Ref. [80].

methylether) copolymer blends. They used a variation of the powerful 2D exchange NMR technique [78] in which ^{129}Xe atoms were allowed to diffuse within the system during the mixing period t_m. In this way, exchange of xenon atoms between different environments, e.g. immiscible phases, could be correlated with guest transport properties within the sample. Figure 7, for example, shows 2D ^{129}Xe diffusion spectra recorded at 294 K, but with different mixing times of 1 s and 8 s, for xenon in a mixture of polystyrene (PS) and poly(vinylmethylether) (PVME). In Fig. 7(a), ^{129}Xe peaks resolved along the diagonal predominate, corresponding to xenon atoms in different phase-separated polymer domains that undergo little or no exchange during the 1-s mixing time. For longer mixing times, however, cross-peaks appear, reflecting ^{129}Xe exchange between

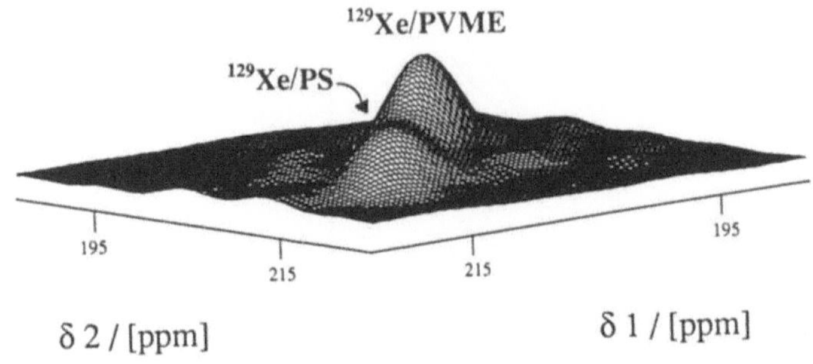

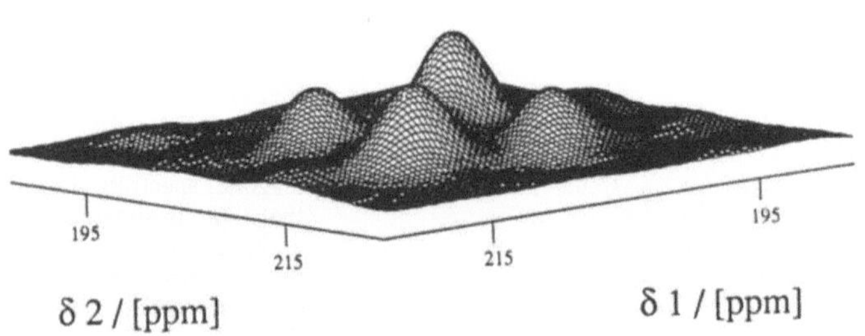

Fig. 7a,b. 2D ^{129}Xe diffusion spectra recorded at 60.5 MHz at 294 K with mixing times of **a** 1 s and **b** 8 s for xenon in a heterogeneous mixture of polystyrene (PS) and poly(vinylmethylether) (PVME). Adapted from Ref. [74].

phase-separated PS and PVME regions [Fig. 7(b)]. By monitoring the relative amplitudes of the ^{129}Xe cross-peaks with respect to the ^{129}Xe/PS or ^{129}Xe/PVME diagonal peaks over a range mixing times (1–20 s), the authors establish effective xenon diffusion coefficients of $D_{eff}^{PS} = 6 \times 10^{-13}\,m^2/s$ and $D_{eff}^{PVME} = 13 \times 10^{-13}\,m^2/s$ in the two media.

3.3 Liquid Crystal Systems

Anisotropic liquid crystal solutions possess properties that place them at the interface between categories of ordered and disordered materials. Studies by Loewenstein and Brenman [81], Bayle et al. [82], and more recently by Diehl, Jokisaari, and coworkers [83–88] have shown that xenon is very sensitive to changes in local order when dissolved in anisotropic solutions. For example,

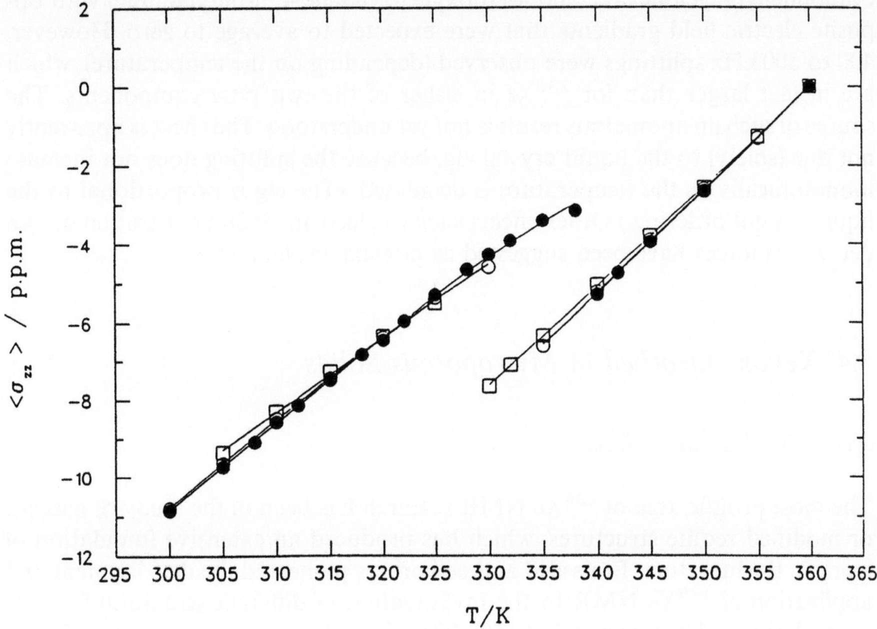

Fig. 8. ^{129}Xe shielding as a function of temperature for xenon dissolved in mixtures of the liquid crystals 4-*n*-alkyl-*trans,trans*-bicyclohexyl-4′-carbonitrile (ZLI1167) and 4-ethoxybenzylidene-4′-*n*-butylaniline (EBBA): (●) 100%, (○) 87.5%, and (□) 78% ZLI1167. Phase changes are clearly seen as discontinuities in the chemical shift behavior as the temperature is changed. Used with permission from Ref. [86].

^{129}Xe and ^{131}Xe NMR show dramatic changes in resonant shifts [82, 86–88] or quadrupolar splittings [81, 83–85], respectively, as thermotropic liquid crystal-xenon solutions undergo transitions from isotropic to ordered nematic phases. As shown in Fig. 8, the ^{129}Xe shielding shows sharp discontinuities across the isotropic-nematic phase boundaries of 4-*n*-alkyl-*trans,trans*-bicyclohexyl-4′-carbonitrile liquid crystal mixtures [86]. The discontinuities reflect changes in the chemical shielding caused by ordering of the liquid crystal director preferentially either along (positive bulk anisotropic susceptibility) or perpendicular (negative bulk anisotropic susceptibility) to the applied magnetic field. These effects can be attributed to interactions of the xenon with a non-spherically symmetric environment that should become more pronounced with increased ordering of the liquid crystal at lower temperatures.

Diehl and coworkers have also examined the electric field gradient experienced by xenon in liquid crystal environments using ^{131}Xe NMR techniques. The quadrupolar splittings observed are proportional to the average efg experienced by the ^{131}Xe nucleus [83]. A surprising result was obtained [84] when xenon was dissolved in a "critical" mixture of two liquid crystals (*trans*-4-*n*-alkyl-(4-

cyanophenyl)-cyclohexane and 4-ethoxybenzylidene-4'-*n*-butylaniline) with opposite electric field gradients that were expected to average to zero. However, 400 to 500 kHz splittings were observed (depending on the temperature), which are in fact larger than for ^{131}Xe in either of the two pure components. The source of such an anomalous result is not yet understood. The effect is apparently not due (solely) to the liquid crystal efg, because the splitting does not increase monotonically as the temperature is decreased. (The efg is proportional to the liquid crystal ordering.) Other effects such as electron cloud polarization or van der Waals forces have been suggested as possible explanations.

3.4 Xenon Adsorbed in Microporous Solids

Xenon in Molecular Sieves

The most prolific area of ^{129}Xe NMR research has been in the study of nascent or modified zeolite structures, which has produced an extensive foundation of work in the literature. Fraissard and coworkers pioneered the development and application of ^{129}Xe NMR to the investigation of different structural features of molecular sieves (e.g. [2, 6, 89–97]), including pore sizes [89, 90] and cation/metal guest influences [91, 92], in a series of papers starting in 1980. Fraissard and his group connected their ^{129}Xe/zeolite studies with the gas phase work of the Jamesons and coworkers [19, 24], empirically correlating xenon–surface interactions with pore sizes in a variety of different zeolites [90]. The ordered zeolitic pore systems, together with high xenon mobility and the sensitive ^{129}Xe nuclear resonance signal, have combined to make these experiments convenient and informative, though care must be taken in establishing general relationships between the data and structural details [98]. As this area has recently been the subject of several thorough reviews [8–11], we touch here only on a few important developments that have recently appeared on this topic. Many of these recent investigations have been linked by their concern for issues related to xenon mobility, including its direct measurement, its modification, and/or its application to the measurement of structural heterogeneities.

Measurements of xenon diffusion in NaCa-A, Na-X, and Na-Y zeolites by Kärger, Pfeifer, and coworkers using pulsed-field gradient methods [99] have proved that xenon mobility in these materials is sufficiently high at room temperature to average xenon interactions over thousands of zeolite cages on the ca. 10^{-3} s time scales of typical 1D ^{129}Xe NMR experiments. In comparison to bulk diffusivities (typically ca. 10^{-5} m^2/s [100]), the loading-dependent Xe diffusion constant in $Na_{4.4}Ca_{3.8}$-A zeolite, for example, has been measured to be 1.5×10^{-9} m^2/s at a bulk concentration of 1.5 xenon atoms per α-cage at room temperature [99]. For similar loadings, xenon diffusivities at room temperature were measured to be 7×10^{-8} m^2/s in Na-X zeolite and 4.0×10^{-9} m^2/s in silicalite. This latter value is supported by a recent molecular dynamics study, which indicates that Xe diffusivity in silicalite is expected to be substantial ($1.9 \times 10^{-9}\,m^2/s$), even at loadings of 4 xenon atoms per unit cell [101]. Such relatively high xenon mobilities have been exploited in a number of instances, notably by Shoemaker and Apple [102] and subsequently by a number of other groups [93, 97, 103–106], to examine macroscopic heterogeneities in collections of molecular sieve particles. Valença and Boudart have made similar observations for xenon adsorbed on Pt/alumina systems [107].

Because xenon atoms in large pore molecular sieves are capable of diffusing through several micron-size zeolite crystallites on the time scale of the NMR experiment [97, 102, 104–106], both inter- and intraparticle diffusional barriers are important to the description of xenon mass transport in molecular sieves [108, 109]. The relevant time scale for the ^{129}Xe NMR measurements is the reciprocal frequency separation of ^{129}Xe signals arising from different xenon adsorption sites. For example, ^{129}Xe resonances in dehydrated Na-Y zeolite environments with and without adsorbed hexamethylbenzene (HMB) are separated by approximately 30 ppm at 9.4 Tesla (^{129}Xe resonant frequency $\cong$ 110 MHz) [105]. Resolution of separate peaks from these two environments in the same sample requires, therefore, that exchange between the different adsorption environments be slow compared to 300 μs; faster exchange between these environments averages the respective contributions and narrows the signals to a single peak in the fast-exchange limit [22]. As a consequence, rapid movement of the xenon probe species through large pore zeolite channels at room temperature limits resolvable structural details to essentially macroscopic dimensions, at least to length scales on the order of microns in faujasite structures and the like. Under these circumstances, Chmelka et al. [108] have exploited slow xenon exchange to measure macroscopic adsorbate distribution heterogeneities in Na-Y zeolite which, when coupled with a crystallite-size scaling analysis, permits diffusivities of coadsorbed organic guest species to be estimated. Structural insight can still be obtained if the length scale of Xe diffusion during the NMR experiment exceeds the dimensions of the structural heterogeneity, though to do so from a single isotropically averaged ^{129}Xe chemical shift in these complex materials requires extra care [98]. In such cases, assistance from other complementary experimental methods is often required to verify interpretations put forward to explain the ^{129}Xe NMR results [110].

To increase the utility of ^{129}Xe NMR to structural issues characterized by shorter length scales, efforts have been made to reduce the mobility of the xenon probe atoms. One method of accomplishing this is through confinement of Xe guest species in molecule-size containers. 0.44-nm-diameter Xe atoms can be introduced, for example, into dehydrated Na-A zeolite cavities at elevated temperature and pressures, conditions that permit the xenon atoms to pass readily through the small apertures separating individual zeolite α-cages and to be distributed throughout the crystallites. The sensitivity of the ^{129}Xe chemical shift to local interactions permits one to resolve signals from α-cages containing different Xe populations [Fig. 9(a)], in this case, because of incremental differences in local xenon density and because xenon exchange

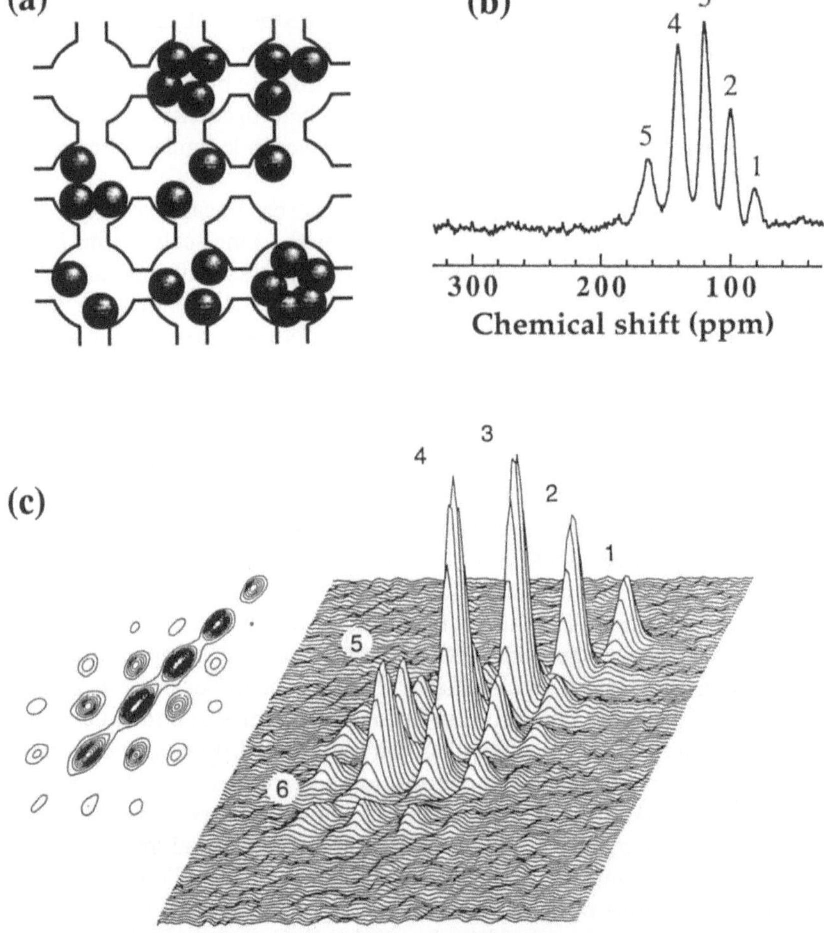

Fig. 9a. Schematic diagram of xenon atoms distributed within the α-cages of microporous Na-A zeolite. **b** Room temperature ^{129}Xe NMR spectrum for xenon occluded in Na-A zeolite at 523 K and at 40 atm. The discrete, chemically-shifted peaks correspond to α-cages containing different numbers of occluded xenon atoms, as indicated by the numbers above the peaks. **c** ^{129}Xe 2D exchange NMR spectrum (contour and stacked plot representations) showing evidence of xenon exchange between different Na-A α-cage environments during the 1-s mixing time. The off-diagonal peaks reflect xenon transport between α-cages with different numbers of xenon guests. The chemical shifts are referenced to xenon gas at very low pressure. Adapted with permission from Refs [113, 121, 122].

between individual Na-A α-cages at room temperature is slow compared to the time scale established by the reciprocal peak splitting (ca. 10^{-3} s).

As a result, ^{129}Xe NMR has been used to study adsorption environments in Na-A cavities with different numbers of xenon guests [111–116]. In Fig. 9(b), for example, a ^{129}Xe spectrum from a xenon/Na-A zeolite sample, prepared at 523 K and 40 atm Xe pressure, is shown where resolved peaks are evident from xenon atoms in α-cages containing from one to five Xe atoms. After normalizing the integrated intensities with respect to cage occupancy, these data and those obtained at higher loading pressures provide insight on the factors relevant to the distribution and packing of atoms in small subvolumes of space. Consistent with the importance of finite atomic volume effects [117], ^{129}Xe NMR data reveal that at low Xe loadings (< 3 Xe atoms/α-cage) the guest distribution is described by binomial statistics, while for higher loadings (at 523 K) a hypergeometric or continuum description is required [113, 118]. Jameson et al. have recently completed a thorough study of temperature and density influences on such Xe/Na-A distributions [115], observing a maximum of 8 Xe/α-cage (at 300 K), which is the highest single-cage occupancy thus far observed for this system. Intracage configurations of such dense assemblies of xenon guests have been shown by van Tassel et al. [119] to depend sensitively on local framework structure, as well as guest loading. For the xenon/Na-A zeolite adsorption temperatures used by Jameson et al. (300 and 360 K), experimental deviations from hypergeometric and continuum statistical distributions were observed, which the authors ascribed to attractive Xe–Xe interactions at low loadings and to disfavored high energy configurations at elevated densities. These findings have been used separately [120] to re-examine xenon adsorption in large pore zeolites, shedding new light on the structural influences of the framework on ^{129}Xe chemical shifts in the fast-exchange limit.

Mass transport of xenon atoms between individual Na-A α-cages has recently been examined by Larsen et al. [122], who have extended significantly studies of slow dynamic processes in molecular sieves through the use of variable temperature ^{129}Xe 2D exchange NMR. As shown in Figure 9(c) for a Xe/Na-A sample prepared at 523 K and 40 atm, numerous off-diagonal peaks appear in a room temperature ^{129}Xe 2D exchange spectrum that was acquired with a mixing time sufficiently long ($t_m = 1.0$ s) to allow movement of some xenon atoms between cavities. For such an exchange process, the transport of a single xenon atom between two α-cages changes the environment of all other xenon atoms in *both* cages, producing up to three off-diagonal peaks in the 2D spectrum. Variable temperature 2D exchange experiments, coupled with careful quantitative analysis of ^{129}Xe peak areas, has permitted loading-dependent rates of Xe exchange and Xe sorption energies to be established [122].

In cases where xenon atoms experience interactions that are anisotropic, local adsorption site symmetries may be reflected by the line shape features of ^{129}Xe NMR spectra. For example, Springuel-Huet and Fraissard have reported [95] that the ellipsoid-shaped channels of SAPO-11 and $AlPO_4$-11 cause chemical shift anisotropy (CSA) effects to be observed in ^{129}Xe NMR spectra of xenon adsorbed on these materials [Fig. 10]. As the concentration of adsorbed xenon is increased, the sign of the anisotropy changes, indicating that asymmetric xenon–xenon interactions along the length of the channel become more important at higher loadings. The authors explained their observations in terms of an axially symmetric chemical shift tensor with components σ_{xx}, σ_{yy}, and σ_{zz} in the principal axis system of the crystallite, as shown in Fig. 10. Xenon–xenon

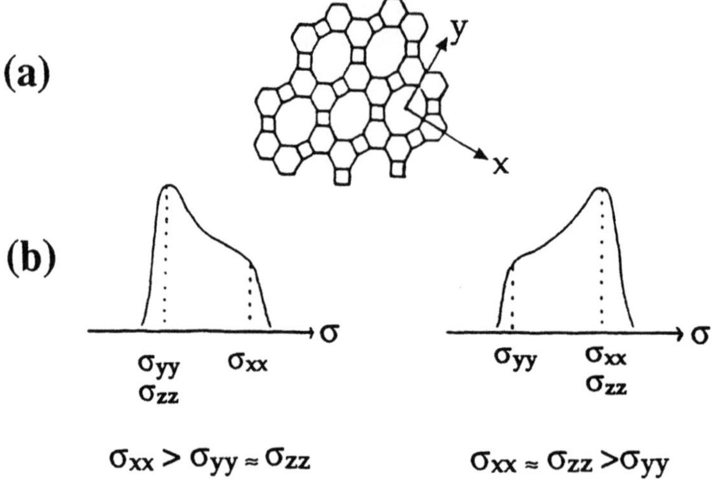

Fig. 10a. Schematic drawing of the SAPO-11 molecular sieve. **b** Representative ^{129}Xe chemical shift anisotropy patterns of xenon adsorbed in SAPO-11 at low and high xenon pressures. At low concentrations of xenon, $\sigma_{xx} > \sigma_{yy}, \sigma_{zz}$ and the line shape has a negative anisotropy, while at high xenon concentrations $\sigma_{xx}, \sigma_{zz} > \sigma_{yy}$ and the sign of the anisotropy changes. In this figure, σ is shown increasing to the right (higher frequency). Used with permission from Ref. [95].

collisions, represented by σ_{zz}, become progressively more important and eventually dominate the surface interactions, represented by σ_{xx} and σ_{yy}, as the loading is increased. The correlation of ^{129}Xe line shapes with anisotropic properties of different host symmetries is closely related to that of Ripmeester et al. [123], whose work on Xe/clathrate systems is discussed below.

Low temperature studies, particularly by Cheung and coworkers [124-128], also exploit diminished mobility of xenon to obtain a better understanding of Xe interactions with zeolites, amorphous silica, and alumina. Cheung et al. [124] have proposed a model that is based on fast chemical exchange between isotropic and adsorbed xenon within the zeolite cages. Their model explains the linear and nonlinear ^{129}Xe chemical shift trends observed in a variety of different xenon-zeolite experiments without resorting to special interactions such as electric fields [91]. Moreover their observation of pore-structure-dependent xenon phase transitions in several zeolites at reduced temperatures [124, 128] indicates promise for examining surface-mediated phase changes at a variety of Xe-solid interfaces. Barrie et al. have recently employed low-temperature ^{129}Xe NMR experiments to investigate similar confinement-induced phase transitions for xenon adsorbed in pillared montmorillonite clays [129]. At reduced temperatures, it is anticipated that lengthened xenon residence times at surface adsorption sites, coupled with well-developed magic-angle spinning or multiple-pulse line narrowing techniques, may extend the utility of solid-state ^{129}Xe NMR to the study of site-specific adsorption phenomena in these complex systems.

Xenon in Clathrates

In 1980, Ripmeester and coworkers showed the applicability of xenon NMR to the study of internal structures, namely xenon clathrate hydrates [3–5]. Confinement of xenon atoms in clathrate and clathrasil cavities of varying symmetries has been shown to produce chemical shift anisotropy patterns in the ^{129}Xe NMR spectra of these inclusion structures as well. For example, Ripmeester and Ratcliffe et al. [5, 123, 130] have studied xenon adsorption and thermodynamics in microporous clathrates and clathrasils using ^{129}Xe NMR line shapes to investigate the symmetry of host cativities. Xenon atoms are occluded in these materials, because they are too large to pass through the small windows separating adjacent cages. Subtle structural transformations in clathrates and clathrasils can be followed by monitoring changes in the ^{129}Xe resonance signals from the occluded xenon guest atoms. As shown in Fig. 11, ^{129}Xe line shapes, corresponding to xenon in the small and large cavities of microporous dodecasil-3C, reflect the symmetry of the clathrasil's cage environments at temperatures above and below the material's phase transition at ≈ 298 K. At 376 K, the ^{129}Xe NMR spectrum in Fig. 11(a) shows an upfield line that is narrow because of the cubic symmetry of the large clathrasil host cavity. By comparison, the small cavities in dodecasil-3C possess lower symmetry at 376 K, manifested by the broadened downfield ^{129}Xe line shape. This downfield signal displays an axially symmetric pattern that is consistent with the three-fold axis of the small clathrasil cages. At 251 K, the isotropic upfield peak is essentially unchanged [Fig. 11(b)], indicating that the dodecasil-3C phase transformation at 298 K produces little modification of the large cage structure. Conversely, the downfield ^{129}Xe line shape at 251 K has broadened further, reflecting non-axially symmetric shielding of the chemical shift tensor. This indicates that the environments of the ^{129}Xe atoms confined in the small cages have departed from axial symmetry, manifesting structural changes produced by the phase transition at 298 K.

In a number of the xenon/clathrate studies [7, 123, 130, 131], proton-xenon cross-polarization (CP) was applied to enhance the signal and thereby reduce the time required for signal averaging. An added advantage of the CP technique is that it discriminates between bulk and clathrate-encapsulated xenon; signals from occluded ^{129}Xe atoms are enhanced because of their close proximity to the framework protons of the clathrate lattice. As in standard CP experiments, polarization is transferred between spin species by matching their respective energies in the rotating frame [132, 133]. One atypical aspect of the proton-xenon CP experiment is that it involves coupling between species that interact while in a comparatively weak physisorbed state. The mobility of a xenon atom trapped in a clathrate cage does not completely average away the dipole–dipole interactions needed for cross-polarization, apparently because the confined xenon does not undergo isotropic motion. The CP contact time needed to effect the transfer, however, can be quite long, on the order of tens of milliseconds [123, 131]. In addition, it was found that the short relaxation time in the rotating

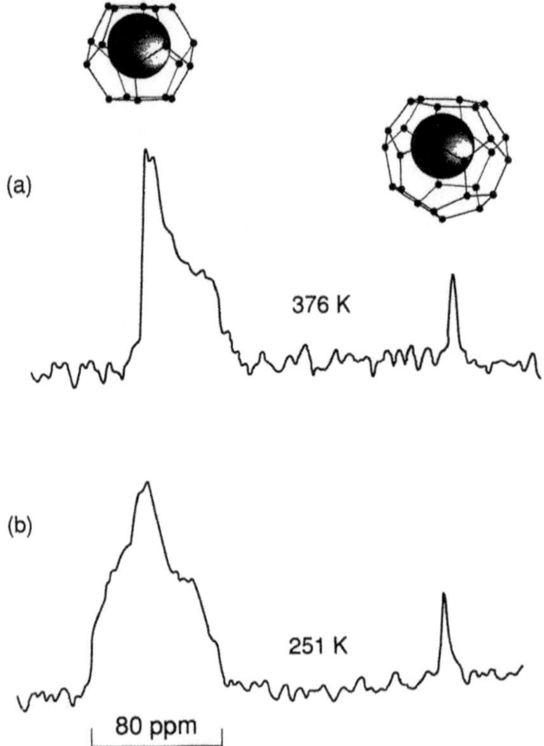

Fig. 11a,b. ^{129}Xe NMR spectra of xenon occluded in the 0.57 nm and 0.75 nm diameter cavities of 136 $SiO_2 \cdot 16(CH_2)_4O \cdot 8Xe$ (dodecasil-3C). Cavity symmetries are reflected in the ^{129}Xe line shapes, which can be used to probe structural transformations across the material's phase transition at 298 K. **a** At 376 K, the ^{129}Xe NMR spectrum shows an upfield line that is narrow because of the cubic symmetry of the large cavity. The downfield ^{129}Xe line shape is dominated by the anisotropic chemical shift, with the axially symmetric pattern consistent with the three-fold axis of the small cage. **b** At 251 K, the isotropic upfield peak is unchanged, reflecting the absence of appreciable structural alteration of the large cages. The downfield ^{129}Xe line shape, however, reflects a chemical shift tensor that is non-axially symmetric; this indicates that, in the small cavities, the environment of the confined ^{129}Xe probe atoms has departed from axial symmetry. Adapted with permission from Ref. [123].

frame, $T_{1\rho}$, caused the apparent ratio of cage occupancies to vary [131]. At 77 K, $T_{1\rho}$ for the protons of the clathrate lattice was much longer, rendering this less troublesome. It is also possible to transfer spin order using an adiabatic process [135] to allow more efficient cross-polarization [136].

Carbonaceous Solids

Xenon NMR has been applied to the study of carbonaceous materials possessing different chemical compositions and varying degrees of order. For example, the problem of coking (the build up of carbonaceous deposits due to incomplete hydrocarbon oxidation) in zeolites has been investigated, drawing upon the known initial structure of the well-defined molecular sieve host. Several studies, including that of de Menorval et al. [137], have shown that ^{129}Xe NMR is sensitive to the identity and density of adsorbed organic molecules inside microporous zeolite channels. Deposition of carbon in the ordered channels of zeolites similarly diminishes the void volume available for xenon adsorption and diffusion, with accompanying influences on the measured shift of the ^{129}Xe resonance frequency. Based on an increase of the ^{129}Xe chemical shift at low xenon concentrations, Ito et al. deduced [94] that up to a 20% reduction in average Na-Y supercage volume occurred following partial oxidation of propylene in Na-Y zeolite under conditions that produced heavy coking. In combination with ^{1}H pulsed-field-gradient methods, ^{129}Xe NMR has indicated that coke is confined largely to the external surface of small-pore NaCa-A zeolite samples, while being deposited essentially homogeneously within more open H-Y zeolite channels [96]. Miller et al. [138] used a variety of techniques including ^{129}Xe NMR to characterize coked H-Y zeolites prepared at different reaction temperatures and with different organic reagents. While argon adsorption experiments indicated the existence of two different-sized pore spaces in their highly coked samples, ^{129}Xe NMR was insensitive to these structural features, indicating fast exchange of xenon between pores that are in close proximity. Dybowski and coworkers investigated ZSM-5 and HZSM-5 zeolites [139], and found that the coking character was radically changed by the addition of hydrogen, which alters the acidity of the material. In the case of ZSM-5, coking blocked the zeolite channels severely, diminishing the accessibility of xenon probe atoms to much of the zeolite pore volume. This was manifested by significantly reduced xenon adsorption, as measured using ^{129}Xe NMR and adsorption isotherms.

For more disordered carbonaceous systems, such as coals and amorphous carbon, ^{129}Xe NMR has been used to characterize internal void structure, though the analyses are complicated by the fact that the signals are invariably broadened inhomogeneously. While these materials share some structural similarities with the amorphous polymer systems discussed above, the rigidity of the carbonaceous 'lattices' has resulted in experimental strategies derived primarily from those used to study internal structure in more ordered porous solids. For example, using ^{129}Xe NMR methods similar to those applied to zeolites, Ryoo and coworkers [140] studied several types of amorphous carbon and found, like in many zeolites, that the xenon chemical shift was a linear function of the loading pressure. Similarly, they found that the value of the chemical shift extrapolated to zero pressure was related to the nature of the carbonaceous surface, though as pointed out by Annen et al. [98], ^{129}Xe chemical shifts at zero pressure are complicated functions of the material composition and structure. Ryoo's group observed no relationship between amorphous carbon pore sizes and the ^{129}Xe shift, possibly because the pore spaces were too large (about 0.8 nm on average) to show a sizable effect.

Nevertheless, differences in pore spaces were noted in Illinois #6 coals [141, 142], where the two peaks in the ^{129}Xe NMR spectrum were observed to exhibit different pressure dependences. Tsiao and Botto [142] attributed the peaks to pores with different degrees of aliphatic or aromatic character. Typically, the inherently inhomogeneous nature of coal materials, reflected by a large distribution of pore sizes and variations in chemical composition, results in broad ^{129}Xe NMR lines. In general, the ^{129}Xe NMR chemical shift provides qualitative information and trends on the structure of amorphous carbon samples, coals, and nascent or coked molecular sieves, though it currently appears risky to rely on existing shift correlations for accurate structural measurements of pore volumes or diameters [98].

For more ordered carbonaceous materials, such as the xenon-graphite system, inhomogeneous structural complications are less troublesome than concerns of sensitivity. The importance of surface interactions with atomic monolayers and the possibility of observing the structure and dynamics of reduced dimensional phenomena [143], has made the xenon-graphite system the focus of many characterization studies, which have included adsorption [144], neutron and X-ray scattering [145], as well as NMR investigations [146, 147]. However, whereas material order and uniformity have placed constraints on ^{129}Xe NMR applications to many of the microporous solids described above, sensitivity limitations have proven the most troublesome for surface NMR investigations of *ordered* low or moderate surface area systems, including Xe/graphite.

Where surface structure is of interest, NMR has had difficulty matching the sensitivity of scattering experiments. Because of this, the two ^{129}Xe NMR studies of the Xe/graphite system to date [146, 147] have made use of high surface area graphitic material, where the basal planes have been expanded by exfoliation with ferric chloride at high temperatures. Nevertheless, in one of the studies, a continuous wave (cw) NMR experiment by Shibanuma et al. [146] showed the possibility of observing two-dimensional phase transitions via chemical shift and line width changes in the ^{129}Xe NMR spectra. The observed line width of the multilayer spectrum is too large to be due solely to bulk dipole–dipole interactions, a fact attributed to multiple adsorption sites and the large diamagnetic susceptibility of the graphitic substrate. The line width may be due to a hexatic (planar hexagonal), commensurate phase, which was recently observed by Hong et al. in X-ray experiments [145]. Similarly, Neue [147] observed asymmetric ^{129}Xe line shapes for one to three monolayer coverages of xenon on graphitized carbon at temperatures between 95 K and 158 K. The line shapes are again indicative of the large magnetic susceptibility of the graphite substrate. At 158 K, Neue observed a small splitting of 180 Hz, which he ascribed to the hexatic phase. Neue also observed two minima in relaxation time measurements of ^{129}Xe. At 160 K, the relaxation is due to rapid motion of xenon, whereas at lower temperatures, relaxation is probably due to the diffusion of a mobile xenon fraction to paramagnetic impurity sites.

For two-dimensional systems, such as graphite and a vast number of other low surface area materials, improvement of signal sensitivity is a prerequisite for extending surface structure studies by NMR. What follows is a discussion of the role optical pumping methods can play in achieving signal enhancements designed to make low surface area xenon NMR studies feasible.

4 Optically Polarized Xenon NMR

4.1 *Optical Pumping Methods*

In a typical NMR experiment for a spin 1/2 nucleus, the signal strength is mainly determined by the difference of the populations P_i of the two energy states, whose ratio is given by the Boltzmann factor:

$$\frac{P_2}{P_1} = e^{-(E_2 - E_1)/kT} = e^{-\gamma\hbar B_0/kT}. \tag{8}$$

At room temperature this population difference is small, about $10^{-5}\,P_i$ (10 ppm) for moderate magnetic field strengths B_o (4 Tesla), so that the accompanying polarization is also small. While such insensitivity generally presents fewer difficulties for abundant nuclei like ^1H, it can be troublesome for dilute species. Many methods have been devised to increase the signal from less abundant or low frequency nuclei, including low temperature experiments or cross polarization [132] from highly polarized nuclei, such as protons, to less polarized or dilute nuclei.

Optical pumping techniques, however, provide alternative means for dramatically increasing nuclear spin polarization. Discovered by Kastler in 1949 [148] (for which he later was awarded the Nobel prize), optical pumping has become important for the study of gas-phase spin interactions and in the production of highly polarized spin reservoirs for use as high energy particle physics targets [164] or for use in NMR experiments. Optical pumping methods have been thoroughly reviewed elsewhere [149–151], so the discussion here will be limited to those aspects pertinent to novel low-surface-area xenon NMR experiments and applications.

In Fig. 12, a typical optical pumping cycle is shown, along with the valence electron energy levels for an alkali metal atom (rubidium is commonly used). Neglecting, for the moment, the hyperfine structure produced by electron interactions with a nuclear spin, rubidium has an $S_{1/2}$ ground state that is split into two levels in a magnetic field. The first two excited states, $P_{1/2}$ and $P_{3/2}$, are also split by the magnetic field into two and four levels, respectively. Right circularly polarized light (σ^+), with a wavelength of 794.7 nm and propagating along the direction of the magnetic field, is absorbed, producing a transition from the ground state to the $P_{1/2}$ excited state according to the selection rule $\Delta m_s = +1$. This has the effect of depopulating the $m_s = -1/2$ ground state. The excited state decays within ≈ 30 ns (due to spontaneous emission) back to the

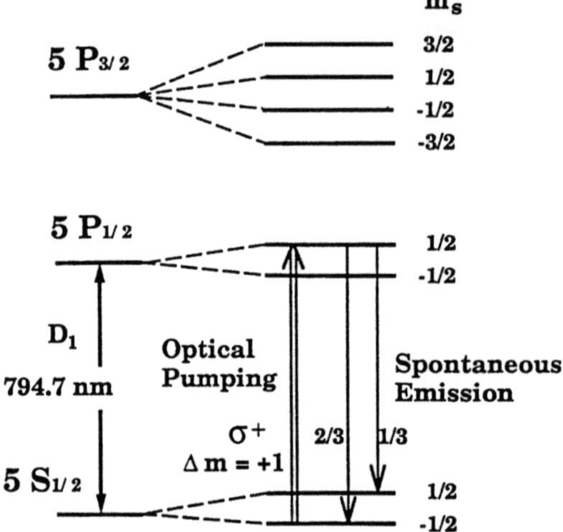

Fig. 12. Electronic energy levels of rubidium, excluding the effects of the Rb nuclear spin. Right circularly polarized light at 794.7 nm causes transitions from the $m_s = -1/2$ ground state to the $m_s = +1/2$ excited state. Emission back to the ground state increases the population of the $m_s = +1/2$ ground state.

two $S_{1/2}$ ground states with relative probabilities 1/3 and 2/3 given by the Clebsch-Gordan coefficients. At the reasonably high pressures used in many experiments (~ 1 atm), excited rubidium atoms collide before de-excitation, causing the excited states to mix with the result that the depolarization rates approach 1/2 and 1/2. Because the relaxation time of the ground state is relatively long (up to 1 s), the net result of the optical pumping cycle is to leave one of the ground state levels highly polarized, e.g. the $m_s = +1/2$ level for right circularly polarized light.

Methods first introduced by Grover at Litton Industries [152] and by Happer and coworkers at Princeton [151 and references therein] showed that optical pumping could be used to increase dramatically the nuclear polarization signal of xenon, and thereby allow its observation under a variety of new and interesting conditions. During collisions with rare gas atoms, spin polarization may be transferred from an alkali metal atom to a rare gas nucleus. This "spin exchange", as it is called, is more efficient for large rare gas atoms like xenon because of their tendency to form long-lived ($10^{-7} - 10^{-9}$ s) van der Waals' molecules [153] and because the spin exchange efficiency varies as the square of the interaction time. A schematic diagram of such a van der Waals complex is shown in Fig. 13. The lifetime of these complexes are short at pressures above one atmosphere, though spin exchange still occurs at a somewhat reduced efficiency. Under such conditions with high powered lasers, dramatic enhancements of xenon nuclear spin polarizations ($> 30\%$) can nevertheless be achieved [154].

The Hamiltonian during the van der Waals collision may be written:

$$\mathscr{H} = A\mathbf{I}\cdot\mathbf{S} + \gamma\mathbf{N}\cdot\mathbf{S} + \alpha\mathbf{K}\cdot\mathbf{S} + \omega_S\mathbf{S}_z + \omega_K\mathbf{K}_z \quad , \tag{9}$$

where A is the rubidium hyperfine interaction between the rubidium electron spin $\mathbf{S}$ and rubidium nuclear spin $\mathbf{I}$, γ is the spin–rotation interaction coupling the rubidium electron spin to the van der Waals complex rotation $\mathbf{N}$, α is the

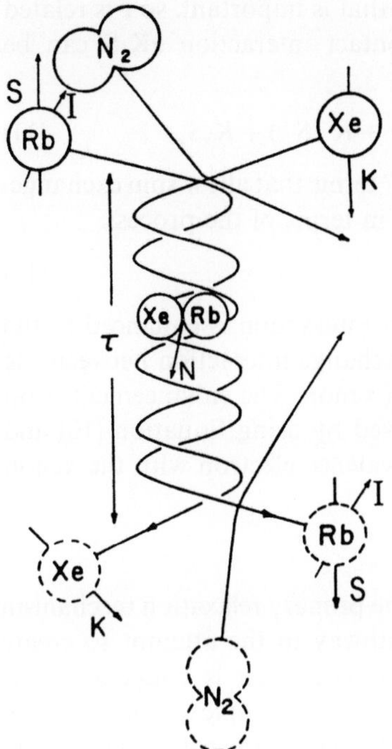

Fig. 13. A schematic diagram of the rubidium-xenon van der Waals' complex in a magnetic field $\mathbf{B}_0$. The rubidium nuclear spin **I** is coupled to the rubidium electron spin **S**, the xenon nuclear spin **K**, and rotational angular momentum **N** of the complex as a whole. The van der Waals' molecules are fragmented at a rate τ^{-1} by collisions with other atoms or molecules, in this case N_2 buffer gas species. Electron-nuclear spin exchange is greatly enhanced by such long-lived molecular complexes. Used with permission from Ref. [162].

Fermi contact interaction coupling the rubidium electron spin with the xenon nuclear spin **K**, and the last two are the Zeeman terms. The hyperfine interaction for ^{85}Rb is approximately 3 GHz, so the rubidium valence electron remains strongly coupled to the nucleus unless the magnetic field is above approximately one Tesla (the Pashen-Back effect [13]). The electron-nuclear spin coupling has no effect on the optical pumping, but reduces the Larmor frequency of the electron in a magnetic field by a factor of $1/(2I+1)$, where I is the nuclear spin. This complicates the rubidium spin level system, but can lead to some interesting effects, such as optically detected sublevel coherence experiments [155] and multiple quantum transitions [156], as well as the possibility of creating higher-order multipole polarizations [151]. These topics have been recently reviewed by Suter and Mlynek [157].

The Fermi contact interaction, given by

$$\alpha = \frac{8\pi}{3} g\beta g_N \beta_N \delta(\mathbf{r}), \tag{10}$$

is responsible for the spin exchange mechanism that transfers polarization from the rubidium electron to the xenon nucleus. The values g and g_N are the electron and nuclear g-factors, β and β_N are the electron and nuclear magnetic moments, and $\delta(\mathbf{r})$ is the probability of finding the electron at a radial coordinate $\mathbf{r}$ from the nucleus. Usually, the electron and nucleus are associated with the same atom, so that $\mathbf{r} = 0$. For Rb–Xe spin exchange, however, it is the probability of finding the Rb electron at the xenon nucleus that is important, so **r** is related to the internuclear separation. The Fermi contact interaction $\alpha\mathbf{K}\cdot\mathbf{S}$ can be rewritten using the relation:

$$\mathbf{K}\cdot\mathbf{S} = (K_xS_x + K_yS_y) + K_zS_z = \tfrac{1}{2}(K^+S^- + K^-S^+) + K_zS_z \quad , \tag{11}$$

where K^+S^- and K^-S^+ constitute the "flip-flop" terms that allow spin exchange. Schematically, one can think of spin exchange in terms of the process:

$$S(\uparrow) + K(\downarrow) \Rightarrow S(\downarrow) + K(\uparrow). \tag{12}$$

As shown by Herman [158], the Fermi contact interaction is enhanced in the rubidium-xenon complex due to the electron exchange interaction between the rubidium 5s electron and the core 1s orbitals of xenon. The enhancement is on the order of 10^5 for Rb-Xe, which can be derived by using Equation (10) and writing a Slater determinant for the rubidium valence electron with the xenon 1s orbitals:

$$\psi = \| \mathrm{Rb(5s)\,Xe(1s^1)\,Xe(1s^2)} \| .$$

The spin–rotation interaction $\gamma\mathbf{N}\cdot\mathbf{S}$ acts as the primary relaxation mechanism for rubidium, and constitutes a major loss pathway in the attempt to create high xenon polarizations. The electron spin polarization is coupled to the rotational angular momentum of the van der Waals' complex, which is then lost upon breakup of the complex. For Rb–Xe complexes, the rotation quantum number **N** has a value of about 70, and therefore can be thought of as a classical vector [159]. At moderate magnetic fields, the rubidium electron may be decoupled from the spin-rotation interaction, as measured by an increase in the rubidium relaxation time [160].

Happer and coworkers have investigated spin exchange between rubidium and ^{129}Xe, ^{131}Xe, or ^{133}Xe, finding that ^{129}Xe polarizes readily with a binary cross section of $7.3 \times 10^{-25}\,\mathrm{m}^2$, and that the xenon polarization rates depend on the rubidium density and polarization [161-163]. They determined that the spin exchange efficiency was about 4%, given approximately by the square of the ratio of the spin-exchange interaction to the spin–rotation interaction. This ratio is about 0.2, so most of the polarized rubidium is relaxed by spin rotation. At high densities, the rubidium polarization can also be reduced by radiation trapping, where de-excited rubidium atoms emit photons with the wrong polarization. Such photons are subsequently absorbed by other rubidium atoms, causing depolarization [164]. The addition of nitrogen or helium gas, however, quenches this mechanism allowing large polarizations of xenon ($>30\%$) to be produced at xenon pressures of about 1 atm using high-power (5 watt) lasers [154]. The spin exchange efficiency is reduced in magnetic fields above 200 Gauss, although recent reports indicate that xenon may be polarized using spin exchange even at high magnetic fields [165].

In Fig. 14(a), a schematic diagram is shown of a simplified apparatus for xenon optical pumping experiments. A laser tuned to the rubidium D1 transition (794.7 nm) illuminates the pumping cell for several minutes. (Rubidium lamps are

(a)

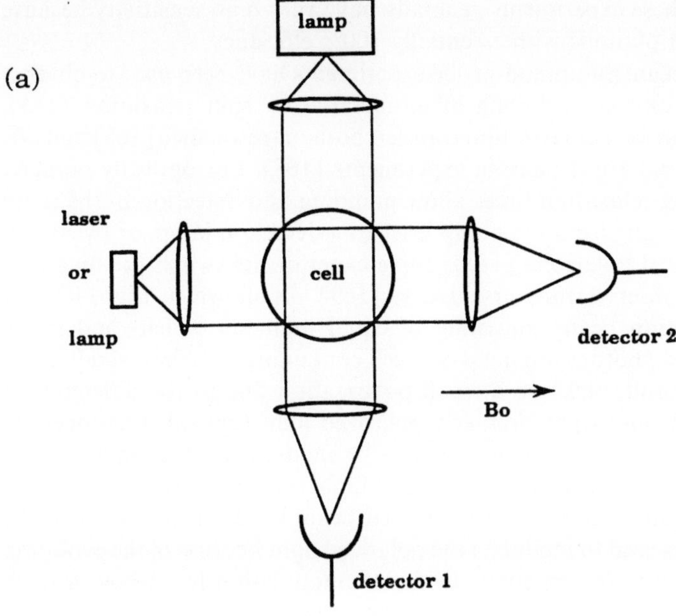

(b)

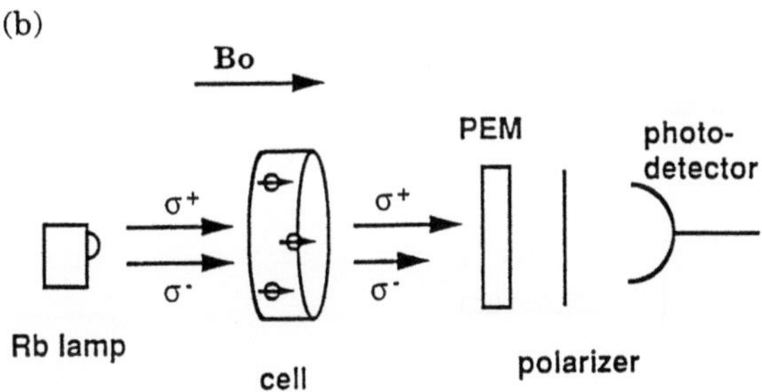

Fig. 14a. A schematic diagram of a typical optical pumping apparatus. The pumping light (a laser) is incident in the *z* direction along the magnetic field $\mathbf{B}_0$, while the detected light (unpolarized rubidium lamp light) can be incident along the *z* direction, or in the transverse direction. **b** A schematic diagram of the optical detection scheme used in low field optical pumping experiments. Unpolarized rubidium light incident on polarized rubidium in the optical cells induces a birefringence in the light. The PEM (photo-elastic modulator) modulates this residual polarization in the light, which may then be detected by a silicon photo detector and lock-in amplifier.

also capable of providing good polarizations.) After pumping, the decay of the polarization can be monitored by simply measuring the change in transmission of rubidium light from a low power probe beam passed through the cell. Highly polarized rubidium vapor does not absorb the circularly polarized photons, so that the detected light initially has a high intensity. As the spin polarization decays, the detected light is increasingly attenuated. The optical detection methods used in these experiments generally have very high sensitivity because they detect optical photons with essentially 100% efficiency.

Such crossed-beam pump-and-probe experiments have been used to observe a variety of phenomena, including alkali-metal-atom spin relaxation [153], rubidium-to-cesium spin polarization transfer, both on-resonance [166] and off-resonance [167], and Hg dephasing experiments [168]. For optically-pumped rare gases, the long relaxation times allow pumping and detection in the same z direction by changing from the pump laser to a detection lamp, or by simply removing the optical polarizers [162]. Some experiments switch the magnetic field direction to observe transverse decays [169]. As shown in Fig. 14(b), an unpolarized detection beam consisting of equal numbers of left- and right-circularly polarized photons impinges on a cell containing polarized alkali metal vapor, and as a result, picks up a small polarization due to the difference in absorption for left- and right-circularly polarized light (optical birefringence). The polarized portion of the light can then be modulated by a photo-elastic modulator (PEM) and detected with very high sensitivity using a lock-in amplifier. This is different from early crossed-beam experiments in which the transverse field was used to modulate the polarized light because of the evolution of the electron spin in the magnetic field. The modulation frequency in these experiments was at the Larmor frequency of the unpaired electron (approximately 700 kHz/Gauss for ^{85}Rb).

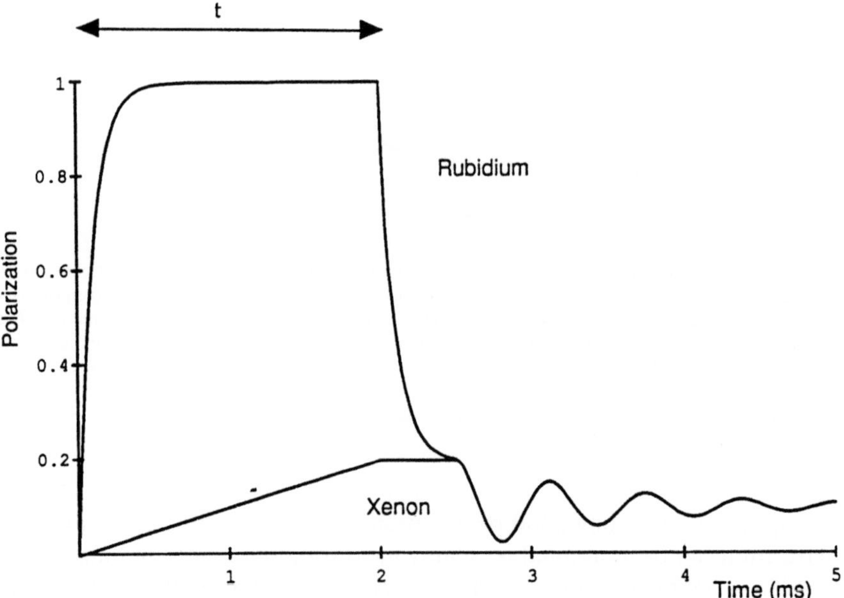

Fig. 15. A typical optical pumping cycle with the pumping occurring during a time t, followed by optical detection of the rubidium polarization, which follows the z component of the xenon magnetization. The xenon pumping rate is exaggerated.

A general polarization cycle is shown in Fig. 15, where the pumping light has been turned on at time $t = 0$. The rubidium polarization P_{Rb} builds up quickly to a value approaching 100% given by [170]:

$$P_{\mathrm{Rb}} = \frac{\Gamma/2}{\Gamma/2 + \gamma_{\mathrm{Xe}} + \gamma_{\mathrm{wall}}} \quad , \tag{13}$$

where Γ is the pumping rate of the laser, which is determined by the incident laser intensity and the rubidium absorption coefficient. The factor of 1/2 arises because only one-half of the absorbed photons end up polarizing the rubidium ground state. The term γ_{Xe} is the relaxation induced by the buffer gas, primarily due to the spin–rotation interaction, and γ_{wall} is the wall relaxation rate. A typical pumping time is of the order 0.1 ms, with the xenon polarization increasing much more slowly due to the smaller spin-exchange cross-section and higher density. The polarization rate per xenon atom is given by:

$$\frac{1}{T_{\mathrm{pump}}} = [\mathrm{Rb}]\bar{v}\sigma_{\mathrm{ex}} \quad , \tag{14}$$

where [Rb] is the rubidium atom concentration, $\bar{v}$ is the average relative velocity between the xenon and the rubidium atoms, and σ_{ex} is the spin-exchange cross-section, which is $7.3 \times 10^{-25}\,\mathrm{m}^2$ [161]. For long pumping times, the final xenon polarization P_{Xe} can be expressed as [170]:

$$P_{\mathrm{Xe}} = \frac{[\mathrm{Rb}]\bar{v}\sigma_{\mathrm{ex}}}{[\mathrm{Rb}]\bar{v}\sigma_{\mathrm{ex}} + \gamma_{\mathrm{wall}}} P_{\mathrm{Rb}} \quad . \tag{15}$$

After the pumping light has been turned off, the rubidium polarization decreases to that of the xenon spin polarization level within a spin-exchange time of about $[\mathrm{Xe}]\bar{v}\sigma_{\mathrm{ex}}$ per rubidium atom (about 1 ms), after which the rubidium polarization then follows the z component of the xenon magnetization. As a consequence, the rubidium polarization can act as a sensitive optical probe of the xenon magnetization and its evolution in these experiments [160, 169].

The primary relaxation mechanism for optically pumped ^{129}Xe in the presence of rubidium vapor is again the Fermi contact interaction, which acts in the reverse direction when the polarized light source is turned off. At low rubidium densities this effect is diminished, so that collisions with the walls of the container then become important, especially for xenon densities below about 20 amagats. Cell wall coatings, such as silicone [171] or paraffinic hydrocarbons, have been found to reduce the wall-induced relaxation rate to about two per hour, probably because the coatings cover paramagnetic centers.

4.2 ^{131}Xe Optical Pumping Experiments

Optical pumping experiments have also been applied to the study of ^{131}Xe, although the reduced spin-exchange efficiency and relaxation time due to quadrupolar interactions are sometimes limiting factors. Volk and coworkers

[172] were the first to observe the effects of collisions of ^{131}Xe with the walls of non-spherical sample cells, which produced a beat pattern in the decay of the transverse ^{131}Xe polarization. A pronounced example of such a beat pattern, recorded by Wu et al. [173], is shown in Fig. 16(a) for ^{131}Xe in a disk-shaped cell. Although Volk et al. were unable to discern the very small quadrupolar splittings because of resolution limitations, they attributed their observations to a quadrupolar dephasing mechanism that had previously been seen in ^{201}Hg (also $I = 3/2$) optical pumping experiments by Cohen-Tannoudji [168, 174]. The dephasing is due to the random fluctuations of the electric field gradient

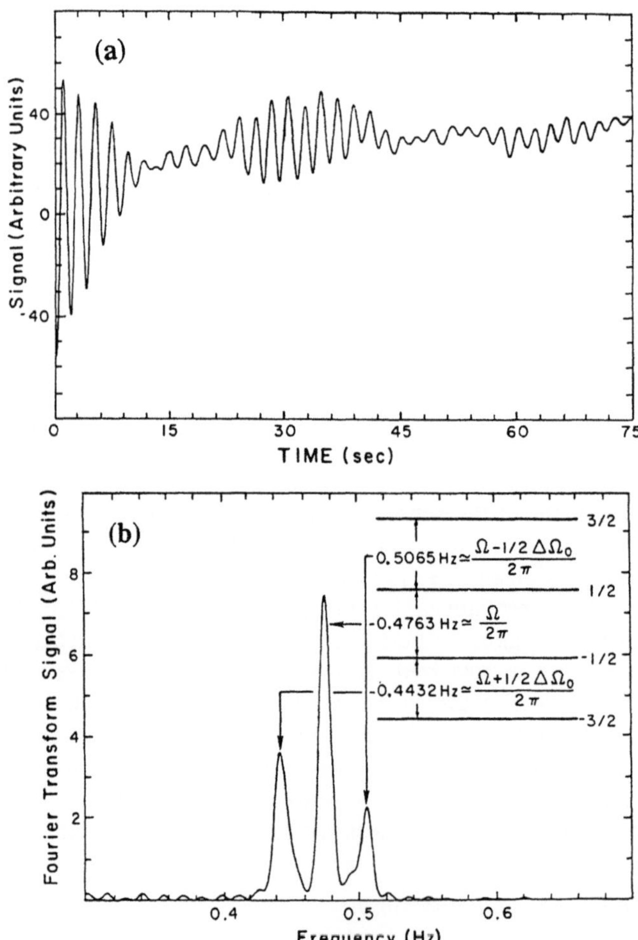

Fig. 16a. ^{131}Xe nutation signal from xenon in a disk-shaped cell 3 mm high and 50 mm in diameter at 373 K in the presence of hydrogen buffer gas. The background decay is due to ^{129}Xe. **b** Fourier transform of **a**. The quadrupolar splitting is 317 mHz. Used with permission from Ref. [173].

experienced by a nucleus at the surface of the cell. The dephasing rate can be expressed as:

$$A(\theta) = \frac{\tau_s}{\tau_c + \tau_s} \frac{e^2 Qq}{4} (3\cos^2\theta - 1) \quad , \tag{16}$$

where τ_s is the sticking time, τ_c is the electric-field-gradient correlation time, and θ is the angle between the quadrupolar interaction principal axis frame and the laboratory or rotating frame (the latter being relevant for nutation experiments). The sticking time may be written as:

$$\tau_s = \tau_o e^{-E_a/kT}, \tag{17}$$

where E_a is a surface activation energy, which has a value of approximately -7 kcal/mol for bare Pyrex glass.

Using the optical pumping and detection techniques described above, Wu et al. [173] have observed splittings in the ^{131}Xe NMR frequencies on the order of 100 mHz in disk-shaped sample cells [Fig. 16(b)]. The quadrupolar splittings scale with the macroscopic asymmetry of the cell such that

$$\Delta\omega_Q = \tfrac{1}{2}\bar{v}\langle\theta\rangle\left(\frac{1}{h} - \frac{1}{d}\right) \quad , \tag{18}$$

where $\bar{v}$ is the average xenon velocity, $(1/h - 1/d)$ is the cell asymmetry, and $\langle\theta\rangle$ is the mean twist angle (per wall collision) of the nuclear polarization produced by the quadrupolar interaction at the cell surface, averaged over the cell surface and all sticking times. The mean twist angle is the product of the sticking time and the quadrupolar interaction strength and can be established by plotting the quadrupolar splitting versus the cell asymmetry [173]. In that study, the splittings displayed a slope $\bar{v}\langle\theta\rangle/4\pi$ equal to 100 mHz-cm. For a mean speed of $\bar{v} = 2.4 \times 10^2$ m/s at 355 K, the mean twist angle was determined to be 38×10^{-6} radians per wall collision.

4.3 *Optically Pumped ^{129}Xe in High Field*

Recently, experiments at Berkeley [12, 175, 176] have combined optical pumping with conventional NMR detection to produce a novel xenon NMR technique for surface investigations. Historically, conventional NMR studies of surface phenomena have been limited by poor sensitivity, unless the temperature is kept extremely low [177], or the spin-lattice relaxation time T_1 is short. Consequently, investigations of surface structure and dynamics with NMR have been constrained primarily to systems with high surface areas, typically in the range of several hundred square meters per gram [15, 178–181]. Pulsed high-field NMR experiments of laser-pumped ^{129}Xe have shown, however, that spectra of xenon adsorbed on powdered solids of moderate surface area (1–10 m^2/g) may be obtained with high sensitivity, opening new possibilities for

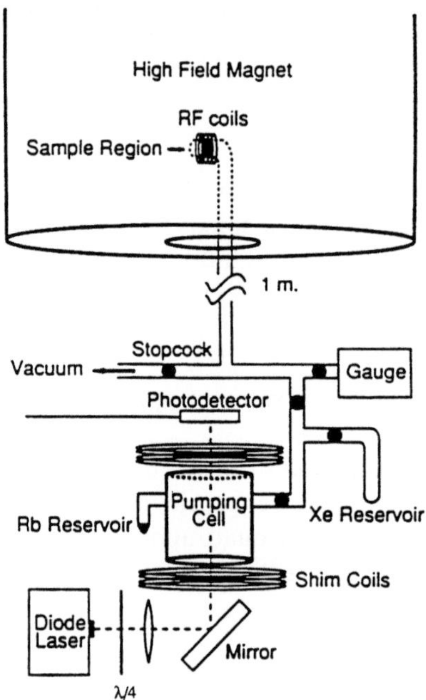

Fig. 17. Schematic diagram of the high-field optical pumping experimental apparatus. Circularly polarized 794.7 nm laser light is focused onto the optical pumping cell. A silicon photo detector is used to detect the rubidium absorption. The oven used to heat the pumping cell and the nitrogen cooling system for the sample region are not shown. Shim coils can be used to cancel gradients and/or reduce the overall magnetic field in the pumping cell. After pumping, the highly polarized xenon is adsorbed on the sample in the bore of the high field magnet, and the ^{129}Xe NMR signal is subsequently detected. Used with permission from Ref. [12].

the study of material surfaces using xenon NMR. These promising results have important implications for the field of surface science, particularly concerning the characterization of adsorbate structures, which are central to numerous issues in catalysis, adhesion, etc.

The apparatus, shown schematically in Fig. 17, consists of an optical pumping cell located in the fringe magnetic field (250 Gauss) beneath a superconducting magnet and a sample/detection cell located in high field in the bore of the magnet. The two regions are connected by a glass tube and separated by a series of stopcocks. Separate connections allow evacuation or measurement of pressure within the sample region. Natural abundance rubidium is contained in a reservoir region connected to the pumping cell, and is heated to produce Rb vapor in an oven with warm flowing nitrogen gas. The rubidium can be optically pumped to a polarization level approaching 100% using circularly polarized light from a semiconductor diode laser operating at 794.7 nm. The entire glass manifold was coated with a siliconizing agent that reduced the wall-induced relaxation rate of ^{129}Xe to less than two per hour. The cylindrical optical pumping cell of approximately 11 cm^3 volume was filled with between 13.3 and 80 kPa of natural abundance or enriched (70%) spin-1/2 ^{129}Xe gas. A Helmholz pair of shim coils were used in the low pressure experiments to reduce the fringe magnetic field to approximately 75 Gauss, which had the effect of increasing the spin exchange efficiency [160]. After optically pumping for approximately 30 minutes, the xenon was transferred to the high field region

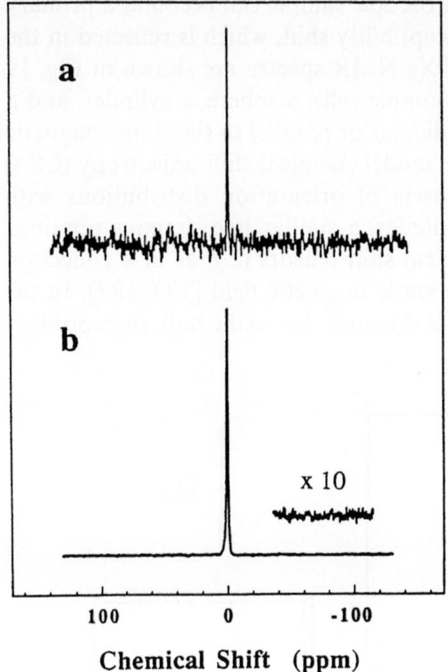

Fig. 18a,b. Room temperature NMR spectra of natural abundance ^{129}Xe acquired at 4.4 T. **a** ^{129}Xe NMR spectrum recorded using a conventional pulse-and-acquire method with 100 acquisitions. (The 3.3 cm^3 sample contained 27.5 kPa xenon and 133 kPa oxygen.) **b** ^{129}Xe NMR spectrum obtained from a single acquisition of optically-pumped xenon (8 cm^3, 4.3 kPa Xe). Used with permission from Ref. [12].

in the magnet bore, where it adsorbed on the surface of the sample at low temperature and the ^{129}Xe NMR signal was subsequently acquired.

Xenon-129 NMR spectra for xenon gas in a 4.4 Tesla magnetic field, obtained with and without optical pumping, are shown in Fig. 18 to estimate the enhancement of xenon nuclear polarization afforded by the optical pumping technique. Comparison of the spectra in Figs. 18(a, b) reveals that the optical pumping method enhances the xenon nuclear polarization by a factor of approximately 750 over conventional ^{129}Xe NMR. The observed NMR signals of the optically pumped xenon gas correspond to a nuclear polarization of about 0.5%. Higher polarizations, and thus larger enhancements, can be achieved with lower xenon gas pressures or through the use of more powerful lasers.

NMR of ^{129}Xe Thin Films

Using optically pumped ^{129}Xe in high field, a variety of effects on the xenon line shape have been observed that correlate with the temperature and macroscopic geometry of micron-thick xenon films. These influences can, in turn, be understood in terms of bulk ^{129}Xe diamagnetic susceptibility shifts. The thinness of the xenon films simplifies analysis of the spectral line shapes by allowing the radial dependence on the susceptibility to be neglected. Under these circumstances, the geometry of the macroscopic sample cell becomes a primary influence on the ^{129}Xe diamagnetic susceptibility shift, which is reflected in the ^{129}Xe line shape. Optically pumped ^{129}Xe NMR spectra are shown in Fig. 19 for xenon in several differently shaped sample cells: a sphere, a cylinder, and a flat rectangular box with planes perpendicular or parallel to the static magnetic field $\mathbf{B}_0$. The observed spectra resemble model chemical shift anisotropy (CSA) line shapes representing restricted subsets of orientation distributions with respect to the applied field. For the cylinder, a similar line shape is obtained as for an uniaxial distribution of chemical shift tensors (e.g. as in a collection of fibers) oriented perpendicular to the static magnetic field [133, 182]. In the experiments on xenon films, however, it is the shifts due to the bulk susceptibility

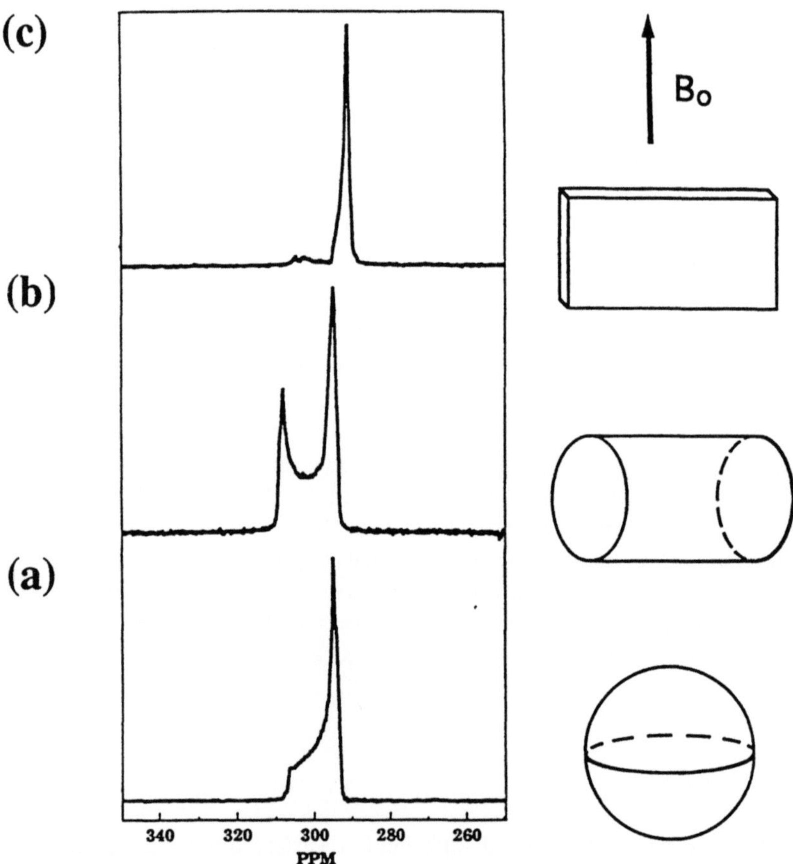

Fig. 19a–c. ^{129}Xe NMR spectra for thin xenon films frozen on the interior surfaces of sample cells with different geometries: **a** a sphere (radius 0.6 cm), **b** a cylinder (radius 0.5 cm, perpendicular to the static 4.4-Tesla $\mathbf{B}_0$ field), and **c** a flat rectangular box (3.0 cm × 1.0 cm × 0.1 cm) with its large planes parallel to $\mathbf{B}_0$. A schematic figure is shown to the right of each spectrum corresponding to the respective vessel and film geometries. Used with permission from Ref. [175].

χ, rather than anisotropic ^{129}Xe chemical shift interactions, that account for the line shapes observed in Fig. 19.

Polyacrylic Acid Studies

As discussed in Sect. 3.2, several NMR studies of xenon adsorbed on or dissolved in polymers have appeared [68, 73, 74, 76, 77, 80], which provide information on polymer structure and phase transitions. Recently, a novel investigation of xenon adsorbed on polyacrylic acid has also been carried out using the new optical pumping techniques described above [176]. The NMR spectra shown in Fig. 20 of optically pumped ^{129}Xe adsorbed on polyacrylic acid are interesting in at least two ways. First, the line widths of these spectra are reasonably narrow (4–18 ppm), indicating a relatively homogeneous interaction of xenon with the surface of the polymer. Second, although some xenon is in the gas phase at room temperature (evidenced by the signal at 0 ppm), there is a large ^{129}Xe peak at about 3 ppm in Fig. 20(a), which is attributed to xenon atoms in contact with the polyacrylic acid surface. This separate downfield xenon peak is due to xenon's significant interaction with the polymer surface, coupled with the moderate polymer surface area of about $15\,m^2/g$. As a consequence, an appreciable concentration of xenon atoms are adsorbed on the polyacrylic acid surface, even at room temperature. Xenon atoms exchange rapidly between the gas phase and a surface-adsorbed phase, so the measured chemical shift of ^{129}Xe is an average over the different environments sampled during the NMR experiment (again, ca. 10^{-3} s). For the xenon atoms that can access the polymer surface, the spectra in Figs. 20(b–d) show large changes in the ^{129}Xe chemical shift as functions of temperature and pressure, which can be understood in terms of the xenon sticking time and the diffusion of xenon atoms on the surface. At lower temperatures (133–173 K), xenon atoms preferentially adsorb on the polymer surface, as evidenced by the diminished relative intensities of the ^{129}Xe gas peaks in Fig. 20(b–d).

The chemical shift of xenon in contact with a surface can be expressed as a virial-type expansion with a pure xenon-surface term $\sigma_{0s}(T)$, a surface xenon–xenon interaction $\sigma_{1s}(T)$, and higher order terms:

$$\delta(T,\theta) = P_s\sigma_{0s}(T) + (P_s)^2\theta\sigma_{1s}(T) + (P_s)^3\theta^2\sigma_{2s}(T) + \cdots, \tag{19}$$

where P_s is the probability of finding a xenon atom at the surface and θ is the coverage in monolayers. The $(P_s)^2\theta\sigma_{1s}(T)$ term is due to binary xenon–xenon collisions that occur at or near the surface [114, 176]. The ^{129}Xe chemical shift observed for xenon adsorbed on polyacrylic acid is a linear function of the coverage over a broad temperature range, and thus only the first two terms need to be considered.

The chemical shift extrapolated to zero xenon pressure δ_S was shown

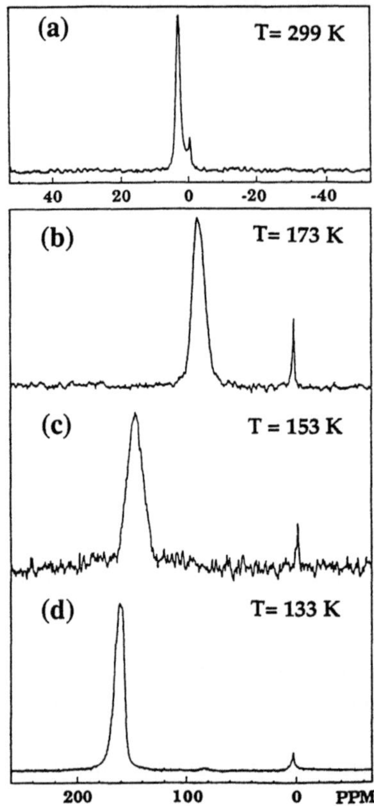

Fig. 20a–d. ^{129}Xe NMR spectra of optically pumped xenon adsorbed on polyacrylic acid at various temperatures: **a** $T = 299\,K$, $P = 4\,kPa$; **b** $T = 173\,K$, $P = 2.6\,kPa$; **c** $T = 153\,K$, $P = 1.3\,kPa$; and **d** $T = 133\,K$, $P = 0.26\,kPa$. Used with permission from Ref. [176].

[Fig. 21a] to fit a function of the form $\delta_S(T) = P_s \sigma_{0s}(T)$ with P_s given by:

$$P_s = \frac{\tau_o e^{H_{ads}/kT}}{\tau_V + \tau_o e^{H_{ads}/kT}} \quad , \tag{20}$$

where τ_o is the pre-exponential factor (ca. 10^{-12} s), τ_V is the reciprocal of the collision rate with the surface, and H_{ads} is the energy of adsorption (about 4.2 kcal/mol for Xe on polyacrylic acid) derived from xenon adsorption isotherms [176]. From the fit in Fig. 21(a), values of $\delta_S = 95$ ppm and $\tau_V = 3 \times 10^{-8}$ s are obtained. The xenon-polyacrylic acid surface interaction is therefore 95 ppm ($P_s = 1$), similar to the 86 ppm value observed for xenon adsorbed on Na-Y zeolite at 144 K [124] and consistent with the similar heats of xenon adsorption on the two materials.

Likewise, analysis of the ^{129}Xe chemical shift slope $d\delta/d\theta$ (i.e., the change in chemical shift as a function of coverage) provides an indication of xenon–xenon interactions at the adsorbent surface. The probability of finding two xenon atoms simultaneously at the surface is P_s^2, which should be related to the slope

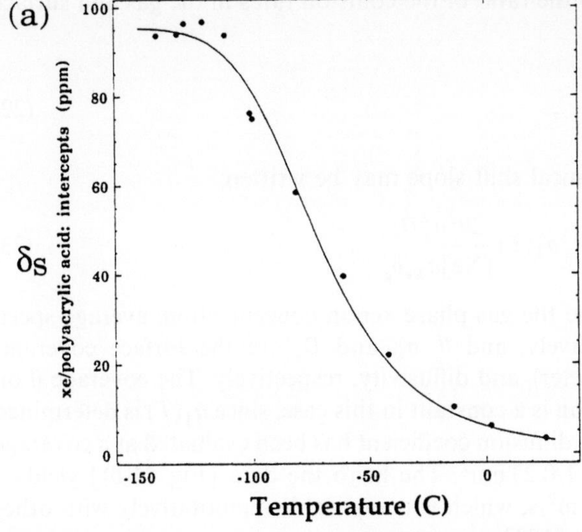

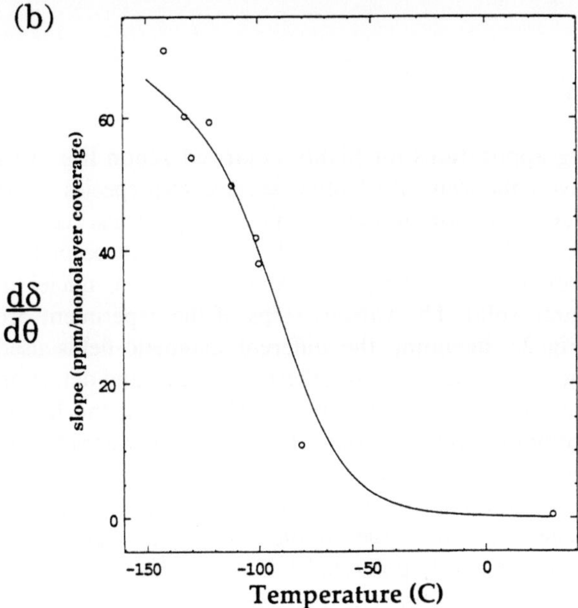

Fig. 21a,b. Plots of **a** the ^{129}Xe chemical shift δ_S and **b** ^{129}Xe chemical shift slopes $d\delta/d\theta$ versus temperature for xenon adsorbed on polyacrylic acid. The fits to the data indicate a large xenon-polyacrylic acid interaction ($\delta_S = 95$ ppm) and a surface diffusion coefficient of D_s of $3.3 \times 10^{-9}\ m^2/s$. Used with permission from Ref. [176].

through the temperature-dependent surface xenon–xenon interaction $\sigma_{1s}(T)$:

$$\frac{d\delta}{d\theta} = P_s^2 \sigma_{1s}(T). \tag{21}$$

$\sigma_{1s}(T)$ can be related to the Jameson second virial coefficient, which has been measured in the gas phase at temperatures between 240 K and 440 K [24]. By scaling $\sigma_1(T)$ according to the ratio of the collision rates in the gas and surface phases:

$$\sigma_{1s}(T) = \sigma_1(T)\frac{\tau_c}{\tau_{sc}} \quad , \tag{22}$$

an expression for the chemical shift slope may be written:

$$\frac{d\delta}{d\theta} = P_s^2\sigma_{1s}(T) = P_s^2\sigma_1(T)\frac{2\theta^2\sigma_s^2 D_s}{[\mathrm{Xe}]\sigma_{\mathrm{Xe}}\bar{v}_g} \quad , \tag{23}$$

where [Xe], $\bar{v}_g$, and σ_{Xe} are the gas-phase xenon concentration, average speed and cross-section, respectively, and θ, σ_s, and D_s are the surface coverage, cross-section (xenon diameter), and diffusivity, respectively. The coverage θ on the right side of the equation is a constant in this case, since $\sigma_1(T)$ is determined at a constant pressure. The diffusion coefficient has been evaluated at a coverage of one monolayer, so $\theta = 1/0.23\,\mathrm{nm}^2$. The fit to the data [Fig. 21(b)] yields a value for D_s of $3.3 \times 10^{-9}\,\mathrm{m}^2/\mathrm{s}$, which is comparable quantitatively with other studies of surface diffusion [183].

4.4 Cross-Polarization

One of the most promising applications for highly polarized xenon lies in its potential utility for selective enhancement of other surface spin species using cross-polarization techniques. A major step in reaching this goal has recently been achieved at Berkeley, where Bowers et al. [184] have succeeded in transferring polarization from optically pumped ^{129}Xe to nearby ^{13}C nuclei (in $^{13}CO_2$ molecules) in a frozen solid. The various steps of the experiment are shown schematically in Fig. 22, including the different magnetic fields used during the optical pumping, polarization storage, thermal mixing, and detection periods. After optically polarizing ^{129}Xe in low field (25 Gauss), the highly polarized xenon gas was combined with gaseous CO_2 (isotopically enriched in ^{13}C) and the mixture then co-condensed in a liquid-nitrogen cooled cell (at 150 Gauss) to form a solid Xe-$^{13}CO_2$ matrix. Reduction of the external magnetic field to B_{mix} allowed the heteronuclear dipole–dipole interaction to evolve via thermal mixing [185], producing highly polarized ^{13}C nuclei that were subsequently detected at 4.2 Tesla using a spin-echo pulse sequence.

Figure 23 shows high-field ^{13}C NMR spectra of the frozen Xe-$^{13}CO_2$ matrix after thermal mixing. The values of the axially symmetric chemical shift tensor, $\sigma_\perp = 232.8 \pm 2.3$ and $\sigma_\parallel = -82.5 \pm 2.3$ ppm, were extracted from the fit that accompanies the spectrum in Fig. 23(a). By changing the helicity of the circularly polarized laser light from right to left, the ^{13}C magnetization was reversed, resulting in the 180°-phase-shifted spectrum of Fig. 23(b). By comparing the integrated signal intensity of Fig. 23(a) with a reference spectrum of $^{13}CO_2$ gas, the enhancement of ^{13}C polarization achieved in this experiment

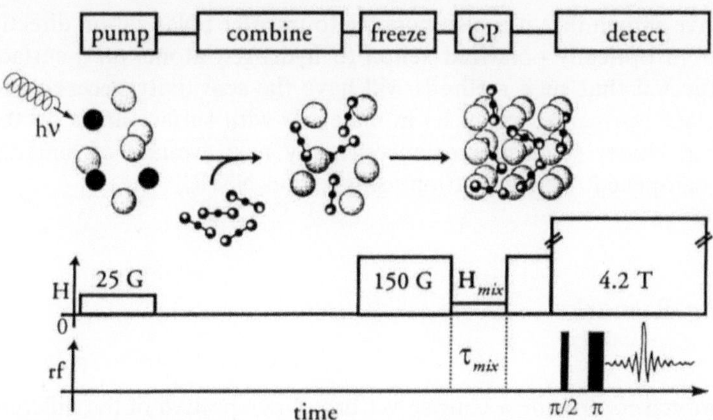

Fig. 22. Timing diagram of the laser-polarized ^{129}Xe-^{13}CO_2 thermal mixing experiment showing the steps of optical pumping at 25 Gauss, and gas mixing and expansion into a precooled sample cell at a field of 150 Gauss. After reducing the field to promote thermal mixing, the sample is subsequently transported to high field where the ^{13}C signal is obtained at 4.2 Tesla. Used with permission from Ref. [184].

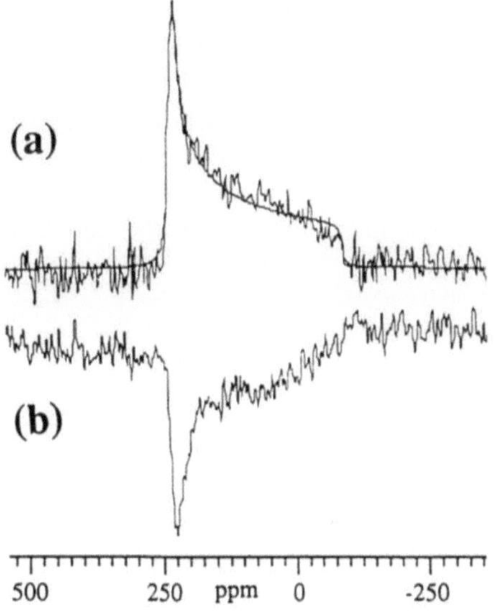

Fig. 23a,b. ^{13}C NMR spectra of approximately 10^{19} molecules of 99%-enriched ^{13}CO_2 in a frozen xenon matrix at high magnetic field, acquired after thermal mixing at low field with optically pumped ^{129}Xe. **a** A ^{13}C NMR spectrum resulting from ^{13}C contact with ^{129}Xe polarized using right-circularly polarized laser light. The signal was obtained in a single scan at 4.2 Tesla using a spin-echo pulse sequence. The solid line represents a fit of the theoretical anisotropic chemical shift line shape to the experimental spectrum. **b** An inverted ^{13}C NMR spectrum obtained using the same procedure as **a**, but with ^{129}Xe polarized using left-circularly polarized light. Used with permission from Ref. [184].

was estimated to be about 200. This is much smaller than enhancements demonstrated for ^{129}Xe alone (up to 2×10^4 using a high powered laser), possibly because of reduced ^{129}Xe-^{13}C polarization transfer efficiency. The efficiency of such polarization transfer depends greatly on the dipolar contact between the two species, the extent of which is not known for the Xe-^{13}CO_2 matrix. Further experiments have shown that it is also possible to transfer polarization directly in high field from optically polarized xenon to hydrogen atoms on a surface [186]. It is expected that such methods will have the sensitivity necessary to investigate surface resonances of nuclei in materials with surface areas on the order of 1 m^2/g. Under such circumstances, many new avenues of material research may be opened to investigation using xenon NMR.

5 Concluding Remarks

We have attempted to provide a concise yet broad perspective of the different types of xenon NMR techniques and data in the literature. Such studies span almost forty years, and include the three bulk phases of xenon, as well as a large variety of interesting and complex chemical systems and host phases. The recent development and application of xenon optical polarization techniques provide particularly creative opportunities for the characterization of surface structures and materials that have previously been inaccessible to NMR study. With the advent of such powerful new methods for increasing sensitivity and resolution, we expect vigorous and novel xenon NMR research to continue, accompanied by an expansion of materials applications in the future.

Acknowledgments. The authors thank their colleagues for the work that appears herein, particularly their co-workers at Berkeley. DR is a NSF Post-Doctoral Fellow in Chemistry. BFC gratefully acknowledges funding support through the NSF Young Investigator program under grant DMR-9257064 and from the Camille and Henry Dreyfus Foundation.

6 References

1. Schrobilgen GJ (1978) In: Harris RK and Mann BE (eds) NMR and the Periodic Table. Academic, London; Reisse J (1986) Nouv. J. Chim. 10: 665
2. Ito T, Fraissard J (1980) In: Rees LVC (ed) Proceedings of the Fifth International Conference on Zeolites, Naples, Italy, Heyden, London, p 510
3. Ripmeester JA (1980) presented at the ISMAR-Ampere International Conference on Magnetic Resonance, Delft, The Netherlands
4. Ripmeester JA (1981) Bullet. Magn. Reson. 79
5. Ripmeester JA, Davidson DW (1981) J. Molec. Struct. 75: 67
6. Ito T, Fraissard J (1982) J. Chem. Phys. 76: 5225
7. Ripmeester JA (1982) J. Am. Chem. Soc. 104: 289
8. Fraissard J, (1987) Zeitsch. Phys. Chem. 152: 159
9. Fraissard J, Ito T (1988) Zeolites 8: 350
10. Barrie PJ, Klinowski J (1992) Prog. in NMR Spectr. 24: 91
11. Dybowski C, Bansal N, Duncan TM (1991) Ann. Rev. Phys. Chem. 42: 433

12. Raftery D, Long H, Meersmann T, Grandinetti PJ, Reven L, Pines A (1991) Phys. Rev. Lett. 66: 584
13. Ramsey NF (1953) Nuclear moments, J. Wiley, New York; Ramsey NF (1950) Phys. Rev. 78: 699
14. Lamb WE (1941) Phys. Rev. 60: 817
15. Slichter CP (1990) Principles of magnetic resonance, 3rd edn. Springer-Verlag Berlin
16. Jameson CJ, Gutowsky HS (1964) J. Chem. Phys. 40: 1714
17. Malli G, Froese C (1967) Int. J. Quant. Chem. 1S: 95
18. Streever RL, Carr HY (1961) Phys. Rev. 121: 20
19. Jameson AK, Jameson CJ, Gutowsky HS (1970) J. Chem. Phys. 53: 2310
20. Grosse R, Burmeister R, Boddenberg B, Gedeon A, Fraissard J (1991) J. Phys. Chem. 95: 2443
21. Bloembergen N (1961) Nuclear magnetic relaxation. WA Benjamin, New York
22. Abragam A (1961) Principles of Nuclear Magnetism. Clarendon, Oxford
23. Proctor WG, Yu FC (1951) Phys. Rev. 81: 20; Brun E, Oeser J, Staub HH, Telschow CG (1954) Phys. Rev. 93: 904; ibid, (1954) Helv. Phys. Acta 27: 174
24. Jameson CJ, Jameson AK, Cohen SM (1973) J. Chem. Phys. 59: 4540
25. Kanegsberg E, Pass B, Carr HY (1969) Phys. Rev. Lett. 23: 572
26. Adrian FJ (1964) Phys. Rev. 136: A980
27. Jameson CJ (1975) J. Chem. Phys. 63: 5296
28. Barker JA, Watts RO, Lee JK, Schafer TP, Lee YT (1974) J. Chem. Phys. 61: 3081
29. Adrian FJ (1970) Chem. Phys. Lett. 7: 201
30. Hunt ER, Carr HY (1963) Phys. Rev. 130: 2302
31. Torrey HC (1963) Phys. Rev. 130: 2306
32. Shizgal B (1974) Chem. Phys. 5: 464
33. Adrian FJ (1965) Phys. Rev. 138: A403
34. Brinkmann D, Brun E, Staub HH (1962) Helv. Phys. Acta 35: 431
35. Jameson CJ, Jameson AK, Cohen SM (1975) J. Chem. Phys. 62: 4224
36. Jameson CJ, Jameson AK, Cohen SM (1976) J. Chem. Phys. 65: 3401
37. Jameson CJ, Jameson AK, Cohen SM (1977) J. Chem. Phys. 66: 5226
38. Jameson CJ, Jameson AK, Parker H (1978) J. Chem. Phys. 68: 3943
39. Jameson CJ, Jameson AK (1971) Mol. Phys. 20: 957
40. Jameson CJ, Jameson AK, Cohen SM (1976) J. Chem. Phys. 65: 3397
41. Jameson CJ, Jameson AK, Hwang JK (1988) J. Chem. Phys. 89: 4074
42. Buckingham AD, Kollman PA (1972) Mol. Phys. 23: 65
43. Jameson CJ, Jameson AK, Cohen SM (1975) Mol. Phys. 29: 1919
44. Yen WM, Norberg RE (1963) Phys. Rev. 131: 269
45. Warren WW Jr, Norberg RE (1966) Phys. Rev. 148: 402
46. Warren WW Jr, Norberg RE (1967) Phys. Rev. 154: 277
47. Cowgill DF, Norberg RE (1972) Phys. Rev. B6: 1636
48. Brinkmann D, Carr HY (1966) Phys. Rev. 150: 174
49. Lurie J, Feldman JL, Horton GK (1964) Phys. Rev. 150: 180
50. Van Kranendonk J (1954) Physica 20: 781
51. Stacey LM, Pass B, Carr HY (1969) Phys. Rev. Lett. 23: 1424; Hayes CE, Carr HY (1977) Phys. Rev. Lett. 39: 1558
52. Le Guillou JC, Zinn-Justin J (1977) Phys. Rev. Lett. 39: 95
53. Bartlett N (1962) Proc. Chem. Soc. 218
54. Brown TH, Whipple EB, Verdier PH (1963) J. Chem. Phys. 38: 3029; ibid, (1963) In: Hyman HH (ed) Noble gas compounds. Univ. of Chicago Press, Chicago
55. Seppelt VK, Rupp HH (1974) Zeitsch. Anorg. Allg. Chem. 409: 331, 338
56. Schrobilgen GJ, Holloway JH, Granger P, Brevard C (1978) Inorg. Chem. 17: 980
57. Holloway JH, Schrobilgen GJ (1980) Inorg. Chem. 19: 2632
58. Sawyer JF, Schrobilgen GJ, Sutherland SJ (1982) J. Chem. Soc., Chem. Commun. 210
59. Naumann D, Tyrra W (1989) J. Chem. Soc., Chem. Commun. 47
60. Turowsky L, Seppelt K (1990) Inorg. Chem. 29: 3226
61. Frohn HJ, Klose A, Henkel G (1993) Angew. Chem. 105: 114
62. Stengle TR, Reo NV, Williamson KL (1984) J. Phys. Chem. 88: 3225
63. Hertz HG (1973) Ber. Bunsenges. Phys. Chem. 77: 531, 688; Hertz HG (1973) 77: 688; Hertz HG, Holz M, Klute R, Stalidis G, Versmold H (1974) Ber. Bunsenges. Phys. Chem., 78: 24
64. Rummens FHA (1975) In: van der Waals' forces and shielding effects, NMR basic principles and progress, vol 10. Springer-Verlag, New York

65. Rummens FHA (1975) Chem. Phys. Lett. 31: 596
66. Stengle TR, Reo NV, Williamson KL (1981) J. Phys. Chem. 85: 3772
67. Stengle TR, Hosseini SM, Basiri HG, Williamson KL (1984) J. Sol. Chem. 13: 779
68. Stengle TR, Williamson KL (1987) Macromolecules 20: 1428
69. Stengle TR, Hosseini SM, Williamson KL (1986) J. Sol. Chem. 15: 777
70. Luhmer M, Dejaegere A, Reisse J (1989) Magn. Reson. Chem. 27: 950
71. Miller KW, Reo NV, Schoot Uiterkamp AJM, Stengle DP, Stengle TR, Williamson KL (1981) Proc. Nat. Acad. Sci. USA 78: 4946
72. Tilton RF Jr, Kuntz ID Jr (1982) Biochemistry 21: 6850
73. Walton JH, Miller JB, Roland CM (1992) J. Polym. Sci., Polym. Phys. 30: 527
74. Tomaselli M, Meier BH, Robyr P, Suter UW, Ernst RR (1993) Chem. Phys. Lett. 205: 145
75. Sefcik MD, Schaefer J, Desá JAE, Yelon WB (1983) Polym. Prepr. 24: 85
76. Kennedy GJ (1990) Polymer Bulletin 23: 605
77. Brownstein SK, Roovers JEL, Worsfold DJ (1988) Magn. Reson. Chem. 26: 392
78. Jeener J, Meier BH, Bachmann P, Ernst RR (1979) J. Chem. Phys. 71: 4546
79. Ernst RR, Bodenhausen G, Wokaun A (1987) Principles of nuclear magnetic resonance in one and two dimensions, Clarendon, Oxford
80. Kentgens APM, van Boxtel HA, Verweel RJ, Veeman WS (1991) Macromolecules 24: 3712
81. Loewenstein A, Brenman M (1978) Chem. Phys. Lett. 58: 435
82. Bayle JP, Courtieu J, Jullien J (1988) J. Chim. Phys. 85: 147
83. Diehl P, Jokisaari J (1990) J. Magn. Reson. 88: 660
84. Diehl P, Jokisaari J (1990) Chem. Phys. Lett. 165: 389
85. Ingman P, Jokisaari J, Diehl P (1991) J. Magn. Reson. 92: 163
86. Jokisaari J, Diehl P (1990) Liq. Cryst. 7: 739
87. Jokisaari J, Diehl P, Muenster O (1990) Mol. Cryst. Liq. Cryst. 188: 189
88. Muenster O, Jokisaari J, Diehl P (1991) Mol. Cryst. Liq. Cryst. 206: 179
89. Ito T, de Menorval LC, Guerrier E, Fraissard J (1984) Chem. Phys. Lett. 111: 271; Chen QJ, Springuel-Huet MA, Fraissard J (1989) Chem. Phys. Lett. 159: 117
90. Demarquay J, Fraissard J (1987) Chem. Phys. Lett. 136: 314
91. Ito T, Fraissard J (1986) J. Chim. Phys. 83: 441
92. de Menorval LC, Fraissard J, Ito T (1982) J. Chem. Soc. Faraday Trans. 1 78: 403; de Menorval LC, Fraissard J, Ito T, Primet M (1985) J. Chem. Soc. Faraday Trans. 1 81: 2855; Petrakis L, Springuel-Huet MA, Ito T, Hughes TR, Chan IY, Fraissard J (1988) In: Phillips MJ, Ternan M (eds) Proceedings of the Ninth International Congress on Catalysis, Calgary, Canada, vol. 1 p. 348
93. Gedeon A, Ito T, Fraissard J (1988) Zeolites 8: 376
94. Ito T, Bonardet JL, Fraissard J, Nagy JB, André C, Gabelica Z, Derouane EG (1988) Appl. Catal. 43: L5
95. Springuel-Huet MA, Fraissard J (1989) Chem. Phys. Lett. 154: 299
96. Fraissard J, Kärger J (1989) Zeolites 9: 351; Ito T, Fraissard J, Kärger J, Pfeifer H (1991) Zeolites 11: 103
97. Chen QJ, Fraissard J (1992) J. Phys. Chem. 96: 1814
98. Annen MJ, Davis ME, Hanson BE (1990) Catal. Lett. 6: 331
99. Heink W, Kärger J, Pfeifer H, Stallmach F (1990) J. Am. Chem. Soc. 112: 2175; Kärger J, Pfeifer H, Stallmach F, Spindler H (1990) Zeolites 10: 288; Kärger J, Pfeifer H (1991) J. Chem. Soc. Faraday Trans. 87: 1989
100. Reid RC, Prausnitz JM, Sherwood TK (1977) The Properties of Gases and Liquids, McGraw-Hill, NY
101. Pickett SD, Nowak AK, Thomas JM, Peterson BK, Swift JFP, Cheetham AK, den Ouden CJJ, Smit B, Post MFM (1990) J. Phys. Chem. 94: 1233
102. Shoemaker R, Apple T (1987) J. Phys. Chem. 91: 4024
103. Bansal N, Dybowski C (1990) J. Magn. Reson. 89: 21
104. Ryoo R, Pak C, Chmelka BF (1990) Zeolites 10: 790
105. Chmelka BF, Pearson JG, Liu SB, Ryoo R, de Menorval LC, Pines A (1991) J. Phys. Chem. 95: 303
106. Tway C, Apple T (1992) J. Catal. 133: 42
107. Valença GP, Boudart M (1991) J. Catal. 128: 447
108. Chmelka BF, Gillis JV, Petersen EE, Radke CJ (1990) AIChE J. 36: 1562
109. Kärger J (1991) J. Magn. Reson. 93: 184

110. Chmelka BF, de Menorval LC, Csencsits R, Ryoo R, Liu SB, Radke CJ, Petersen EE, Pines A (1989) Stud. Surf. Sci. Catal. 48: 269; Chmelka BF, Rosin RR, Went GT, Bell AT, Radke CJ, Petersen EE (1989) Stud. Surf. Sci. Catal. 49: 995; Seo G, Ryoo R (1990) J. Catal. 124: 224; Ahn DH, Lee JS, Nomura M, Sachtler WMH, Moretti G, Woo SI, Ryoo R (1992) J. Catal. 133: 191; Yang OB, Woo SI, Ryoo R (1992) J. Catal. 137: 357
111. Samant MG, de Menorval LC, Dalla Betta RA, Boudart M (1988) J. Phys. Chem. 92: 3937
112. Ripmeester JA, Ratcliffe CI (1990) J. Phys. Chem. 94: 7652
113. Chmelka BF, Raftery D, McCormick AV, de Menorval LC, Levine RD, Pines A (1991) Phys. Rev. Lett. 66: 580; ibid (1991) Phys. Rev. Lett. 67: 931
114. McCormick AV, Chmelka BF (1991) Molec. Phys. 73: 603
115. Jameson CJ, Jameson AK, Gerald R, de Dios AC (1992) J. Chem. Phys. 96: 1676
116. Ryoo R, de Menorval LC (1992) In: Higgins JB, von Ballmoos R, Treacy MMJ (eds) Ninth International Zeolites Conference Extended Abstracts and Program. Butterworth-Heinemann, Stoneham, MA
117. Güémez J, Velasco S (1987) Am. J. Phys. 55: 154
118. Rowlinson JS, Woods GB (1990) Physica A 164: 117
119. van Tassel PR, Davis HT, McCormick AV (1991) Molec. Phys. 73: 1107; van Tassel PR, Davis HT, McCormick AV (1992) Molec. Phys. 76: 411; van Tassel PR, Hall C, Davis HT, McCormick AV (1992) Pure App. Chem. 64: 1629
120. Jameson CJ, Jameson AK, Gerald R, de Dios AC (1992) J. Chem. Phys 96: 1690
121. Shore JS, Emsley L, Larsen RG, Chmelka BF, Pines A (1993) presented at 34th Experimental Nuclear Magnetic Conference, St. Louis, Abstract P94, pp. 119
122. Larsen RG, Shore J, Schmidt-Rohr K, Emsley L, Long H, Pines A, Janicke M, Chmelka BF (1993) Chem. Phys. Lett. 214: 220
123. Ripmeester JA, Ratcliffe CI, Tse JS (1988) J. Chem. Soc., Faraday Trans. 1 84: 3731
124. Cheung TTP, Fu CM, Wharry S (1988) J. Phys. Chem. 92: 5170
125. Cheung TTP, Fu CM (1989) J. Phys. Chem. 93: 3740
126. Cheung TTP (1989) J. Phys. Chem. 93: 7549
127. Cheung TTP (1990) J. Catal. 124: 511
128. Cheung TTP (1990) J. Phys. Chem. 94: 376
129. Barrie PJ, McCann GF, Gameson I, Rayment T, Klinowski J (1991) J. Phys. Chem. 95: 9416
130. Ripmeester JA, Ratcliffe CI (1990) J. Phys. Chem. 94: 8773
131. Davidson DW, Handa YP, Ripmeester JA (1986) J. Phys. Chem. 90: 6549
132. Hartmann SR, Hahn EL (1962) Phys. Rev. 128: 2042; Pines A, Gibby MG, Waugh JS (1972) J. Chem. Phys. 56: 1776; Pines A, Gibby MG, Waugh JS (1973) J. Chem. Phys. 59: 569
133. Mehring M (1983) Principles of high resolution NMR in solids, 2nd edn. Springer, Berlin
134. Ripmeester JA, Tse JS, Ratcliffe CI, Powell BM (1987) Nature 325: 135
135. Slichter CP, Holton WC (1961) Phys. Rev. 122: 1701
136. Pines A, Shattuck TW (1973) Chem. Phys. Lett. 23: 614
137. de Menorval LC, Raftery D, Liu S-B, Takegoshi K, Ryoo R, Pines A (1990) J. Phys. Chem. 94: 27
138. Miller JT, Meyers BL, Ray GJ (1991) J. Catal. 128: 436
139. Tsiao C, Dybowski C, Gaffney AM, Sofranko JA (1990) J. Catal. 128: 520
140. Suh DJ, Park T-J, Ihm S-K, Ryoo R (1991) J. Phys. Chem. 95: 3767
141. Wernett PC, Larsen JW, Yamada O, Yue HJ (1990) Energy and Fuels 4: 412
142. Tsiao C, Botto RE (1991) Energy and Fuels 5: 87
143. Birgeneau RJ, Horn PM (1986) Science 232: 329, and Refs. therein
144. Thomy A, Duval X (1970) J. Chim. Phys. 67: 1101; Régnier J, Thomy A, Duval X (1977) J. Chim. Phys. 74: 926
145. Hong H, Peters CJ, Mak A, Birgeneau RJ, Horn PM, Suematsu H (1989) Phys. Rev. B 40: 4797; Hong H, Birgeneau RJ (1989) Zeitsch. Phys. B-Cond. Matt. 77: 413
146. Shibanuma T, Asada H, Ishi S, Matsui T (1983) Jap. J. Appl. Phys., (Eng.) 22: 1656
147. Neue G (1987) Zeitsch. Phys. Chem. Neue Folge, Bd. (Eng.) 152: 13
148. Brossel J, Kastler A (1949) Comp. Rend. 229: 1213; Kastler A (1950) J. Phys. Radium 11: 255
149. Carver TR (1963) Science 141: 599
150. Bernheim R (1965) Optical pumping. W. A. Benjamin, New York
151. Happer W (1972) Rev. Mod. Phys. 44: 169; Happer W (1985) Ann. Phys. Fr. 10: 645; Knize RJ, Wu Z, Happer W (1988) In: Advances in Atomic and Molecular Physics, vol 24. Academic, New York

152. Grover BC (1978) Phys. Rev. Lett. 40: 391
153. Bouchiat CC, Bouchiat MA, Pottier LCL (1969) Phys. Rev. 181: 144; Bouchiat MA, Brossel J, Pottier LC (1972) J. Chem. Phys. 56: 3703
154. Cates GD, Benton DR, Gatzke M, Happer W, Hasson KC, Newbury NR (1990) Phys. Rev. Lett. 65: 2591
155. Appelt S, Scheufler P, Mehring M (1989) Opt. Comm. 74: 110
156. Suter D, Rosatzin M, Mlynek J (1991) Phys. Rev. Lett. 67: 34
157. Suter D, Mlynek J (1991) In: Advances in Magnetic and Optical Resonance, vol 16 Academic, New York
158. Herman, RM (1964) Phys. Rev. 136A: 1576; Herman RM (1965) Phys. Rev. 137A: 1062
159. Happer W, Miron E, Schaefer S, van Wijngaarden WA, Zeng X (1984) Phys. Rev. A 29: 3092
160. Bhaskar ND, Happer W, Larsson M, Zeng X (1983) Phys. Rev. Lett. 50: 105
161. Bhaskar ND, Happer W, McClelland T (1982) Phys. Rev. Lett. 49: 25
162. Zeng X, Wu Z, Call T, Miron E, Schreiber D, Happer W (1985) Phys. Rev. A 31: 260
163. Calaprice FP, Happer W, Schreiber DF, Lowry MM, Miron E, Zeng X (1985) Phys. Rev. Lett. 54: 174
164. Chupp TE, Wagshul ME, Coulter KP, McDonald AB, Happer W (1987) Phys. Rev. C 36: 2244
165. Zeng X, Wu C, Zhao M, Li S, Li L, Zhang X, Liu Z, Liu W (1991) Chem. Phys. Lett. 182: 538; Liu Z, Zhao M, Wu C, Li L, Li S, Zeng X, Xiong W (1992) Chem. Phys. Lett. 194: 440
166. Haroche S, Cohen-Tannoudji C (1970) Phys. Rev. Lett. 24: 974
167. Haroche S, Cohen-Tannoudji C, Audoin C, Schermann JP (1970) Phys. Rev. Lett. 24: 861
168. Cohen-Tannoudji C (1963) J. Phys. (Paris) 24: 653
169. Volk CH, Kwon TM, Mark JG (1980) Phys. Rev. A 21: 1549
170. Redsun SG, Knize RJ, Cates GD, Happer W (1990) Phys. Rev. A 42: 1293; Redsun SG (1990) Ph.D. Thesis, Princeton University
171. Zeng X, Miron E, van Wijngaarden WA, Schreiber D, Happer W (1983) Phys. Lett. 96A: 191
172. Volk CH, Kwon TM, Mark JG, Kim YB, Woo JC (1980) Phys. Rev. Lett. 44: 136; Kwon TM, Mark JG, Volk CH (1981) Phys. Rev. A 24: 1894
173. Wu Z, Happer W, Daniels JM (1987) Phys. Rev. Lett. 59: 1480; Wu Z, Schaefer S, Cates GD, Happer W (1988) Phys. Rev. A 37: 1161
174. Cohen-Tannoudji C (1962) Ph. D. Thesis, In: Ann. Phys. (Paris) 7: 423–460, 469–504
175. Raftery D, Long H, Reven L, Tang P, Pines A (1992) Chem. Phys. Lett. 191: 385
176. Raftery D, Reven L, Long H, Pines A, Tang P, Reimer JA (1993) J. Phys. Chem. 97: 1649
177. Gonen O, Kuhns PL, Waugh JS, Fraissard JP (1989) J. Phys. Chem. 93: 504
178. Derouane EG, Fraissard J, Fripiat JJ, Stone WEE (1972) Catal. Rev. 7: 121
179. Lechert H (1976) Catal. Rev.-Sci. Eng. 14: 1
180. Duncan TM, Dybowski C (1981) Surf. Sci. Rep. 1: 157
181. Slichter CP (1986) Ann. Rev. Phys. Chem. 37: 25; Ansermet J-Ph, Slichter CP, Sinfelt JH (1990) Prog. In NMR Spectr. 22: 401
182. Opella SJ, Waugh JS (1977) J. Chem. Phys. 66: 4919
183. Tabony J, Cosgrove T (1979) Chem. Phys. Lett. 67: 103
184. Bowers CR, Long HW, Pietrass T, Gaede HC, Pines A (1993) Chem. Phys. Lett. 205: 168
185. Goldman M (1970) Spin temperature and nuclear magnetic resonance in solids, Oxford, London
186. Long HW, Gaede HC, Shore J, Reven L, Bowers CR, Kritzenberger J, Pietrass T, Pines A, Tang P, Reimer JA (1993) J. Am. Chem. Soc. 115: 8491

NMR as a Generalized Incoherent Scattering Experiment

Gerald Fleischer[1] **and Franz Fujara**[2]

[1] Universität Leipzig, Fachbereich Physik, Linnéstraße 5, D-04103 Leipzig, FRG
[2] Universität Dortmund, Experimentelle Physik III, Postfach 500500, D-44221 Dortmund, FRG

Table of Contents

NMR Basic Principles and Progress, Vol. 30

Two types of stimulated NMR echo experiments, ^{2}H-spin alignment and field gradient NMR, are formulated in terms of a "generalized dynamic scattering function". The analogies to incoherent quasielastic neutron scattering are discussed. The concept is illustrated by selected examples covering molecular reorientations and self diffusion in molecular crystals and supercooled liquids, anomalous diffusion in linear chain polymers and restricted diffusion of molecules in confined geometries.

1 Motivation

When dealing with molecular motion in condensed phases we look for experimental methods which allow us to identify a dynamic process with respect to its *geometry* and its *temporal* behaviour. A given type of motion defines a dynamic structure in space. It has therefore been the privilege of scattering methods to study the geometry of molecular motions. Every scattering experiment leads to a dynamic structure factor $S(\boldsymbol{Q}, \omega)$ or its spatial and temporal Fourier transform, the so called van Hove correlation function $G(\boldsymbol{r}, t)$. $G(\boldsymbol{r}, t)$ denotes the probability—averaged over the whole sample—to find a particle at time t at position $\boldsymbol{r}$ if at an earlier time $t = 0$ some other (or the same) particle had been at $\boldsymbol{r} = 0$. Let us, according to common use, roughly distinguish between microscopic ($\leqq 10^{-12}$ s), mesoscopic ($10^{-12} \cdots 10^{-8}$ s) and macroscopic ($\geqq 10^{-8}$ s) times and between microscopic ($\leqq 10$ Å), mesoscopic ($10 \cdots 10^4$ Å) and macroscopic ($\geqq 10^4$ Å) distances.

It is generally accepted that *neutron scattering* is the outstanding experimental method. It allows us to obtain $G(\boldsymbol{r}, t)$ at microscopic and mesoscopic times and microscopic distances. Although there has been a lot of effort to extend the temporal window toward longer times (i.e. higher energy resolution), at present, processes slower than 10^{-8} s remain unaccessible with neutron scattering. Other scattering techniques like Mössbauer-Rayleigh scattering allow to go beyond this limit but still do not provide sufficient intensity for practical applications.

A different approach to extend the dynamic range toward macroscopic times without loosing the information about the microscopic geometry of the process is *NMR*. For many years, solid state NMR has been used to detect and characterize slow motions via line shape analysis, stimulated echoes and multidimensional spectra. Mostly, however, the relationship to scattering methods remained unexplored. It is only since the recent onset of NMR imaging and NMR microscopy that NMR spectroscopists have started to be more interested in looking at their results in terms of a dynamic structure factor. Therefore, it is the purpose of this contribution to analyse NMR in terms of a generalized scattering experiment. There is no doubt that one would wish to obtain a unified description of NMR spectroscopy and scattering methods. One would be able to expect progress in both, theoretical concepts and practical applications.

To be specific: It will be shown in the following that the concept of the *incoherent structure factor* known from neutron scattering can be generalized and thus be applied to stimulated NMR echoes. Note that this generalization actually concerns only the *in*coherent scattering, i.e. the tagged particle motion. We are far from being able to generalize the coherent structure factor, although approaches are thinkable.

The work is that of experimentalists and will thus focus on the basic conceptional ideas in an illustrative manner rather than in terms of mathematical rigidity.

Emphasis is also put on applications taken from recent studies about polymer dynamics, the glass transition, molecular reorientation in crystals and defect migration. The contribution is organized as follows: In Sect. 2, the correlation functions of neutron scattering and the NMR stimulated echo are pointed out. It will be shown that there are two domains where the analogies between *n*-scattering and NMR allow closely complementary experiments in the study of molecular reorientations (NMR spin alignment) and the study of mesoscopically ranged diffusion phenomena (field gradient NMR). In Sect. 3, a few experimental examples will be given for both types of applications. The examples are chosen with the aim to point out the wealth of combined NMR-neutron scattering studies, to illustrate the wide range of their close analogies but also discuss the limitations. Finally (Sect. 4), among other concluding remarks, we will comment on a new technology in gradient NMR which allows an enormous increase in the spatial resolution.

2 Analogies Between Incoherent Quasielastic Neutron Scattering and NMR Stimulated Echoes

In this section, it will be shown that both incoherent quasielastic neutron scattering (IQNS) and NMR stimulated echoes (NMR-SE) yield direct information about molecular correlation functions in a comparable way. Therefore, let us formulate the main results of both methods in a—for this purpose—suitable way.

2.1 Quasielastic Neutron Scattering

The significant quantity in neutron scattering is the double differential cross section per atom $\mathrm{d}^2\sigma/\mathrm{d}\Omega\,\mathrm{d}E$. If, for simplicity, the system is assumed to consist only of one atomic species, we get [1–4]

$$\frac{\mathrm{d}^2\sigma}{\mathrm{d}\Omega\,\mathrm{d}E}=\frac{1}{\hbar}\frac{k}{k_0}\left[\sigma_{\mathrm{coh}}S_{\mathrm{coh}}(Q,\omega)+\sigma_{\mathrm{inc}}S_{\mathrm{inc}}(Q,\omega)\right] \tag{1}$$

with $\boldsymbol{k}_0$ and $\boldsymbol{k}$ being the wavevectors of the incoming and scattered neutrons, respectively, the scattering vector $\boldsymbol{Q}=\boldsymbol{k}-\boldsymbol{k}_0$, the energy transfer $E=\hbar\omega=\hbar^2Q^2/2m$, the neutron mass m, the coherent scattering cross section σ_{coh}, which weights the contribution from interferences of the neutron waves scattered at different centers, and the incoherent scattering cross section σ_{inc} due to scattering at individual particles. The coherent and incoherent dynamic structure factors $S_{\mathrm{coh}}(Q,\omega)$ and $S_{\mathrm{inc}}(Q,\omega)$, respectively, are quantities which contain information on physical properties of the sample only. They are linked with the intermediate (coherent and incoherent) scattering functions via Fourier transformation

$$S_{\mathrm{coh}}(Q,\omega)=\frac{1}{2\pi}\int S_{\mathrm{coh}}(Q,t)\mathrm{e}^{-\mathrm{i}\omega t}\mathrm{d}t$$
$$S_{\mathrm{inc}}(Q,\omega)=\frac{1}{2\pi}\int S_{\mathrm{inc}}(Q,t)\mathrm{e}^{-\mathrm{i}\omega t}\mathrm{d}t \tag{2}$$

where the intermediate scattering functions are defined as [5, 6]

$$S_{\mathrm{coh}}(Q,t)=1/N\sum_{\mathrm{i,j}}\langle \mathrm{e}^{-\mathrm{i}QR_{\mathrm{i}}(0)}\cdot \mathrm{e}^{\mathrm{i}QR_{\mathrm{j}}(t)}\rangle$$
$$S_{\mathrm{inc}}(Q,t)=1/N\sum_{\mathrm{i}}\langle \mathrm{e}^{-\mathrm{i}QR_{\mathrm{i}}(0)}\cdot \mathrm{e}^{\mathrm{i}QR_{\mathrm{i}}(t)}\rangle \tag{3}$$

with N being the number of particles and $\boldsymbol{R}_{\mathrm{i}}(t)$ denoting the position of the i-th particle at time t. The brackets $\langle \cdots \rangle$ represent the ensemble average. When we deal with powders, which is the case for all of our experimental examples to be treated in Sect. 3, we can agree that the brackets also include the proper powder average. Correspondingly, it is only the absolute value Q—instead of the full vector $\boldsymbol{Q}$—which appears in Eqs. (2) and (3). Applying another Fourier transformation

$$S_{\mathrm{coh}}(Q,\omega)=(2\pi\hbar)^{-1}\int \mathrm{d}t\int \mathrm{d}r\mathrm{e}^{-\mathrm{i}(\omega t-Qr)}G_{\mathrm{coh}}(r,t)$$
$$S_{\mathrm{inc}}(Q,\omega)=(2\pi\hbar)^{-1}\int \mathrm{d}t\int \mathrm{d}r\mathrm{e}^{-\mathrm{i}(\omega t-Qr)}G_{\mathrm{s}}(r,t) \tag{4}$$

we obtain the "distinct" and the "self" part of the dynamic pair correlation function, $G_{\mathrm{coh}}(r,t)$ and $G_{\mathrm{s}}(r,t)$, respectively, also named van Hove correlation function. A comparison with Eqs. (2) and (3) leads to the following results for G_{coh} and G_{s}:

$$G_{\mathrm{coh}}(r,t)=1/N\sum_{\mathrm{i,j}}\int\langle[\delta(\boldsymbol{r}'-\boldsymbol{R}_{\mathrm{i}}(0)]\delta[\boldsymbol{r}'+\boldsymbol{r}-\boldsymbol{R}_{\mathrm{j}}(t)]\rangle\mathrm{d}\boldsymbol{r}'$$
$$G_{\mathrm{s}}(r,t)=1/N\sum_{\mathrm{i}}\int\langle[\delta(\boldsymbol{r}'-\boldsymbol{R}_{\mathrm{i}}(0)]\delta[\boldsymbol{r}'+\boldsymbol{r}-\boldsymbol{R}_{\mathrm{i}}(t)]\rangle\mathrm{d}\boldsymbol{r}' \tag{5}$$

In the classical limit $kT \gg h\omega$, Eq. (5) simplifies to

$$G_{\mathrm{coh}}^{\mathrm{class}}(r,t)=\sum_{\mathrm{i}}\langle\delta[\boldsymbol{r}-\boldsymbol{R}_{\mathrm{i}}(t)+\boldsymbol{R}_0(0)]\rangle$$
$$G_{\mathrm{s}}^{\mathrm{class}}(r,t)=\langle\delta[\boldsymbol{r}-\boldsymbol{R}_0(t)+\boldsymbol{R}_0(0)]\rangle. \tag{6}$$

Equation (6) clarifies, that $G_{\mathrm{coh}}^{\mathrm{class}}(r,t)$ describes the conditional probability of finding a particle at time t at distance r if another particle has started at time $t=0$ at the origin $r=0$. $G_{\mathrm{s}}^{\mathrm{class}}(r,t)$ is the probability of finding a particle at time t at distance r if *itself* has started at time $t=0$ at $r=0$. Since we will exclusively deal with cases in the classical limit we leave away the superscripts "class" in the following. Note, however, that the analogy between NMR and neutron scattering to be formulated in this work, is expected to hold in the nonclassical regime as well.

In the following we further restrict ourselves to *in*coherent scattering. With Eq. (6) we also see the context of the powder averaged $G_s(r,t)$ and the probability density $P(\boldsymbol{r},\boldsymbol{r}_0,t)$ for a particle jump from $(\boldsymbol{r}_0, t=0)$ to $(\boldsymbol{r},t)$:

$$G_s(r,t) = \langle \int P(\boldsymbol{r},\boldsymbol{r}_0,t)p(\boldsymbol{r}_0)\mathrm{d}\boldsymbol{r}_0 \rangle_{\text{powder}} \tag{7}$$

with the a priori probability distribution $p(\boldsymbol{r}_0)$ of origins. If we split up $G_s(r,t)$ into a time independent and a (to zero decaying) time dependent part

$$G_s(r,t) = G_s(r,\infty) + G_s'(r,t) \tag{8}$$

we obtain e.g. for the intermediate incoherent scattering function

$$\begin{aligned} S_{\text{inc}}(Q,t) &= \langle \int e^{i\boldsymbol{Q}\boldsymbol{r}} \int P(\boldsymbol{r},\boldsymbol{r}_0,t)p(\boldsymbol{r}_0)\mathrm{d}\boldsymbol{r}_0\mathrm{d}\boldsymbol{r} \rangle_{\text{powder}} \\ &= S_{\text{inc}}'(Q,t) + A_0(Q) \end{aligned} \tag{9}$$

with a plateau value $A_0(Q)$. If we calculate $S_{\text{inc}}(Q,\omega)$ by applying Eq. (2), $A_0(Q)$ transforms into the elastic amplitude

$$S_{\text{inc}}(Q,\omega) = A_0(Q)\delta(\omega) + S_{\text{inc}}'(Q,\omega) \tag{10}$$

and is therefore called *e*lastic *i*ncoherent *s*tructure *f*actor (EISF). From Eq. (9) we get

$$A_0(Q) = \text{“EISF”} = \langle |\int e^{i\boldsymbol{Q}\boldsymbol{r}} p(\boldsymbol{r})\mathrm{d}\boldsymbol{r}|^2 \rangle_{\text{powder}} \tag{11}$$

The EISF is a central quantity in incoherent neutron scattering since via Eq. (11) it contains information about the “structure” which is spanned by all positions which can be reached by the tagged particle in the run of the time.

In incoherent neutron scattering there are no selection rules. Therefore principally all types of motion contribute to the scattered intensity. It is essentially the Q-dependence which allows to distinguish between different types of motion. To give an example: If we study protonated molecular powders, the differential cross section is well approximated by the incoherent proton scattering. [It happens that $\sigma_{\text{inc}}(^1\text{H}) \gg \sigma_{\text{coh}}(^1\text{H})$ and also much larger than the cross sections of practically all other nuclei [7]]. If we factorize the molecular motion in a center of mass part (T), a reorientational part (R) and a vibrational part (V)

$$S_{\text{inc}}(Q,t) = S_{\text{inc}}^{\text{T}}(Q,t) \cdot S_{\text{inc}}^{\text{R}}(Q,t) \cdot S_{\text{inc}}^{\text{V}}(Q,t) \tag{12}$$

we can give explicit expressions for some basic modes of motion [8]. Perhaps the most important models are the hydrodynamic translational diffusion

$$\vartheta/\vartheta t P(\boldsymbol{r},\boldsymbol{r}_0,t) = D\Delta P(\boldsymbol{r},\boldsymbol{r}_0,t) \tag{13}$$

(D being the self diffusion constant) with the solution

$$S_{\text{inc}}^{\text{T}}(Q,t) = e^{-DQ^2 t} \tag{14}$$

and the rotational diffusion for the angular part of the probability density

$$\vartheta/\vartheta t P(\Omega,\Omega_0,t) = D_{\text{R}}\Delta_{\Omega} P(\Omega,\Omega_0,t) \tag{15}$$

(solid angle Ω, rotational diffusion constant D_R) with the solution

$$S_{inc}^{R}(Q,t) = \sum_{l=0}^{\infty} (2l+1) j_l^2(Qa) e^{-l(l+1)D_R t} \tag{16}$$

(a being the radius of the spherical surface of diffusion; j_l being the Bessel function of lth order). The characteristic feature of these diffusional models is the strong decay of the intermediate scattering functon with increasing Q. In the case of translational diffusion the decay is monoexponential with a time constant which is proportional to Q^{-2}. For the rotational diffusion the decay is multiexponential such that with higher Q the weight of "partial" time constants $\tau_l = [l(l+1)D_R]^{-1}$ with large values of l increases. There is another distinction between translational and rotational diffusion: The EISF of translational diffusion is zero, whereas that of rotational diffusion is given by

$$\mathrm{EISF}_{\mathrm{rot.diff.}} = j_0(Qa). \tag{17}$$

Diffusional models are correct only in the hydrodynamic regime, i.e. as long as the jump length d of the elementary process is sufficiently smaller than Q^{-1}. If this condition is not fulfilled, we deal with jump diffusion. For $d \geqq Q^{-1}$ the intermediate scattering function of both, translational and rotational jump diffusion, tends toward a Q-independent temporal behaviour ("plateau regime"). An important and often [9] applied model is that of Chudley and Elliott [10].

Another type of molecular reorientational motion which often occurs in crystalline rotator phases is the N-site jump (N finite). The intermediate scattering function can often be calculated analytically [8, 11]. For N-site jumps the temporal decay contains N time constants at most. It is most important for the following to note that for $Q \to \infty$ the EISF of N-site jumps converges *always* toward $1/N$. An example [11]: For rotational jumps between N equivalent positions on a circle of radius a the powder averaged EISF is

$$\mathrm{EISF}_{N\text{-site}} = 1/N \sum_{j=1}^{N} \frac{\sin[2Qa\sin(\pi j/N)]}{2Qa\sin(\pi j/N)} \tag{18}$$

One may infer from this concise description of several motional models that in favourable cases S_{inc}^{T} and S_{inc}^{R} can be measured and interpreted in detail if translational and/or rotational dynamics is much slower than vibrations, which in fact is often the case.

The determination of S_{inc}^{V} due to vibrations allows less precise assignments since many molecular and lattice modes occur at microscopic times and are coupled with each other. Only in well selected crystalline systems has it been possible to determine the dispersion rules using coherent neutron scattering [12]. In crystalline powders, amorphous solids and liquids one usually contents oneself with the determination of the Debye-Waller factor or at best with the vibrational density of states. The (in-coherent) Debye-Waller factor f_Q^{inc} (sometimes also called Mössbauer-Lamb factor) can be interpreted as long time limit

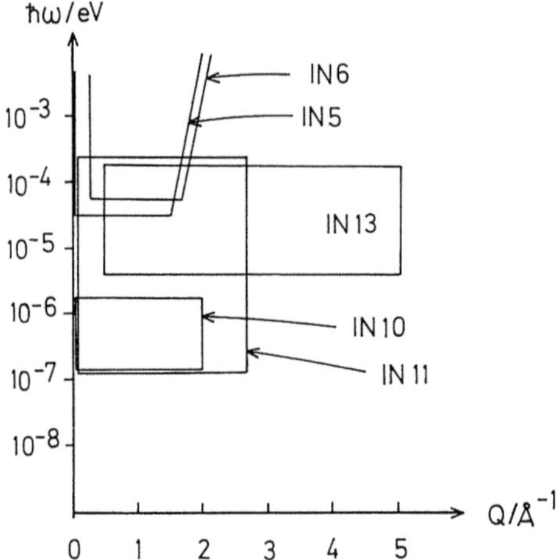

Fig. 1. Typical "working" regimes of the high resolution neutron scattering spectrometers at the ILL

and thus as EISF of the vibrational part of the intermediate scattering function

$$f_Q^{inc} = \lim_{t \to \infty} S_{inc}^{V}(Q, t) \tag{19}$$

For lattice vibrations of a classical harmonic crystal the Debye-Waller factor is of Gaussian shape

$$f_Q^{inc} = \exp[-\tfrac{1}{3}\langle r^2 \rangle Q^2] \tag{20}$$

The Debye-Waller factor can be interpreted as the "form factor" of the vibrations, the EISF as form factor of the jump process. If the jumps occur on a time scale comparable to that of vibrations, the Debye-Waller factor is the product of a Gaussian due to the vibrational excitations and an EISF due to the jumps.

The Mecca of neutron scattering is Grenoble with its Institute Laue-Langevin. Let us, without going into any further details, shortly mention those ILL spectrometers which are currently used for the study of molecular dynamics. For further information we refer to Ref. [8] which contains an excellent description of all relevant spectrometers. There are three types of spectrometers, the time of flight (tof) machines IN5 and IN6, the backscattering spectrometers IN10 and IN13 and the spin echo spectrometer IN11. The "working" regime of all machines is depicted in Fig. 1. [1 meV $\cong$ 0.0965 kJ mol^{-1} $\cong$ 2.418 $\times$ 10^{11} Hz $\cong$ 11.605 K.]

2.2 NMR Spin Alignment

Deuterons are spin $I = 1$ nuclei with a quadrupolar interaction which is much larger than the dipole–dipole interaction but still small enough for NMR pulse experiments. Therefore they represent an ideal probe for the study of molecular motions in organic substances with C-^{2}H or O-^{2}H bonds [13]. Since the electrical field gradient tensor is nearly axially symmetric with respect to the bond direction, there is a simple relationship between the NMR-quadrupolar frequency ω_Q and the angle Θ between the external magnetic field $\boldsymbol{B}_0$ and the bond direction:

$$\omega_Q = (\delta_Q/2)(3\cos^2\Theta - 1) \tag{21}$$

Here the parameter $\delta_Q/2\pi$ is equal to 3/4 of the deuteron quadrupolar coupling constant e^2Qq/h. Molecular motions which involve a time dependence of the bond axis orientation $\Theta(t)$ thus lead to a time dependent quadrupolar frequency $\omega_Q(t)$. This frequency coding of geometrical information is the basis of ^{2}H echo experiments [14] and the ^{2}H spin alignment [15]. Especially, it has been shown that a modified Jeener-Broekaert pulse sequence of the "spin alignment 3 pulse experiment" $(\pi/2)_x - \tau - (\pi/4)_y - t - (\pi/4)_y$

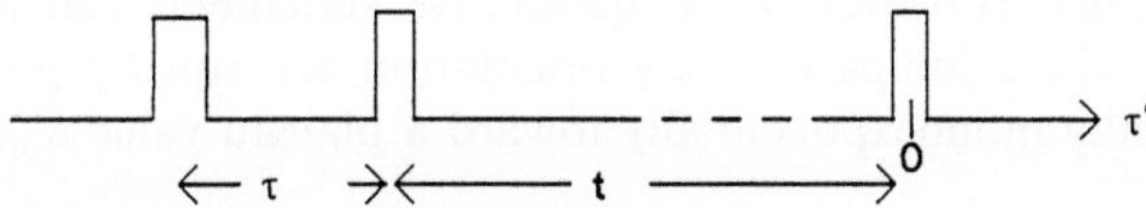

leads to the single particle correlation function

$$S_{\sin}(\tau, t, \tau') = \langle \sin[\tau\omega_Q(t=0)] \sin[\tau'\omega_Q(t)] \rangle \tag{22}$$

and the "stimulated 3 pulse echo experiment" $(\pi/2)_x - \tau - (\pi/2)_x - t - (\pi/2)_x$ to the single particle correlation function

$$S_{\cos}(\tau, t, \tau') = \langle \cos[\tau\omega_Q(t=0)] \cos[\tau'\omega_Q(t)] \rangle. \tag{23}$$

The range of validity of Eqs. (22) and (23) is limited to $\tau \leqq T_2 < t \leqq T_1$, where T_2 and T_1 are the spin–spin- and the spin–lattice relaxation times, respectively. In practice this means typically $\tau \leqq 300\,\mu s < t \leqq$ min. The correlation functions Eqs. (22) and (23) can be read in several variations. An extremely successful application is the 2-dimensional exchange experiment [16]. For the purpose of this work we set $\tau' = \tau$. This means experimentally that we measure only the amplitude of the echo after the third pulse. Thereby we are left with the correlation functions

$$\begin{aligned} S_{\sin}(\tau, t) &= \langle \sin[\tau\omega_Q(0)] \sin[\tau\omega_Q(t)] \rangle \\ S_{\cos}(\tau, t) &= \langle \cos[\tau\omega_Q(0)] \cos[\tau\omega_Q(t)] \rangle. \end{aligned} \tag{24}$$

$S_{\sin}$ and $S_{\cos}$ can be measured separately and therefore also their sum

$$\begin{aligned} S(\tau, t) &= S_{\cos}(\tau, t) + S_{\sin}(\tau, t) \\ &= \langle \exp[-i\tau\omega_Q(0)] \exp[i\tau\omega_Q(t)] \rangle \end{aligned} \tag{25}$$

This is correct since in $\langle e^{-i}\cdots e^{i}\cdots\rangle = \langle\cos\cdots\cos\cdots\rangle + \langle\sin\cdots\sin\cdots\rangle + i\langle\cos\cdots\sin\rangle - i\langle\sin\cdots\cos\cdots\rangle$ the two last terms cancel each other since the time arguments can be interchanged in thermal equilibrium.

As for neutron scattering we will restrict ourselves to powder samples. Thus, as remarked already in connection with Eq. (3), it is understood that also here the brackets $\langle \cdots \rangle$ include the powder average. $S(\tau, t)$ is a single particle correlation function which can be calculated with a given motional model. Since the spin alignment experiment detects reorientations only, calculation of $S(\tau, t)$ requires specification of the conditional probability density $P(\Omega_1, \Omega_2, t)$ that a C-^{2}H (or a O-^{2}H) bond points at time t into the solid angle $d\Omega_2$ about Ω_2 if its starting (i.e. $t = 0$) orientation was in the interval $d\Omega_1$ about Ω_1 and the a priori probability distribution $p(\Omega_1)$. Then $S(\tau, t)$ is given by

$$S(\tau, t) = \langle \int d\Omega_1 \int d\Omega_2 e^{-i\tau\omega_Q(0)} e^{-i\tau\omega_Q(t)} P(\Omega_1, \Omega_2, t) p(\Omega_1) \rangle \quad (26)$$

Because of the nontrivial relationship between ω_Q and Θ, Eq. (21), the spin alignment function $S(\tau, t)$ can mostly not be calculated analytically, in contrast to the intermediate scattering function $S_{inc}(Q, t)$. Therefore we have to evaluate $S(\tau, t)$ numerically. However, a few qualitative statements can be made: For N-site jumps, where jumps to every orientation are equally probable, $S(\tau, t)$ decays essentially monoexponentially toward a plateau value $S_\infty(\tau)$

$$S(\tau, t)/S(\tau, t \to 0) = [1 - S_\infty(\tau)] e^{-t/\tau_c} + S_\infty(\tau) \quad (27)$$

In general, however, we get a multiexponential decay with time constants and weights which depend on τ. It has been shown [17–19] that the correlation function $S(\tau, t)$ can be written as a convolution

$$S(\tau, t) = \int_0^{\pi/2} d\beta K(\tau, \beta) \Gamma_g(\beta, t) \quad (28)$$

with a screening function $K(\tau, \beta)$ which is specific for the experimental method and analytically given in Refs. [17, 18] and a reduced probability density $\Gamma_g(\beta, t)$

$$\Gamma_g(\beta, t) = \Gamma(\beta, t) + \Gamma(\pi - \beta, t) \quad (29)$$

which depends according to Eq. (29) on the probability density $\Gamma(\beta, t)$, that a C-^{2}H (or O-^{2}H) bond reorients during the time interval t by the angle β. Note that $\Gamma(\beta, t)$, which contains the maximum information on the jump process, is a *one*-dimensional jump angle distribution. $K(\tau, \beta)$ describes the weight, by which the spin alignment experiment in a powder is sensitive to the jump angle β. This is illustrated in Fig. 2, where $K(\tau, \beta)$ is plotted for two different τ-values. Note the increased sensitivity of the experiment for small jump angles when τ is increased.

As far as our own spin alignment experiments are concerned, they have been performed at the University of Mainz on a specialized ^{2}H spectrometer at a deuteron frequency of 55 MHz. Several probeheads (furnace, He-cryostate) are available such that a temperature regime from 4 to 1000 K can be covered. Care has been taken to achieve very high long term temperature stability of about 0.1 K/day. The ^{2}H 90° pulse length of about 2.4 µs is reached.

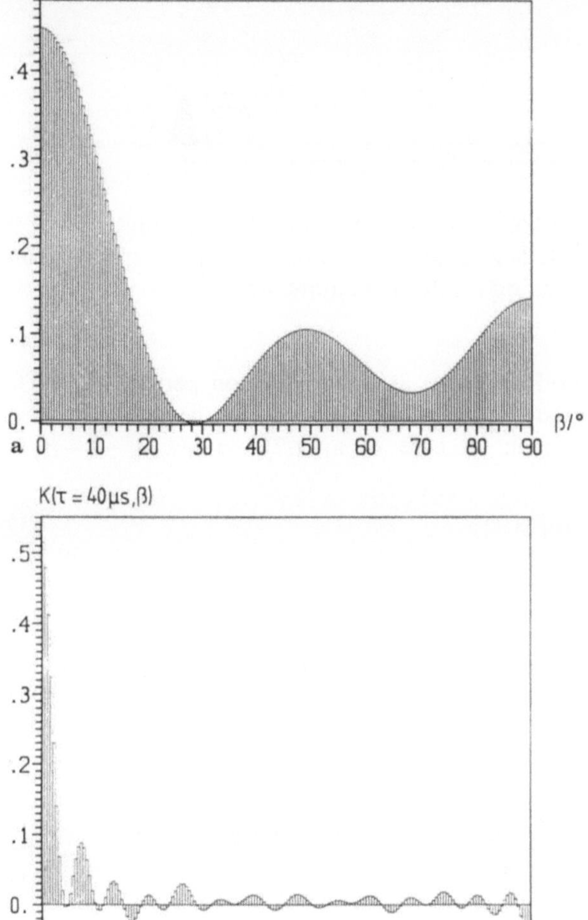

Fig. 2a, b. Kernels $K(\tau, \beta)$ of the integrals Eq. (28) as a function of jump angle β for two different values of the geometry parameter $\tau = 6\,\mu s$ (**a**) and $40\,\mu s$ (**b**) assuming $e^2qQ/h = 215\,kHz$. Taken from Ref. [19]

2.3 *Field Gradient NMR*

Magnetic field gradient NMR has been established for more than 20 years [20–22] as a method for studying self diffusion processes in a large variety of systems, whenever their species are sufficiently mobile (liquids, polymers, biological systems etc.).

Field gradient NMR is based on the possibility of producing a spatial dependence of the Larmor frequency $\omega(\boldsymbol{r})$ of a spin by application of a magnetic field gradient $\boldsymbol{g}$ in addition to the background field $\boldsymbol{B}_0$

$$\omega(\boldsymbol{r}) = \gamma \boldsymbol{B}_0 + \gamma \boldsymbol{g}\boldsymbol{r}. \tag{30}$$

Let us, for the purpose of this paper, consider a stimulated echo experiment of type

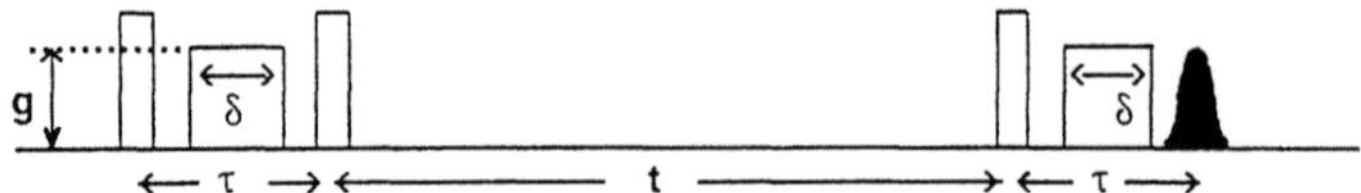

$(\pi/2)-\tau-(\pi/2)-t-(\pi/2)$ in the limit $t \gg \tau$ and where in the evolution and in the detection period magnetic field gradients of strength g and duration δ are applied. A tagged spin at position $\boldsymbol{r}$ will then aquire a phase

$$\varphi=\gamma \boldsymbol{g}\boldsymbol{r}\delta \tag{31}$$

in both, the evolution period ($\varphi(t=0)$) and the detection period ($\varphi(t)$). The measured echo height correlates these two phases in quite the same way as in the spin alignment experiment: In case of a $(\pi/2)_x-\tau-(\pi/2)_y-t-(\pi/2)_y$ sequence we get

$$S^{\sin} \sim \langle \sin[\varphi(0)]\cdot\sin[\varphi(t)]\rangle \tag{32}$$

and for a $(\pi/2)_x-\tau-(\pi/2)_x-t-(\pi/2)_x$ sequence we get

$$S^{\cos} \sim \langle \cos[\varphi(0)]\cdot\cos[\varphi(t)]\rangle \tag{33}$$

As in the case of the ^{2}H spin alignment experiment, either function $S^{\cos}$ or $S^{\sin}$ can be measured separately, and the sum is given by

$$\begin{aligned} S &\sim S^{\cos}+S^{\sin} \\ &\sim \langle e^{-i\varphi(0)}e^{+i\varphi(t)}\rangle. \end{aligned} \tag{34}$$

If we define a generalized momentum transfer or scattering vector

$$\boldsymbol{Q}=\gamma \boldsymbol{g}\delta, \tag{35}$$

the phase writes

$$\varphi(t)=\gamma \boldsymbol{g}\delta \boldsymbol{r}(t)=\boldsymbol{Q}\boldsymbol{r}(t) \tag{36}$$

yielding

$$S \rightarrow S(\boldsymbol{Q},t) \sim \langle e^{-i\boldsymbol{Q}\boldsymbol{r}(0)}e^{+i\boldsymbol{Q}\boldsymbol{r}(t)}\rangle \tag{37}$$

which is nothing else than the incoherent or self part of the intermediate scattering function, of Eq. (3). We may therefore apply—without any restriction—the interpretation of the neutron scattering function $S_{\mathrm{inc}}(Q,t)$, Eqs. (4–18), also to field gradient NMR.

The experimentally accessible dynamic range is determined by spin–lattice relaxation which produces an additional decay channel $\sim \exp(-t/T_1)$ for the correlation function S, Eq. (37). The accessible Q-range is limited by spin–spin relaxation in the evolution and detection period giving rise to a factor $\sim \exp(-2\tau/T_2)$ in the correlation function S, Eq. (37).

There are two different ways of performing the experiment which differ in the way the gradients are applied:

a) Pulsed gradients: The gradients are applied as depicted above and assumed in Eqs. (35) and (36). Large pulsed gradients of up to 50 T/m can be realized. The method has the advantage that because of the absence of the gradient during the rf pulses the whole sample is excited. Further, according to Eq. (35), Q can easily be varied by changing the gradient strength g or duration δ. However, the condition $\delta < \tau$ and the technical difficulty of producing large gradients and fast gradient switching imposes a serious limitation on the experiment, that is, samples with $T_2 \leqq 1$ ms are practically inaccessible.

b) Static gradients: The use of cryomagnets with their high stray field gradients ($g_{max} \approx 80$ T/m in a 8.5 T magnet of 54 mm bore) have led to a "renaissance" of the old fashioned static gradient diffusion experiment [23, 24]. The whole derivation of the generalized scattering law, Eq. (37), stays valid except that the pulsed gradient width δ is replaced by the spacing τ between the first two pulses, i.e.

$$Q = \gamma g \tau. \tag{35a}$$

This method has some obvious disadvantages. Since the gradient persists during the rf pulses, only a small slice of the sample can be excited. This problem can be reduced by the use of very strong rf pulses (90°-pulse length ≈ 0.3–0.4 μs), thus enlarging the excitation bandwidth. Another problem is that, according to Eq. (35a), Q can only be changed via τ, which in turn also changes one of the relaxation prefactors. On the other hand, static gradient NMR allows us to extend the experiments to very small τ-values, the only limitation being the spectrometer dead time after the rf-pulses (a few μs). Consequently, there should be no limitation in T_2 if only the diffusion is fast enough. In future work, static gradients of more than 100 T/m will be reached, a value which is beyond the technical possibilities for pulsed gradients. In our concluding remarks (Sect. 4) we will comment on a design of such an ultrahigh static field gradient.

2.4 Analogies

Both single particle correlation functions, Eq. (3)

$$S_{inc}(Q,t) \sim \langle e^{-i\boldsymbol{QR}(0)} e^{i\boldsymbol{QR}(t)} \rangle$$

and Eq. (25)

$$S(\tau,t) \sim \langle e^{-i\tau\omega_Q(0)} e^{i\tau\omega_Q(t)} \rangle$$

have the same structure. Numerical calculations [25] show that for given motional models $S(\tau,t)$ behaves quite similar to $S_{inc}(Q,t)$. It is of special interest to compare the final states

$$A_0(Q) = S_{inc}(Q, t \to \infty)/S_{inc}(Q, t \to 0) \tag{38}$$

$$S_\infty(\tau) = S(\tau, t \to \infty)/S(\tau, t \to 0) \tag{39}$$

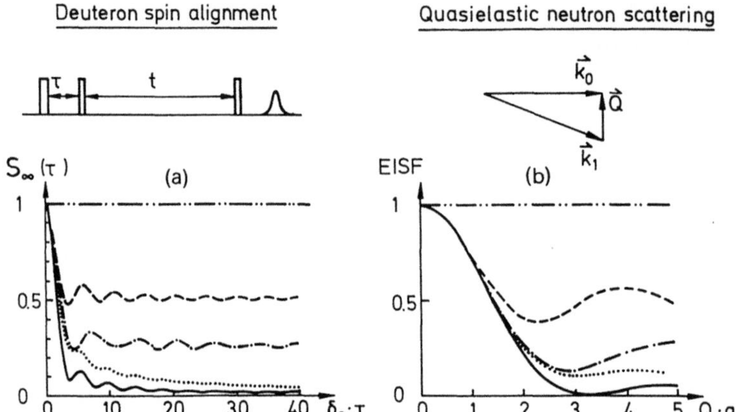

Fig. 3a, b. Powder calculations of final states (**a**) $S_\infty(\tau)$ and (**b**) $A_0(Q)$ according to Eqs. (39) and (38), respectively, for various molecular reorientational models (—··—··—no motion; ——— two-site jump; —·—·—four-site jump; ······ rotational diffusion on a circle; —— isotropic rotational diffusion). The semiangle of the cone models is assumed to be 30°. The radius of the circle (sphere) which defines the assumed particle positions is a. Taken from Ref. [26]

for given models. For this purpose Fig. 3 shows for a set of models the final states $A_0(Q)$ [= EISF] and $S_\infty(\tau)$. Some similarities are obvious.

Let us emphasize a few observations:

a) The NMR experiment is sensitive on *reorientation* of C-^{2}H or O-^{2}H bonds only. Because of Eq. (21), a pure translational mode will not led to a decay of $S(\tau, t)$ and will thus not be detected. Therefore, $A_0(Q)$ and $S_\infty(\tau)$ can be directly compared with each other only for reorientational models.

b) Because of Eq. (21), the NMR experiment cannot distinguish between an orientation and its inverse. We will come back to this point in the context of some experiments.

c) For N-site jump models the final states of both experiments allow a direct determination of N, namely

$$\lim_{\tau \to \infty} S_\infty(\tau) = 1/N \tag{40}$$

$$\lim_{Q \to \infty} A_0(Q) = 1/N \tag{41}$$

[Exception: $\lim_{\tau \to \infty} S_\infty(\tau) = 2/N$ if the point group of the model contains the inversion operation [25].]

d) For $\tau \to 0$ and $Q \to 0$ the correlation functions contain no information since for all motional models $S_\infty(\tau \to 0) = A_0(Q \to 0) = 1$.

In contrast to neutron scattering NMR measures both parts $S_{\cos}(\tau, t)$ and $S_{\sin}(\tau, t)$

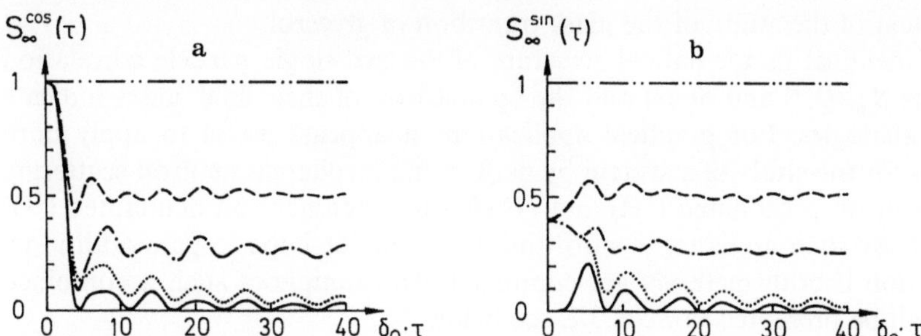

Fig. 4a, b. Phase selected parts (**a**) $S_\infty^{\cos}(\tau)$ and (**b**) $S_\infty^{\sin}(\tau)$ Eq. (42), of the final state geometries as chosen in Fig. 3. Taken from Ref. [26]

separately, Fig. 4. Therefore it may be useful also to discuss both parts

$$S^{i}_{\infty}(\tau) = S_i(\tau, t \to \infty)/S_i(\tau, t \to 0) \quad [i = \cos, \sin] \tag{42}$$

separately.

This comparison shows that $S_\infty^{\sin}(\tau \to 0) < 1$ contains more information than $S_\infty^{\cos}(\tau \to 0) = 1$. Some simple models of reorientation even allow to give analytical expressions for $S_\infty^{\sin}(\tau \to 0)$ [17, 25]. For reorientations between N equivalent sites on a cone with semiangle α we get

$$\lim_{\tau \to 0} S_\infty^{\sin}(\tau_1) = \begin{cases} 3/2 \sin^4 \alpha + 3/2 \cos^4 \alpha - 1/2 & [\text{for } N = 2] \\ 3/4 \sin^4 \alpha + 3/2 \cos^4 \alpha - 1/2 & [\text{for } N > 2] \end{cases} \tag{43}$$

For demonstrating these properties, we will present in Sect. 3.1 some experimental results on selected model systems, see below. For N-site jump models with at least cubic point symmetry we obtain [25]

$$\lim_{\tau \to 0} S_\infty^{\sin}(\tau) = 0 \tag{44}$$

which will also be demonstrated in Sect. 3.1 by several experimental examples (tetrahedral jumps, isotropic reorientation). On the whole it can be stated that the limiting value $\lim_{\tau \to \infty} S_\infty^{\sin}(\tau)$ tells about the *number* of reachable final state orientations, whereas $\lim_{\tau \to 0} S_\infty^{\sin}(\tau)$ tells about the *arrangement* of the orientations.

On the other hand measuring $S^{\sin}$ has also a disadvantage compared to measuring $S^{\cos}$. Because of spin-lattice relaxation $S^{\sin}$ and $S^{\cos}$ decay with additional time constants T_{1Q} and T_1, respectively. In general $T_{1Q} < T_1$ [26]. Whereas T_1 can be measured independently, T_{1Q} cannot. Thus, whenever the time scale of the reorientational process under study is not sufficiently separated from T_{1Q}, the analysis of $S^{\sin}(\tau, t)$ can become difficult. In this case one may prefer to measure $S^{\cos}(\tau, t)$, especially in cases when one does not need the information $\lim_{\tau\to 0} S^{\sin}(\tau, t)$. In Sect. 3.5, see below, we will present this case in the context of the study of the glass transition of glycerol.

The identical mathematical structure of the two single particle correlation functions $S_{inc}(Q, t)$ and $S(\tau, t)$ and the comparison of their final states indicate certain analogies. For practical applications it appears useful to apply both methods for the study of a system by performing incoherent neutron scattering experiment on protonated (^1H) and NMR spin alignment on deuterated (^2H) material. By such an isotopic substitution one might hope to gain additional information if both methods are combined. An example of such a combined study will be presented in Sect. 3.2, see below.

As already noted above, a basic difference between both methods is the fact that the intermediate scattering function $S_{inc}(Q, t)$ is a full *positional* correlation function and $S(\tau, t)$ a pure *orientational* correlation function. Both methods have rather different and thus complementary dynamic regimes, neutron scattering works within 10^{-13}–10^{-8} s, NMR spin alignment within 10^{-4}–10^{+2} s. The most important analogy is that in NMR spin alignment the pulse spacing τ plays, as a *geometry parameter*, essentially the same role as the momentum transfer Q in neutron scattering. As has been discussed in connection with Fig. 2, the sensitivity of the experiment on a given jump process can be changed by variation of τ just as by variation of Q in neutron scattering. This is also illustrated in Fig. 5; for instance, when large values of the geometry parameter τ (or Q) are chosen, the NMR (neutron scattering) experiment examines small amplitude reorientations (displacements). The reason for this analogy is the fact that both methods measure—strictly speaking—a phase correlation function. In NMR spin alignment the phase $\tau \cdot \omega_Q(t)$ is the product of two conjugated quantities as well as in neutron scattering where the phase is given by the product $Q \cdot R(t)$.

Because of the existence of a geometry parameter τ and its role as "generalized momentum transfer" we may truely consider the NMR spin alignment correla-

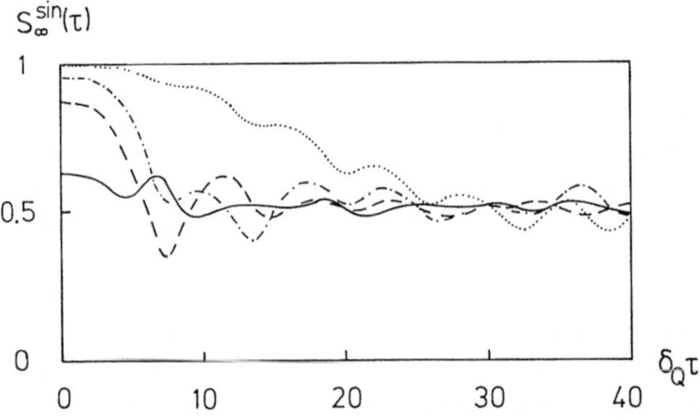

Fig. 5. Comparison of powder averaged final states $S_\infty^{sin}(\tau)$ for 2-site jumps on cones with different semiangles $\alpha = 22.5°$ (——), 12.5° (— — —), 7.5° (—·—·—), and 2.5° (······)

tion function as a *generalized dynamic structure factor*. The spin alignment final states can then be considered as a generalized EISF.

Whereas neutron scattering and spin alignment may be termed "analogous", the field gradient NMR experiment measures a single particle correlation function which is *identical* to that accessible in incoherent neutron scattering. This was pointed out in Sect. 2.3. It is obvious, however, that both methods differ strongly in their dynamic ranges and Q-windows. The dynamic regimes differ by many orders of magnitude, neutron scattering reaching as high as $\approx 10^{-8}$ s at best, gradient NMR only going down to $\approx 10^{-3}$ s typically. The Q-ranges, however, approach each other: Neutron scattering reaches down to $\approx 10^{-2}$ Å^{-1}, gradient NMR may reach up to $Q_{\max} = \gamma g_{\max} \tau \approx (2.68 \times 10^8\, \mathrm{T}^{-1}\mathrm{s}^{-1})$ $(100\, \mathrm{Tm}^{-1})\,(4 \cdot 10^{-3}\, \mathrm{s}) \approx 10^{-2}$ Å^{-1} as well if static gradients of about 100 T/m (or more) become applicable. The typical working regimes of $S(Q, t)$ in neutron scattering and in gradient NMR are plotted in Fig. 6, which suggests that both methods may be applied advantageously in concert. This may be exemplified as follows: Self diffusion leads in the hydrodynamic regime to an exponential decay of $S(Q, t) \sim \exp(-Q^2 D t)$ (Eq. 14). For this decay to be measurable, the product $Q^2 t$ must be of the order of D^{-1}. This may be achieved either in a neutron scattering experiment at, say, 1 Å^{-1} and in the ps $\cdots$ ns range or, equivalently, in the gradient NMR working regime at 10^{-4} Å^{-1} and in the ms $\cdots$ s range. In other words: The trajectory (in (Q, t)-space) of a diffusing particle crosses both the neutron scattering *and* the field gradient NMR working regimes as indicated in Fig. 6. Neutron scattering measures the diffusion at short diffusion times, NMR at long diffusion times. Figure 6 also indicates that field gradient NMR allows us to measure smaller diffusion coefficients than neutron scattering. [A smaller diffusion coefficient corresponds to an upward parallel shift of the arrow]. This fact suggests that gradient NMR might be well suited for studying self diffusion in supercooled liquids (see e.g. Sect. 3.4.2) and generally

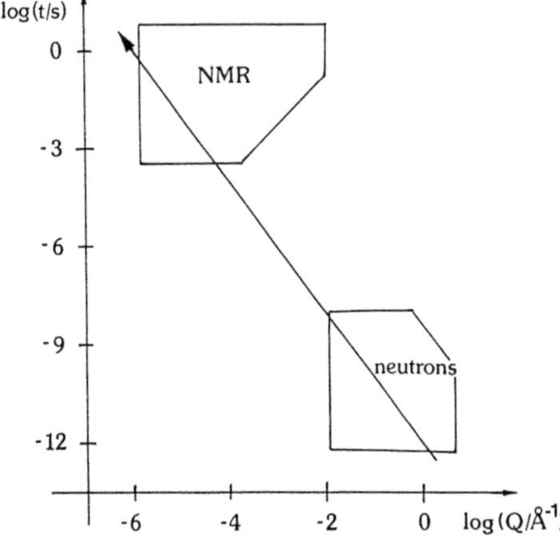

Fig. 6. Working regimes of quasielastic neutron scatering and gradient NMR. The *arrow* represents the trajectory of a free diffusing particle

in polymer systems where diffusion is slow because of the restrictions in mobility of the chain segments (see e.g. Sect. 3.5).

We would emphasize once again an essential feature on which the analogy between NMR stimulated echoes and neutron scattering is based: In both methods the measured quantity is a phase correlation function of type

$$S \sim \langle e^{-i\varphi(0)} e^{+i\varphi(t)} \rangle. \tag{45}$$

In NMR, the phases are acquired in the *e*volution and *re*volution (detection) period of a 3-pulse experiment and are given by

$$\varphi(t) = \omega_Q(t) \cdot \tau \quad (^2\text{H-NMR spin alignment}) \tag{46}$$

or

$$\varphi(t) = \gamma g \delta \boldsymbol{r}(t) \quad (\text{gradient NMR}). \tag{47}$$

In neutron scattering, the phases

$$\varphi(t) = (\boldsymbol{k}_1 - \boldsymbol{k}_0)\boldsymbol{r}(t) \tag{48}$$

stem from the incoming and outgoing neutron waves being shifted by the scattering event. Whereas the analogy between NMR-SA and neutron scattering is restricted to reorientations only, gradient NMR can be truely called a generalized incoherent scattering experiment at a (generalized) $\boldsymbol{Q} = \gamma g \delta \cong \boldsymbol{k}_1 - \boldsymbol{k}_0$.

Let us finally summarize the comparison of the three methods in the following table:

	Incoh. *n*-scattering	NMR-SA	Field gradient NMR
nucleus	^{1}H	^{2}H	^{1}H
measured quantity	$S_{inc}(Q, t)$	$S(\tau, t)$	$S_{inc}(Q, t)$
correlation function	orientation + translation	orientation	orientation + translation
time window	$10^{-13}\,\text{s} \cdots 10^{-8}\,\text{s}$	$10^{-4}\,\text{s} \cdots 10^{1}\,\text{s}$	$10^{-3}\,\text{s} \cdots 10^{1}\,\text{s}$
geometry parameter	$\boldsymbol{Q} = \boldsymbol{k}_1 - \boldsymbol{k}_0$	τ	$\boldsymbol{Q} = \gamma \mathbf{g} \delta$
phase	$\boldsymbol{Q} \cdot \boldsymbol{r}(t)$	$\tau \cdot \omega_Q(t)$	$\boldsymbol{Q} \cdot \boldsymbol{r}(t)$

3 Experiments

3.1 Tests: Spin Alignment Final States

In this section a few test experiments on easily accessible model systems with well known types of molecular motion are summarized. These tests are meant to demonstrate the validity and beauty of the concept of the generalized elastic incoherent structure factor $S_\infty(\tau)$.

3.1.1 2-Site Jumps

It is know from the literature [27] that, in addition to a fast CH_3-rotation, crystalline dimethylsulfone (DMS) performs a thermally activated rotational jump about its C_2-axis (see Fig. 8, below) between two equivalent orientations. This jump process is characterized by an activation energy of 56 kJ mol^{-1} and a correlation time of a few ms at room temperature, which falls ideally into the spin alignment dynamic window. Thus DMS is well suited for a first demonstration of the behaviour of the spin alignment correlation function. In Fig. 7, spin alignment decay curves $S^{sin}(\tau, t)$ of deuterated polycrystalline DMS are shown for two τ-values. The plot shows a fast decay which is due to the 2-site jump (time constant τ_c) and a slow decay due to spin lattice relaxation (time constant T_{1Q}). It is apparent that the idealization of a constant $S_{\infty}^{sin}(\tau)$ holds only in the limit $T_{1Q} \gg \tau_c$ (vanishing spin–lattice relaxation). For small T_{1Q} the condition $T_2 < \tau_c < T_{1Q}$ can render the time window quite narrow. In the case of DMS at 310 K, the values $T_2 \approx 100\,\mu s$, $\tau_c \approx 3$ ms, $T_{1Q} = 36$ ms fulfill this relationship. When the temperature is lowered, however, T_{1Q} is lowered and τ_c increased (e.g. at 288 K $\tau_c \approx 4$ ms, $T_{1Q} = 22$ ms) such that the two quantities become less well separated. At temperatures above 310 K τ_c becomes smaller and soon approaches T_2. For DMS, these considerations confine the applicability of the spin alignment technique to the temperature range of 280 K < T < 340 K. From extrapolation of the slow T_{1Q} decay to $t \to 0$, the value $S_{\infty}^{sin}(\tau)$ can be obtained. The two decay curves shown in Fig. 7 approximately represent the two limiting cases

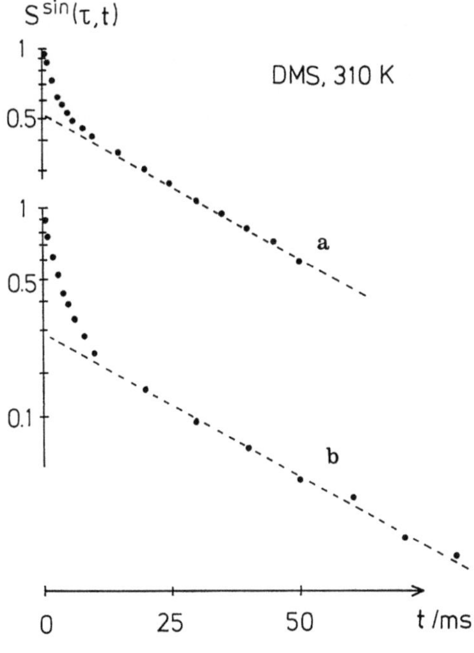

Fig. 7. Spin alignment decay curves $S^{sin}(\tau, t)/S^{sin}(\tau, t \to 0)$ for dimethylsulfone at 310 K. *a*) $\tau = 40\,\mu s$, *b*) $\tau = 5\,\mu s$. The *dashed lines* fit the relaxation induced long time tail and yield for $t \to 0$ the plateau values $S_{\infty}^{sin}(\tau)$

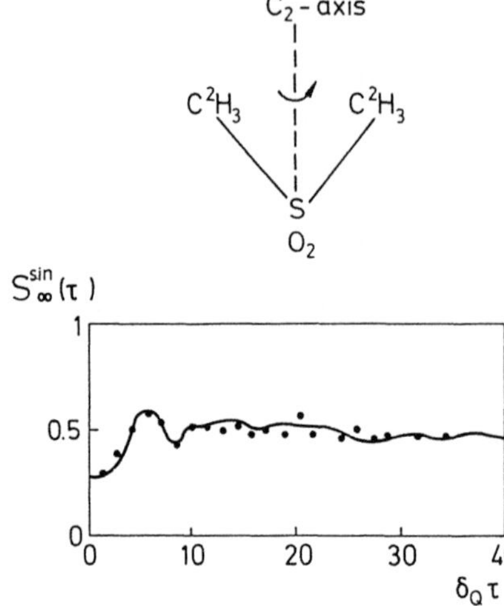

Fig. 8. Measured spin alignment final state $S_{\infty}^{\sin}(\tau)$ of polycrystalline dimethylsulfone at 310 K. The asymptotic value of 0.5 indicates a two site jump. The *drawn curves* has been calculated for a 180° flip about the C_2 axis assuming a C—S—C angle of 103° ($\delta_Q = 2\pi \cdot 46$ kHz). Taken from Ref [25]

(a) $\lim_{\tau\to\infty} S_{\infty}^{\sin}(\tau) = 0.5$ (Eq. 40) and (b) $\lim_{\tau\to 0} S_{\infty}^{\sin}(\tau) = 3/2\sin^4\alpha + 3/2\cos^4\alpha - 1/2$ (Eq. 43) and thus allow us to obtain the angle α between the C^2H_3-S bond and the C_2 axis. The full final state curve $S_{\infty}^{\sin}(\tau)$ of DMS is shown in Fig. 8 as a function of the reduced geometry parameter $\delta_Q\tau$ with a quadrupolar coupling strength $\delta_Q = 2\pi \cdot 46$ kHz of the fast rotating methyl groups measured independently. The drawn curve represents a model calculation using Eq. (26) assuming $\alpha = 51.5°$ [27]. The agreement of the completely independent model calculation with the experimental data is obvious. Further experiments on DMS using ^{2}H 2D-NMR [28] and the ^{13}C stimulated echo [29] are fully consistent with our result.

3.1.2 4-Site Jumps

Another nice model system is (poly)crystalline hexamethylenetetramine (HMT). As known from earlier spin alignment measurements [30], HMT performs tetrahedral jumps with an activation energy of 73 kJ mol^{-1} over about 8 decades ranging from 10^{-6}–10^2 s. Using the same procedure discussed above, the ^{2}H spin alignment final state $S_{\infty}^{\sin}(\tau)$ has been measured at 316 K (Fig. 9). The decay curves yield $\tau_c = 2$ ms.

An independent measurement of the spectrum yields $\delta_Q = 2\pi \cdot 121.5$ kHz. Because of the large spin–lattice relaxation time (several minutes) only a moderate statistical precision but a very good separation of the motional and the relaxational timescales is achieved. The most important characteristics of $S_{\infty}^{\sin}(\tau)$ become apparent from inspection of Fig. 9: $\lim_{\tau\to\infty} S_{\infty}^{\sin}(\tau) = 1/4$, cf. Eq. (40) and $\lim_{\tau\to 0} S_{\infty}^{\sin}(\tau) = 0$ because of the cubic symmetry of the motion, cf. Eq. (44). A

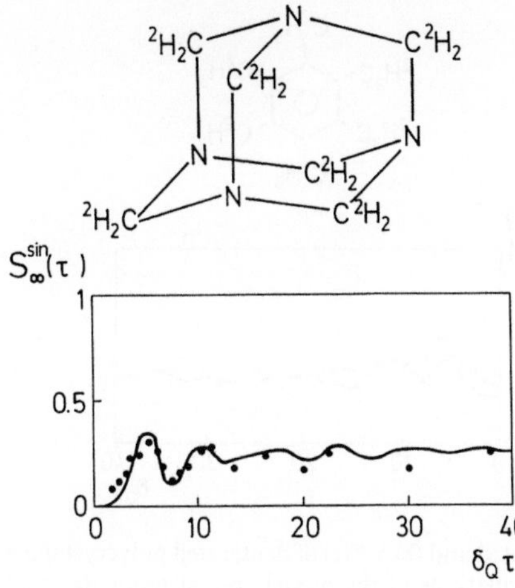

Fig. 9. Measured spin alignment final state $S_\infty^{sin}(\tau)$ of polycrystalline hexamethylene-tetramine at 316 K. The *drawn curve* represents the calculated final state for the model "tetrahedral jumps" ($\delta_Q = 2\pi \cdot 121.5$ kHz). Taken from Ref. [25]

reasonable agreement with the model "tetrahedral jump" is noted. Deviations at small τ_1 and at the marked oscillations of $S_\infty(\tau)$ are due to finite pulse lengths (≈ 4 μs for a ^{2}H 90° pulse) which average $S_\infty^{sin}(\tau)$ over a corresponding τ-interval. (Remember that the point to point distance at small τ is $\Delta\tau \approx 1$ μs. In case of DMS the finite pulse lengths did not lead to a smearing out since $\Delta\tau = 5$ μs could be chosen because of the much smaller quadrupole coupling constant).

3.1.3 6-Site Jumps

A third test system is hexamethylbenzene (HMB) in its (poly)crystalline phase II [31], where in addition to a CH_3-rotation the molecules perform rotations about their pseudo 6-fold axes. This process has been described by Allen and Cowking [31] by an activation energy of 28.1 kJ mol^{-1} and a preexponential factor of 1.3×10^{-14} s. Here, this process is used for a first experimental comparison of *both* methods discussed, quasielastic neutron scattering and spin alignment. (The presentation is kept short since an analogous problem will be presented in great detail in Sect. 3.2). The EISF of (protonated) HMB at 310 K, measured with the ILL backscattering spectrometer IN10, and the spin alignment final state $S_\infty^{sin}(\tau)$ of (deuterated) HMB at 135 K are compared in Fig. 10. Again, reduced abscissa scales are chosen, using the "proton radius" $a = 3.293$ Å [32] and a coupling parameter $\delta_Q = 2\pi \cdot 39$ kHz (from a separate measurement of the ^{2}H spectrum). The drawn curves represent the generalized elastic incoherent structure factors for the 6-site rotational jumps of the HMB molecules about their molecular axes. This case appears to be instructive because at the NMR side it represents a special case which has already been pointed out: Since the molecular reorientational process includes with each orientation also its inverse, $\lim_{\tau\to\infty} S_\infty^{sin}(\tau) = 1/3$ instead of 1/6 (an exception to Eq. 40). In this case, the NMR experiment cannot distinguish between a 3- and a 6-site jump;

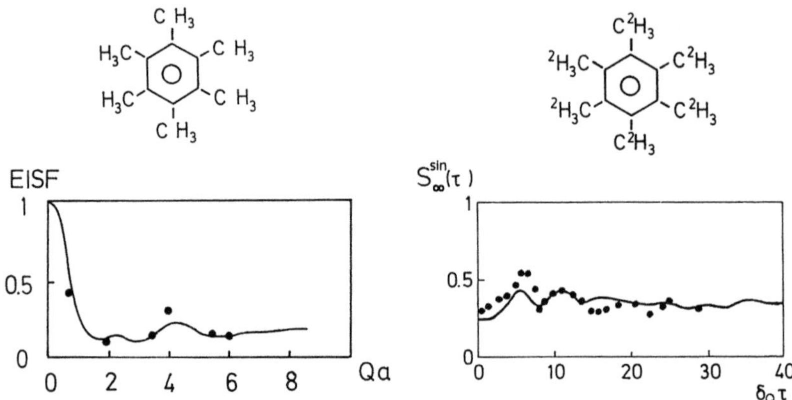

Fig. 10a, b. Measured final states (**a**) $A_0(Q)$ of protonated and (**b**) $S_\infty^{\sin}(\tau)$ of deuterated polycrystalline hexamethylbenzene. The *drawn curves* represent calculations of the model "6-fold reorientational jumps" about the molecular C_6-axis. (The proton radius is $a = 3.292$ Å, the quadrupole coupling parameter $\delta_Q = 2\pi \cdot 39$ kHz)

distinguishing between them can only be achieved by neutron scattering. For a detailed discussion of this type of motion we refer to Sect. 3.2. As in the case of the 2-site jump of DMS (see Sect. 3.1.1) an analogous study of the HMB jump rotation is given in Ref. 29 using the ^{13}C-NMR stimulated echo and by 2D-NMR [28].

3.1.4 *Isotropic Reorientation*

The final state of an isotropic reorientation has been experimentally demonstrated by the example of the primary glass transition (α-process) of chain deuterated polystyrene (PS-d_3) [33] at 395 K, just above the caloric glass

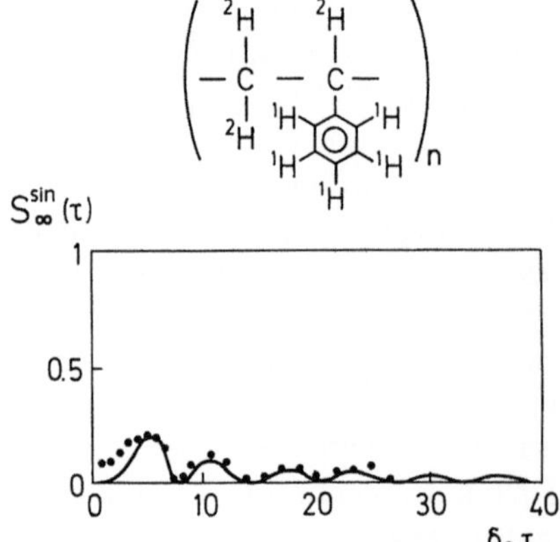

Fig. 11. Measured spin alignment final state $S_\infty^{\sin}(\tau)$ of chain deuterated amorphous polystyrene (PS-d_3) at 395 K. The *drawn curve* represents the calculated final state for the model "isotropic reorientation" ($\delta_Q = 2\pi \cdot 125.8$ kHz). Taken from Ref. [25]

transition temperature of $T_g = 373\,\text{K}$. Figure 11 shows the comparison with the corresponding model calculation. The agreement, especially with the characteristic zeros of $S_{\infty}^{\sin}(\tau)$, is reasonable. Deviations for small τ can again be attributed to the finite pulse lengths. The rotational dynamics of the primary glass transition of polystyrene has been extensively studied by ^{2}H-2D-NMR [34].

A more detailed discussion of the α-process (isotropic reorientation) of molecular glasses will be given in Sect. 3.4.

3.2 Reorientation of Benzene [26]

After the above test experiments we now proceed by a detailed discussion of a selected example which is meant to illustrate the full potential of the generalized dynamic structure factor obtained by a combined NMR and neutron scattering experiment.

As model system we choose polycrystalline benzene C_6H_6 (Fig. 12). It is well known that in this orthorhombic crystal lattice the molecules perform rotational jumps about their molecular C_6 axes. NMR relaxation data show that this molecular rotation is termally activated in a wide temperature range and obeys the Arrhenius law $\tau_c = \tau_0 \exp(E_A/RT)$ [35] with an activation energy of $E_A = 17.6\,\text{kJ}\,\text{mol}^{-1}$ and a pre-exponential factor $\tau_0 = 9.2 \times 10^{-15}\,\text{s}$ [36, 37]

3.2.1 NMR Spin Alignment

The NMR alignment experiment on deuterated benzene is performed at 84 K, where the condition $T_2 = 100\,\mu\text{s} < \tau_c = 340\,\mu\text{s}$ is fulfilled and where T_1 is still acceptably short. In Fig. 13 the measured $S_{\infty}^{\sin}(\tau)$ is shown. The data are compared with the calculated final states of two models, the 6-fold jump model (which as an exception to Eq. 40 cannot be distinguished from the 3-fold jump model) and the rotational diffusion on the circle. The data clearly show that the molecules perform rotational jumps and not rotational diffusion, a result which

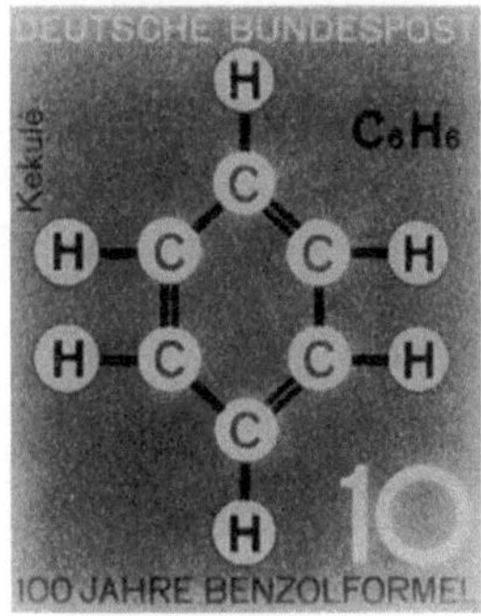

Fig. 12. Special issue—the postage stamp "100 Jahre Benzolformel" (1964) from the Deutsche Bundespost as part of a three stamp memorial edition dedicated to progress in science and technology (benzene, nuclear energy, the combustion engine). This selection reflects the attitude of the 1960s. Today, carcinogenic benzene is a symbol of the tragic ambivalence of science and technology, which are responsible for the pollution of the biosphere

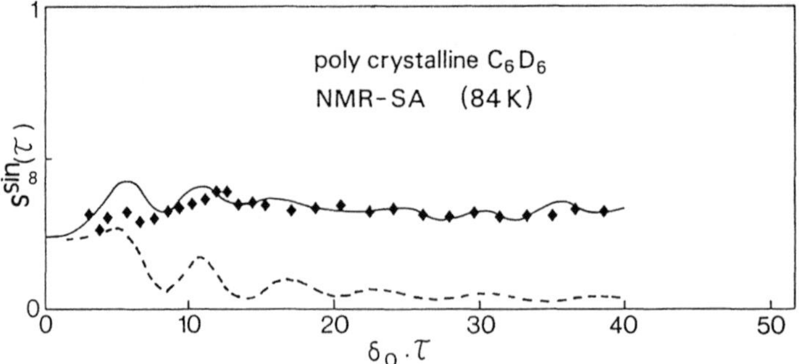

Fig. 13. Measured final state $S^{sin}_{\infty}(\tau)$ of polycrystalline benzene C_6D_6 at 84 K. The *drawn curve* represents the calculated final state for 6-site rotational jumps, the *dashed curve* uniaxial rotational diffusion ($\delta_Q = 2\pi \cdot 138$ kHz)

is physically reasonable and also supported by X-ray data [38]. Furthermore, the $S^{sin}(\tau, t)$ decays monoexponentially toward the plateau value since each molecule attains only three distinguishable orientations among which it jumps with equal probabilities, see Eq. (27).

3.2.2 *Quasielastic Neutron Scattering*

Neutron scattering experiments on (protonated) benzene have been performed on several ILL high resolution spectrometers, the time of flight spectrometer IN5 at 270 K, the "cold" backscattering spectrometer IN10 at 210 K and on the "thermal" backscattering spectrometer IN13 at 202, 227 and 246 K. With all the mentioned spectrometers the incoherent part of the dynamic structure factor $S_{inc}(Q, \omega)$ has been obtained. Some typical spectra are shown in Fig. 14. The spectra consist of an elastic contribution which is somewhat broadened because of finite instrumental resolution and a quasielastic broadening due to the molecular motion, see Eq. (10). It is evident from the figure that the determination of the EISF by separation of the elastic and quasielastic scattering is a nontrivial task and will depend sensitively on the chosen lineshape for the fits. This difficulty does not appear or at least less frequently in NMR spin alignment since a timescale separation $\tau_c \ll T_{1Q}$ is often well-realized.

In the following we will analyze the data in terms of two reorientational models. Since we already know from NMR spin alignment that the molecules perform 6-fold rotational jumps, we can impose the corresponding EISF as a boundary condition and thus reduce the number of variables.

Model I: Random jumps among the 6 molecular orientations

This model allows jumps among all six equivalent molecular orientations such that starting from any initial orientation each other one is reached with equal probability. In the calculus of stochastical processes this jump model leads to an exchange matrix [39] with only one eigenvalue τ_c^{-1} since all off

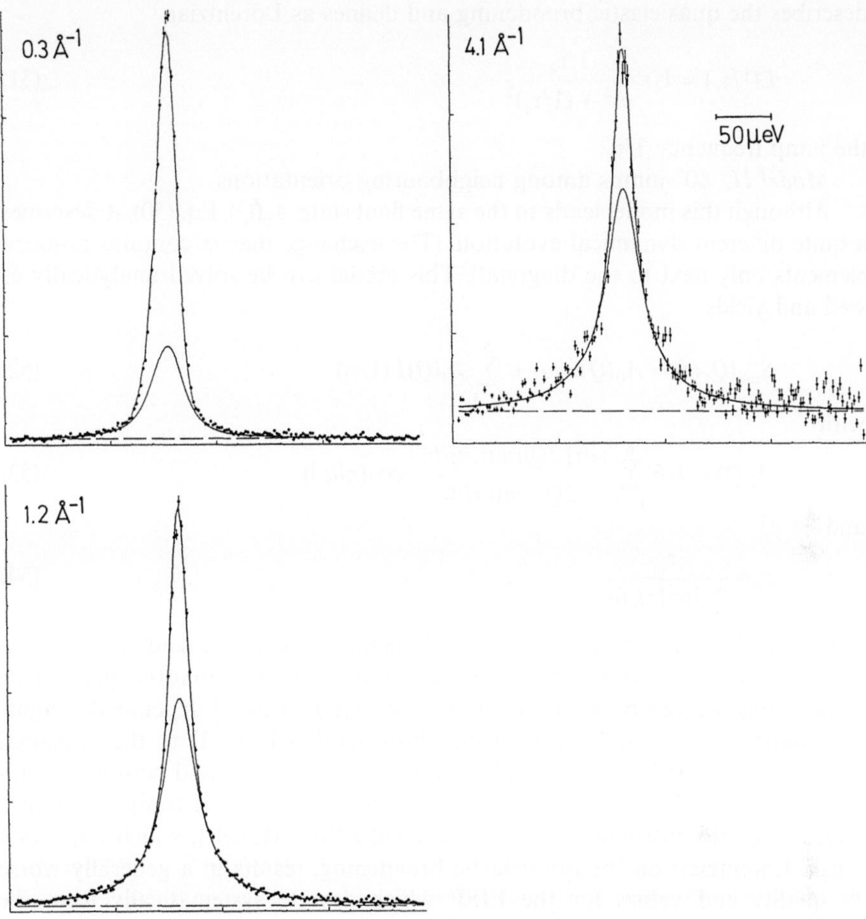

Fig. 14. Typical benzene neutron spectra taken with the backscattering spectrometer IN13 at 246 K and three different Q-values. The *outer drawn curve* represents a fit of Eq. (52) folded with the instrumental resolution function, the *inner curve* shows the quasielastic part only

diagonal elements are nonzero and equal to each other. This eigenvalue appears as inverse time constant in the conditional probability $P(\Omega_1, \Omega_2, t)$, see Eq. (26), and leads to an analytical solution of the incoherent dynamical structure factor in the (powder averaged) problem

$$S_{\text{inc}}(Q, \omega) = A_0(Q)\delta(\omega) + [1 - A_0(Q)]L(1/\tau_c) \tag{49}$$

with the EISF as already given in Eq. (18) for $N = 6$

$$A_0(Q) = 1/6 \sum_{j=1}^{6} \frac{\sin[2Qa\sin(\pi j/6)]}{2Qa\sin(\pi j/6)} \tag{50}$$

Here, $a = 2.479$ Å [32] is the proton radius of the benzene molecule. $L(1/\tau_c)$

describes the quasielastic broadening and defines as Lorentzian

$$L(1/\tau_c) = 1/\pi \frac{1/\tau_c}{\omega^2 + (1/\tau_c)^2} \tag{51}$$

the jump frequency $1/\tau_c$.

Model II: 60°-jumps among neighbouring orientations

Although this model leads to the same final state $A_0(Q)$, Eq. (50), it describes a quite different dynamical evolution. (The exchange matrix contains nonzero elements only next to the diagonal). This model can be solved analytically as well and yields

$$S_{inc}(Q,\omega) = A_0(Q)\delta(\omega) + \sum_{l=1}^{5} A_l(Q)L(1/\tau_l) \tag{52}$$

with

$$A_l(Q) = 1/6 \sum_{j=1}^{6} \frac{\sin[2Qa\sin(\pi j/6)]}{2Qa\sin(\pi j/6)} \cos(\pi lj/3) \tag{53}$$

and

$$\tau_l = \frac{\tau_c}{2\sin^2(\pi l/6)} \tag{54}$$

The $A_l(Q)$ $(l > 0)$ are called quasielastic structure factors. We read $A_2 = A_4$ and $A_1 = A_5$ because of symmetry reasons. Equation (52) shows that the quasielastic broadening is given by a sum of three Lorentzians with Q-dependent weights as is illustrated in Fig. 15. Fits of the above models I and II to the measured spectra yield EISF's which are plotted in Figs. 16 and 17 and compared with the calculated EISF of Eq. (50). It is obvious that model II (only 60°-jumps) yields a result consistent with the calculated EISF. Model I, which imposes a single Lorentzian on the quasielastic broadening, results in a generally worse fit quality and values for the EISF which deviate systematically from the calculated EISF.

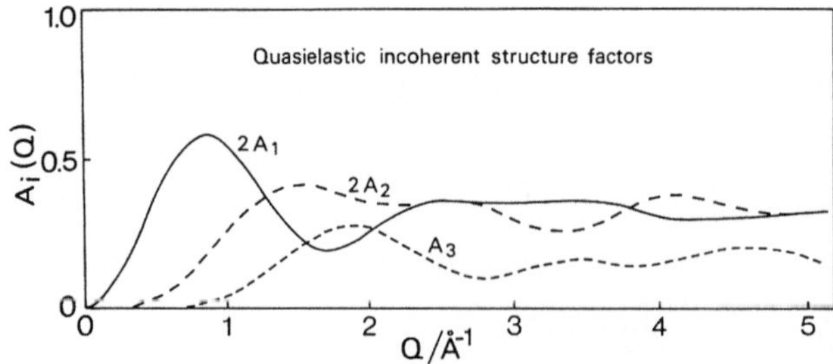

Fig. 15. Quasielastic structure factors for powder averaged 60°-jumps of benzene, calculated with Eq. (53)

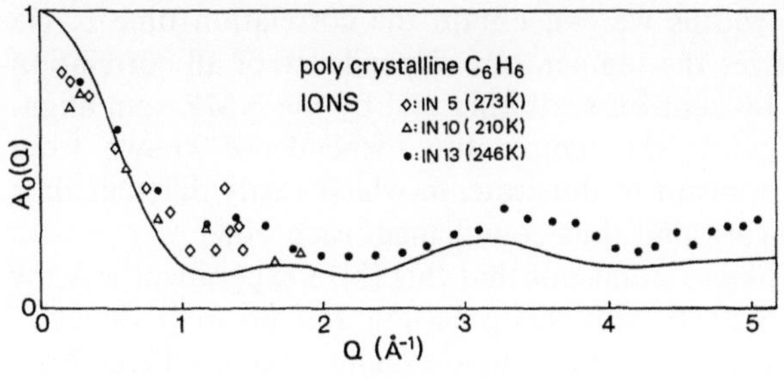

Fig. 16. EISF from the fit of model I (random jumps, Eq. (49)) to the measured neutron scattering spectra of polycrystalline benzene. A comparison with the calculated final state of "6-fold reorientational jumps" (*drawn curve*) shows a systematic deviation. Taken from Ref. [26]

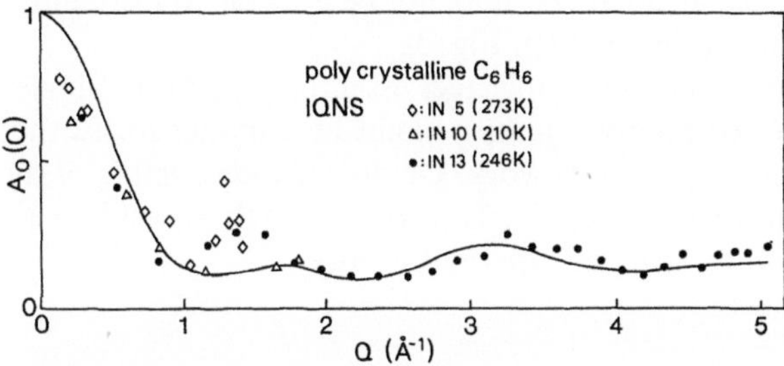

Fig. 17. EISF from the fit of model II (60°-jumps, Eq. (52)) to the measured neutron spectra of polycrystalline benzene. The data are consistent with the calculated final state of "6-fold reorientational jumps" (*drawn curve*). Taken from Ref. [26]

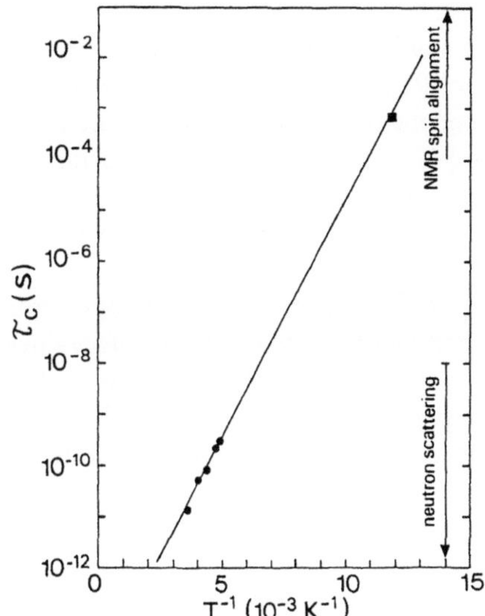

Fig. 18. Arrhenius plot of $\ln \tau_c$ vs T^{-1} for the temperature dependence of the measured rotational correlation times of benzene (●: quasi-elastic neutron scattering; ■: ^{2}H-NMR spin alignment). The *drawn curve* represents literature data [36]. Taken from Ref. [26]

Some artifacts of Figs. 16 and 17 necessitate a comment:

—A finite sample thickness and a resulting transmission of only $\approx 80\%$ implies some multiple scattering which primarily reduces the EISF at small Q-values [8].

—A strong Bragg reflection at $Q = 1.4\,\text{Å}^{-1}$ leads to some additional coherent intensity. All other reflections are comparatively weak and uniformly distributed such that apart from $\approx 1.4\,\text{Å}^{-1}$ the incoherent approximation is well satisfied.

From the quasielastic line widths we can obtain the correlation time τ_c via Eq. (54). Figure 18 summarizes the temperature dependence of all correlation times obtained in this way by neutron scattering and by the NMR spin alignment. The solid line represents the temperature dependence known from literature [39]. The figure is meant to illustrate, in which vastly different time domains the two methods work and thus complement each other.

At this point we would like to point out that this IN13 experiment was the first attempt to measure an EISF with this precision and up to Q-values of about $5\,Å^{-1}$. Earlier studies [8] of the molecular dynamics often suffered from the limitation in Q-range, since at small Q it may not be possible to distinguish between two models but only at large Q (cf. Fig. 3).

The physical result, that the benzene molecules essentially perform "single" (i.e. 60°) and not "multiple" (or random) jumps, might be a matter for further discussion but is beyond the scope of this work. Obviously the coupling of the molcular benzene molecule to the phonons is strong enough to lead to an instantaneous dissipation of its "trapped" thermal energy.

3.3 *Vacancy Diffusion in Crystalline Benzene*

Using ^{1}H- and ^{13}C-NMR [40], self diffusion of benzene molecules in the crystalline state with an activation energy of $95.5\,\mathrm{kJ\,mol^{-1}}$ has been reported. This result has also been confirmed by tracer measurements [41]. On the basis of ^{13}C-NMR stimulated echo experiments [42] on polycrystalline and monocrystalline benzene a detailed picture of this diffusion process has been given, which includes some anisotropy and is essentially attributed to vacancy migration.

Let us take up this example with ^{2}H spin alignment, since (1) it represents a most beautiful application of the generalized incoherent structure factor, since (2) it shows that the experiment provides complete information even in a polycrystal and since (3) some remaining discrepancy of the earlier ^{13}C stimulated echo with tracer measurements concerning activation parameters are left.

To this end, ^{2}H spin alignment measurements of $S^{\sin}(\tau, t)$ have been performed [43] in the (poly)crystalline state between 250 and 273 K. As in the case of DMS (see Sect. 3.1.1) the lower limit of this temperature is caused by the fact that with lower temperatures the condition $\tau_c < T_{1Q}$ cases to be fulfilled. At high temperature the limit is given by the melting point (278 K). The measured data are well fitted by an exponential (time constant τ_c) which decays toward a plateau value $S^{\sin}_{\infty}(\tau)$ (Fig. 19) multiplied by another overall decay due to spin–lattice relaxation. For the reduced representation of $S^{\sin}_{\infty}(\tau)$ the coupling parameter of $\delta_Q = 2\pi \cdot 69\,\mathrm{kHz}$ is used. The reduction by the factor of 2 compared to the low temperature value of δ_Q (cf. Fig. 13) is explained by the fast rotation of the benzene molecules about their C_6 axes. Because of this rotation the principle field gradient tensor axis is orthogonal to the molecular plane. If the

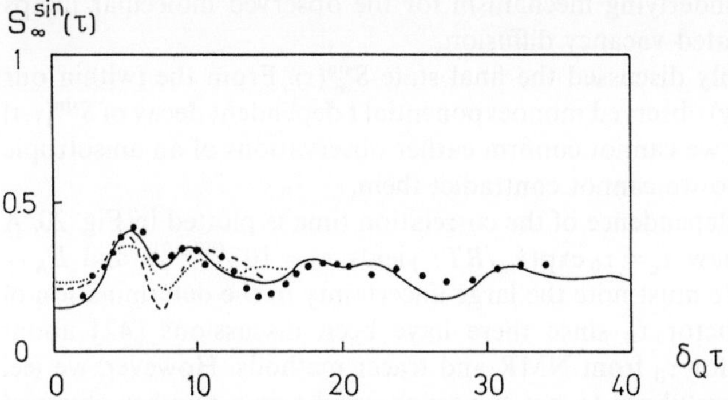

Fig. 19. Measured spin alignment final state $S_\infty^{sin}(\tau)$ of polycrystalline deuterated benzene at 266 K. The data are compared with curves calculated for vacancy diffusion which implies during each elementary jump step (for details see text) a 4-site jump on a cone with semiangles $\alpha = 50.55°$ (*drawn curve*), $\alpha = 40°$ (———) and $\alpha = 60°$ (····). [$\delta_Q = 2\pi \cdot 69$ kHz]

decay of the spin alignment is explained by jumps of this molecular axis among the four orientations which exist in the crystalline unit cell, we can calculate $S_\infty^{sin}(\tau)$ using known crystal data of benzene [32]. It is easy to show that the molecular reorientation can be mapped onto 4-fold jumps on a cone with a semiangle of $\alpha = 50.55°$. As can be seen from Fig. 19, the model calculation and the experimental results show excellent agreement. In order to get a feeling for the sensitivity of this method we have included some more calculated final states with somewhat different cone angles in the figure. It turns out that the discrepancies between the experiment and the "wrong" jump geometries are large especially for small τ while the asymptotic value $S_\infty^{sin}(\tau \to \infty)$ is 1/4 in any case, Eq. (40). This discussion indicates that the experiment corresponds within $\approx \pm 5°$ to the expectations from the crystal structure and thereby supports the

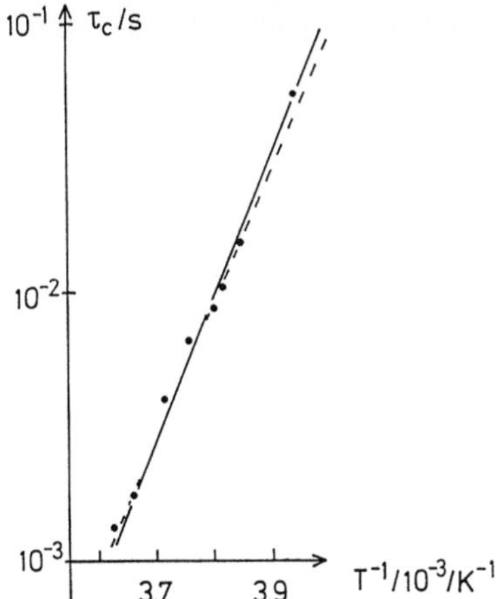

Fig. 20. Arrhenius plot of the correlation times for the vacancy diffusion in crystalline benzene. The *drawn solid line* represents the fit (see text) which considers all experimental points whereas the *dashed line* ignores the data point at the lowest temperature

hypothesis that the underlying mechanism for the observed molecular jumps is the thermally activated vacancy diffusion.

So far we have only discussed the final state $S_{\infty}^{\text{sin}}(\tau)$. From the (within our experimental accuracy) observed monoexponential t dependent decay of $S^{\text{sin}}(\tau, t)$ toward the final state we cannot confirm earlier observations of an anisotropic diffusion [42] but also we cannot contradict them.

The temperature dependence of the correlation time is plotted in Fig. 20. A fit of the Arrhenius law $\tau_c = \tau_0 \exp(E_A/RT)$ yields $\tau_0 = 10^{-(22 \pm 2)}$s and $E_A = (103 \pm 7)\,\text{kJ mol}^{-1}$. We must note the large uncertainty in the determination of the preexponential factor τ_0 since there have been discussions [42] about contradicting results for τ_0 from NMR and tracer methods. However, we feel that one should be careful not to put too much weight on a number obtained by a measurement over just two decades which extrapolates to a numerical value about 20 orders of magnitude away.

3.4 On the Glass Transition Dynamics

Another wide field of application of both spin alignment and field gradient NMR is the study of the glass transition of supercooled organic molecular liquids. A first example has already been discussed briefly in Sect. 3.1.4. The phenomenon "glass transition" [44–47] is a topic of high current interest. Besides the traditional interest in the role of chemical constituents, side groups, mixtures etc., which is a major motivation in polymer research and material science to study the glass transition, there are several fundamental questions about the physics involved. One of these questions is due to the fact that in the frozen glass there are microscopic excitations (phonons) and typical solid like transport processes which do not exist in the liquid state. These findings suggest that glass dynamics is similar to that of a solid, although the glass structure is undistinguishable from that of a liquid. Then the following question might be raised: If there is a transition from liquidlike to solidlike dynamics, at which temperature does this transition occur? Such a change of mechanism has recently been proposed to take place *above* T_g by mode coupling theory [47], the discussion of which is beyond the scope of this work. In any case, present experimental research work on the glass transition is largely motivated by this problem and the aim to test corresponding theoretical predictions.

Since ^{2}H spin alignment allows us to look at molecular reorientations in the ms$\cdots$s range this technique is well suited for studies just above T_g. This will be demonstrated in Sect. 3.4.1.

As has been discussed in Sect. 2.4, field gradient NMR also works in the ms$\cdots$s regime. In connection with Fig. 6, we suggested that the experiment might be suited to measure small self diffusion coefficients in the supercooled regime. This will be shown in Sect. 3.4.2.

3.4.1 *Reorientational Motion in Viscous Glycerol near Its Glass Transition Temperature [48]*

Glycerol $C_3H_8O_3$ ($T_g = 185$ K) is one of the best studied glass forming H-bridged liquids (dielectric relaxation [49], photon correlation spectroscopy [50] NMR [51–53], Mössbauer-Rayleigh scattering [54], neutron scattering [55], NMR + neutron scattering [56], frequency dependent calorimetry [57]). It is expected that molecular reorientation in the highly viscous state is isotropic. However, it is unclear which *path* the reorienting molecules take. Proper spin alignment experiments should shed some light on just this point.

In contrast to the earlier reported experiments we chose to measure $S^{\cos}(\tau, t) \sim \langle \cos[\tau\omega_Q(0)] \cos[\tau\omega_Q(t)] \rangle \exp[-t/T_1]$ since T_1 can be measured independently [56]. We performed measurements on two selectively deuterated glycerols, namely glycerol-d_5 ($C_3{}^2H_5O_3H_3$) and glycerol-d_3 ($C_3H_5O_3{}^2H_3$). Figure 21 shows a representative set of decay curves on glycerol-d_5 for a number of τ-values. (Note that in our original publication [48] we chose to show the corresponding curves on glycerol-d_3). The data are compared with three reorientational models [17, 18] for the rotation angle distribution $\Gamma_g(\beta, t)$, Eq. (29), of the C-^{2}H (glycerol-d_5) and O-^{2}H (glycerol-d_3) bonds:

(i) random reorientational jumps;
(ii) rotational diffusion;
(iii) rotational fixed angle jumps.

In model (iii) the elementary step consists of a fixed angle jump performed around arbitrary axes. The limiting case of infinitesimal small angles is model (ii). Thus, the cross-over from diffusive motion to large angular jumps can be monitored continuously. The models also allow for the inclusion of a log-Gaussian distribution of correlation times. Since the whole set of curves is fitted simultaneously, the mean correlation time $\langle \tau_c \rangle$ and the distribution width are the only relevant fit parameters. The results of the data analysis for glycerol-d_5 at 206 K ($= T_g + 21$ K) are included in the figure as full curves.

The model "random reorientational jumps" (i) does not reproduce the strong τ-dependence of the time constant of the decay curves, whereas the model "rotational diffusion" (ii) overestimates the τ-dependence. The model "rotational fixed angle jumps" leads to the best agreement with the entire data set. The average rotational jump size $\Delta\beta$ is found to be 10 degrees, the mean correlation time $\langle \tau_c \rangle = 40$ ms and the width of the log-Gaussian distribution is 1 decade.

The same analysis has been performed at 199 K ($= T_g + 14$ K) yielding for glycerol-d_5 $\langle \tau_c \rangle = 1$ s, an unchanged distribution width of 1 decade but a tendency toward smaller jump size $\Delta\beta$ of about 6 degrees. Analogous and quantitatively the same results are also found for the reorientation of the hydroxyl deuterated glycerol-d_5 at 206 and 199 K, the only difference being a tendency toward a broader distribution width of about 1.5 decades of correlation times at 199 K.

The fact that we find for both glycerol isotopes the same value of reorientational step sizes is not surprising. From structure analysis by neutron diffraction it is known [58] that glycerol is a highly associated liquid. A reorientation of a C—H bond can occur only if a O—H—O bond is broken. On the other hand every reorientation of a C—H bond should also be manifested in the O—H motion. It is conceivable that the highly viscous liquid consists of extended clusters which restrict the jump possibilities of the hydroxylic protons. Therefore, the same values for $\Delta\beta$ for both bonds indicate the highly cooperative behaviour. The tendency that $\Delta\beta$ becomes smaller with

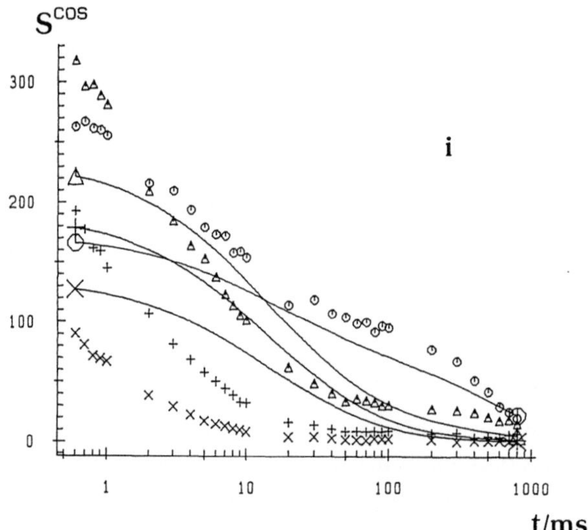

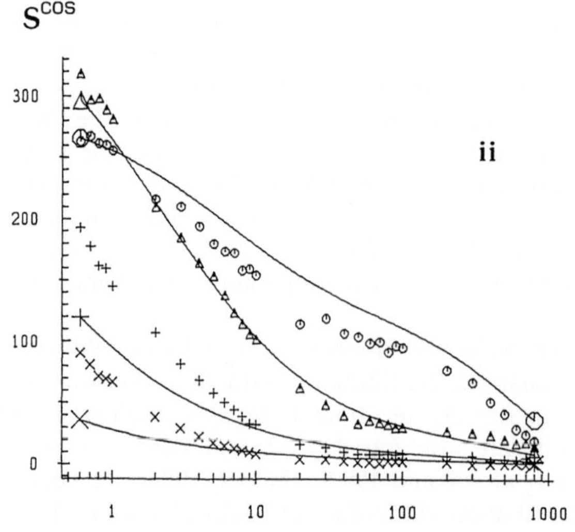

Fig. 21. (*Cont.*)

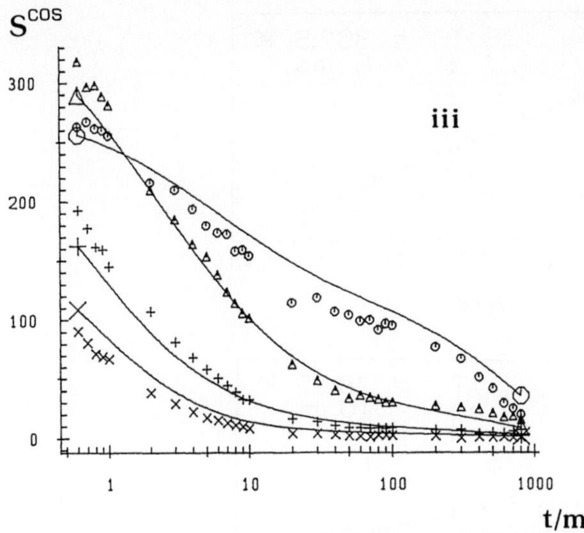

Fig. 21. Zeeman echo decay curves S^{cos} for glycerol-d_5 at 206 K at $\tau = 3.4\,\mu s$ (○), 6.4 μs (Δ), 20.4 μs (+) and 31.4 μs (×). The data are fitted with three motional models i–iii as described in the text

decreasing temperature may indicate a change of the reorientational mechanism when approaching the glass transition temperature.

3.4.2 *"Decoupling" of Self Diffusion and Viscosity in the Supercooled Liquid [24]*

The molecular van der Waals liquid orthoterphenyl (OTP) is meant to be a good model system for testing the mode coupling theory of the glass transition [47, 59–61]. As mentioned above, one of the predictions is a crossover from liquidlike to solidlike behaviour at some temperature $T_c > T_g$. It has been suggested [62] that such a crossover may be reflected in a decoupling of self diffusion and viscosity: Whereas above T_c the (translational) self-diffusion coefficient D and the shear viscosity η scale according to the Stokes-Einstein relation $D \sim T/\eta$, there should be some additional mechanism below T_c which contributes to D. For this purpose the known $\eta(T)$-data [63] have been compared with $D(T)$ in the liquid and supercooled regime of OTP.

$D(T)$ have been measured using the stimulated echo in both, the pulsed and the static field gradient technique. For measuring very small D values at temperatures as low as possible very high field gradients should be used, see Eqs. (14) and (35). Since in the deep supercooled regime $T_2 < 1$ ms is expected, the static field gradient technique is the preferred method. Figure 22 shows a set of measured diffusion curves at two temperatures, one of which is at the lower edge of the accessible regime where the T_2-decay interferes, cf. discussion after Eq. (37). The gradient in the stray field of a 360 MHz cryomagnet was chosen to be 53 T/m corresponding to a ^1H frequency of 101 MHz. Whereas at higher temperatures (Fig. 22a) the T_2-decay can be safely neglected, the lower temperature curves (Fig. 22b) are strongly influenced by the short T_2. The two

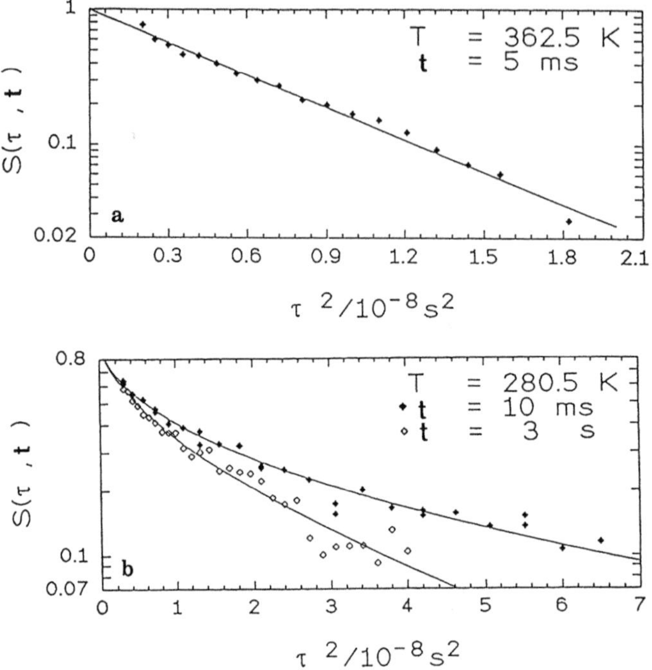

Fig. 22a, b. Diffusion curves for OPT-h measured at a static gradient $g = 53\,\text{T/m}$ at (**a**) 362.5 K yielding $D = 1.83 \times 10^{-10}\,\text{m}^2\text{s}^{-1}$ and at (**b**) 280.5 K yielding $D = 2.6 \times 10^{-14}\,\text{m}^2\text{s}^{-1}$. For the discussion of the fit curves see text. Taken from Ref. [24]

curves in Fig. 22b are fitted simultaneously by

$$S(\tau, t) \sim \exp[-\gamma^2\tau^2 g^2 Dt]\exp[-2\tau/T_2] \tag{55}$$

yielding $T_2 = 0.22\,\text{ms}$ and $D = 2.6 \times 10^{-14}\,\text{m}^2\text{s}^{-1}$. Figure 23 summarizes the T-dependence of our field gradient NMR self diffusion data, a beautiful set of tracer diffusion data from forced Rayleigh scattering [64] and compares them to interpolated viscosity data [63]. It is seen from the figure that the diffusion coefficient D_t decouples from η^{-1} as the glass transition is approached. The decoupling is illustrated in a more pronounced manner by the insert, where the product $D_t \cdot \eta \cdot T^{-1}$ ceases to be T-independent below $\approx 300\,\text{K}$. It is a most remarkable result that this temperature roughly coincides with the "critical" temperature $T_c = (290 \pm 5)\,\text{K}$ found in neutron scattering [66–69] where a number of anomalies predicted by mode coupling theory do in fact occur.

3.5 *Diffusion of Linear Chain Molecules*

For a long time, the diffusion of chain molecules in the melt was a scarcely investigated phenomenon since no suitable measuring methods were available and no well-founded theoretical models existed. About 20 years ago vigorous

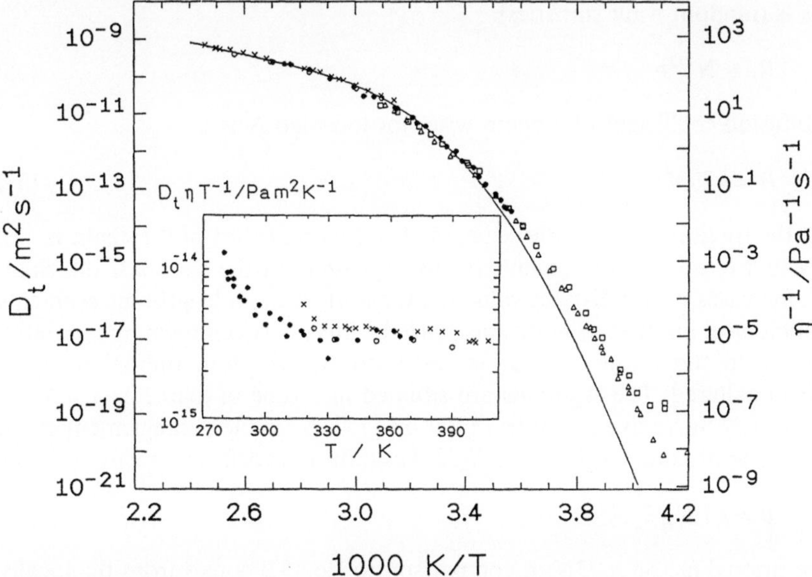

Fig. 23. Logarithmic plot of the NMR self diffusion coefficient [stimulated echoes in a static (●) and a pulsed (○) gradient; Hahn echoes (X) [65]] and the tracer diffusion coefficient (Δ, □) [64] as compared to the inverse viscosity [63]. The *insert* shows the product $D\cdot\eta\cdot T^{-1}$ containing only NMR self diffusion data. Taken from Ref. [24]

development in this field started because of new theoretical concepts: reptation mechanism [70], tube picture [71] and new measuring methods: pulsed field gradient NMR [72] and optical labelling (IR-microdensitometry [73], forced Rayleigh scattering [74]). The pulsed field gradient (PFG) NMR has provided important contributions to the understanding of diffusion processes of polymer molecules in solution and melt. We will report in this chapter some basic experiments.

3.5.1 Chain-Length Dependence of Self Diffusion of Polymers in the Melt

In the context of this chapter, self diffusion means the center-of-mass diffusion of coiled polymer chains in a monodisperse homogeneous melt. Monodispersity is practically never realized, we always have a distribution of chain lengths in polymer samples but macromolecular chemistry today provides samples with such small distribution widths that one can neglect the influence of the chain length distribution or easily eliminate it numerically.

A polymer chain in the melt can be described as consisting of *N* free linked statistical or Kuhn's segments of length *b*. The end-to-end distance or Flory radius is (random walk statistics)

$$R_F = N^{1/2} b \tag{56}$$

The diffusion coefficient of a chain with not too high N is

$$D = kT/N\varsigma. \tag{57}$$

with ς the friction factor of the segment. The friction factor of the chain is $N\varsigma$, simply that of all linked segments (Rouse-dynamics of the chain). If the chain length increases over a distinct value N_c, the critical chain length, the segments feel topological hindrances additionally to the links between them, the mobility transverse to the chain contour is restricted but the longitudinal mobility remains unaltered. The segments are situated in a tube of diameter $a = N_e^{1/2} b$ in which they move like a reptile: reptation [75]. N_e is the entanglement chain length and approximately equal to $N_c/2$. The diffusion coefficient turns out to be

$$D = kTN_e/3\varsigma N^2 \tag{58}$$

The additional factor $N_e/3N$ in comparison to Eq. (57) comes from the locally one-dimensional motion of the segments along the Gaussian tube consisting of $Z = N/N_e$ chain parts (primitive path steps, [76]).

A temporary stable tube is only established for a chain length $N \gg N_e$ (about $N \approx 10\,N_e$). Up to this chain length the chain also diffuses due to lateral fluctuations of the tube because the constraints formed by the neighboring chains are not yet stable. We have "constraint release". With the most applied model of Graessley [77] we obtain for D

$$D = kT/\varsigma N(N_e/3N + \alpha N_e^3/N^3) \tag{59}$$

The second term, the constraint release contribution, quickly decreases with increasing N/N_e. α is in the order of 2–5 and characterizes the number of constraints of a chain part. If an isolated test chain diffuses in a matrix of chain length P we have to change N by P in the second term in Eq. (59).

The experimental verification of center-of-mass self diffusion, Eqs. (57) and (59), with field gradient NMR implies a measurement under the condition $QR_F \ll 1$. On the other hand, in the experimentally accessible time $t \leqq 1$ s (see Sect. 2.3) the spin echo amplitude must have decreased on the value, say, 1/e, i.e. $Q^2Dt = 1$. Since D strongly decreases with increasing N, one has to use the maximum Q in the experiment to expand the measured range of molecular weights as large as possible. Therefore, R_F should be limited to about 100 Å, for flexible polymers this corresponds to a polymerization index of about 350.

Only very flexible polymers with low glass transition temperatures have a sufficiently high mobility (and, hence, a sufficiently long T_1 and T_2) that a broad range of chain lengths can be measured. Three examples are reviewed: poly(dimethylsiloxane), PDMS [78, 79], poly(ethylene oxide), PEO [79–82] and poly(ethylene), PE [83–85]. In Fig. 24 the dependence of the self-diffusion

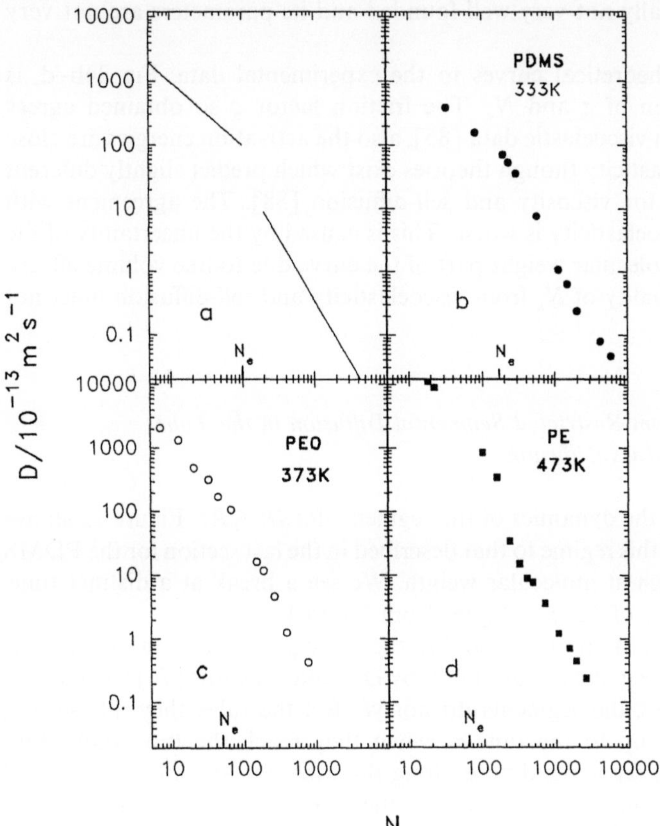

Fig. 24. Dependence of the self diffusion coefficient D of linear chains on the polymerization index N. *a)* theoretically according to Eqs. (57) and (58) with $N_e = 100$ and kT/ς arbitrary, *b)* PDMS at 303 K, $N_e = 145$ [79], *c)* PEO at 373 K, $N_e = 50$ [79], *d)* PE at 473 K, $N_e = 62$ [85]

coefficient on the chain length together with the theoretical expected dependence, Eqs. (57) and (58), are shown.

Looking at the figures, one will realize that $D \sim N^{-2}$ is well fulfilled, best seen for PE which has the smallest N_e and the highest measured N. The dependence $D \sim N^{-1}$ is best fulfilled for PDMS. Here one has to take into consideration that the friction factor for small molecular weights is no longer constant but decreases due to the markedly increasing influence of the chain ends in the system. They raise the free volume and consequently the molecular mobility. D is additionally increased with decreasing N. This effect is less pronounced for the highly mobile PDMS with CH_3-substituents and with a low glass transition temperature. It is possible to take into account the free volume of the chain ends in the data evaluation and to verify the linear dependence of D on N for small molecular weights [86, 87], but the free volume concept is theoretically not very well founded and its parameters are not very reliable.

The fit of the theoretical curves to the experimental data, Fig. 24b–d, is achieved by variaion of ς and N_e. The friction factor ς so obtained agrees sufficiently well with viscoelastic data [85], also the activation energies are close to that from viscoelasticity though theories exist which predict slightly different activation energies for viscosity and self-diffusion [88]. The agreement with N_e-values from viscoelasticity is worse. This is caused by the uncertainty of the height of the low molecular weight part of the curve due to free volume effects. In addition, the equality of N_e from viscoelasticity and self-diffusion must not be true a priori.

3.5.2 Cross-Over from Restricted Segmental Diffusion in the Tube to Center-of-Mass Diffusion

More interesting is the dynamics of the segments for $Dt \leqq R_F^2$. Figure 25 shows the cross-over from this regime to that described in the last section for the PDMS sample with the highest molecular weigth. We see a break at a distinct time. The time exponent α of $\langle z^2 \rangle \sim t^\alpha$ goes from 1/2 to 1.

At first, we will look in more detail at the principles of motion of segments within the tube (see, e.g. the famous book by Doi and Edwards [76]). Beginning at the shortest times, the segments do not yet feel the tube, they diffuse in a Rouse-like manner up to the time τ_e when they reach the tube wall. With increasing time, the segments diffuse along the tube but are still uncorrelated at different parts of the chain. At time T_R, the Rouse time, the chain begins to move as a whole within the tube. Caused by this diffusive motion, the tube loses its initial conformation more and more, the life time of this initial

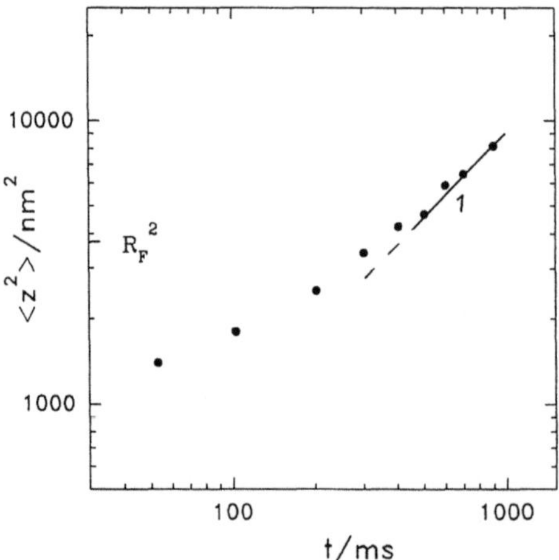

Fig. 25. Mean Square displacement $\langle z^2 \rangle$ vs t for PDMS with $M_W = 717000$. $T = 333$K

conformation is T_d, the tube disengagement time. For times much longer than T_d, all the segments diffuse with the center-of-mass diffusion coefficient, Eq. (58). The relations relevant for our purposes are [89]

$$\begin{array}{lccccccc} D \simeq & & & a(kT/N\varsigma t)^{-1/2} & & & & kTN_e/\varsigma N^2 \\ \langle r^2\rangle \simeq R_F^2(t/Z^2T_R)^{1/4} & & & R_F^2(t/T_d)^{1/2} & & & & R_F^2(t/T_d) \\ t & \longrightarrow & T_R \sim \tau_0 N^2 & & \longrightarrow & T_d \sim \tau_0 N^3/N_e & \longrightarrow & \end{array}$$

$Z = N/N_e$, $\tau_0 \simeq b^2\varsigma/kT$, numerical prefactors are omitted, they are to the order of one. In Fig. 26, the dependence of $\langle r^2\rangle$ on t according to the given relations is shown. (In the field gradient NMR the rms displacement in one dimension $\langle z^2\rangle$ is measured which is $\langle r^2\rangle/3$ in the always isotropic samples.)

What can we expect in the experiments? If the time t crosses the tube disengagement time T_d and becomes shorter, we should measure an apparent diffusion coefficient with the time dependence $\sim t^{-1/2}$. $\langle r^2\rangle \sim t$ changes into $\langle r^2\rangle \sim t^{1/2}$. The crossover time T_d should scale with N^3, the long-time D with N^{-2}. The apparent activation energy of D, which (apart from a possibly weak temperature dependence of a or N_e) is controlled by the temperature dependence of ς, should be for $T_R < t < T_d$ 50% of the activation energy for $t > T_d$.

Experiments to detect the segmental diffusion for $t < T_d$ are a challenge for field gradient NMR because very large gradients are necessary. Here the use of the stray field in a cryomagnet is very advantageous. The first experiment in this direction was done by Kimmich and coworkers [23]. Using extremely large field gradients, there is a real chance to connect the Q-regions of field gradient NMR and neutron scattering, see Fig. 6. Here we can still see that the trajectory shown in Fig. 6 is no longer a straight line with slope -1 for the polymer chains. Due to the previously mentioned complicated diffusion mechanism and two characteristic mesoscopic space scales (tube diameter, Flory radius) the trajectory will look similar to the trajectory in Fig. 26.

Let us first deal with some basic results obtained with QENS experiments with the IN11 spectrometer in Grenoble for polymer diffusion. QENS is, similar to field gradient NMR experiments, only possible with highly flexible polymers to exploit the accessible Q-range of $0.02 < Q < 0.15\,\text{Å}^{-1}$ in the experimentally

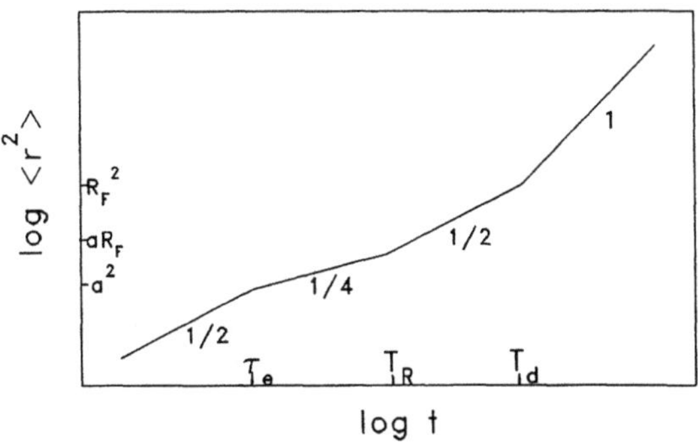

Fig. 26. Mean square displacement of segments $\langle r^2\rangle$ in dependence on time, after Ref. [89]

given time window in the order of ns. Experiments were carried out with PDMS [90] and hydrogenated poly(isoprene) (h-PI), a polymer related to poly(ethylene) [91]. The incoherent scattering function showed Rouse dynamics of the segments for $t < \tau_e$ according to

$$S(Q,t) = \exp(-Q^2(kTb^2t/3\pi\varsigma)^{1/2}). \tag{60}$$

For h-PI an EISF for the lowest Q was seen indicating the constraints of the tube imposed to the segments ($a \approx 50$ Å). The value of $3\,kTb^2/\varsigma$ was determined to be 1.75×10^{13} Å^4/s for PDMS at 373 K and to $3 \cdots 4 \times 10^{13}$ Å^4/s for h-PI at 473 K. From the self-diffusion measurements for $t > T_d$ reported in the last section we obtain for $3\,kTb^2/\varsigma$ 3.5×10^{13} Å^4/s (PDMS, recalculated for 373 K with $E_A = 15$ kJ/mol) and 6.5×10^{13} Å^4/s (PE). Taking into account possible uncertainties of the data necessary to calculate these values from the experimental

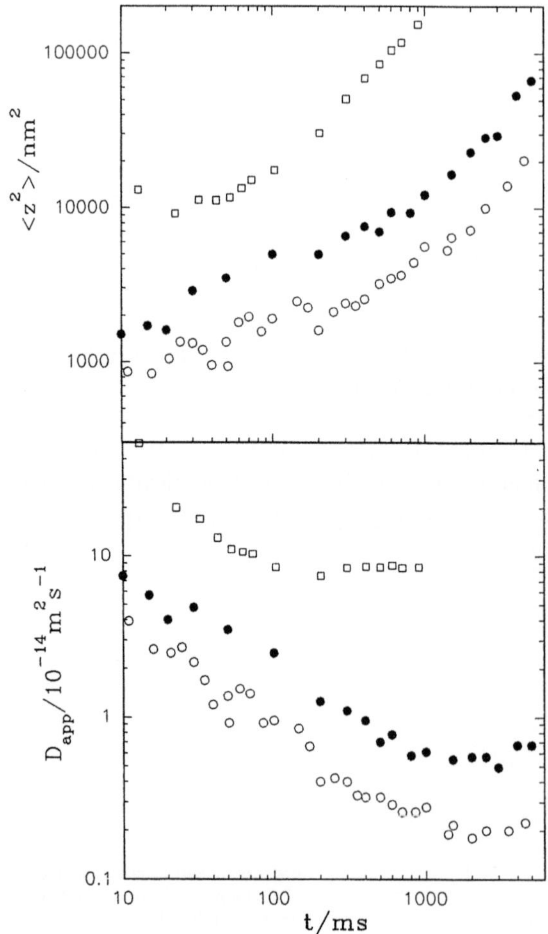

Fig. 27. Log-log-plot of D_{app} and $\langle z^2 \rangle$ on time t for PEO 5Mio 50% in C_6D_6 (○) and 5% in D_2O (□) and PEO 2Mio 50% in C_6D_6 (●), $T = 333$ K

data (N_e and R_F), the agreement is not bad, but there seems to be a systematic deviation of the friction factor obtained from NMR data to lower values.

In the field gradient NMR experiments one has to work with $QR_F > 1$ in order to see the segmental motion, i.e. samples of high molecular weight must be used and large field gradients applied. Simultaneously T_d must be within the time window of the experiment. This latter condition is equivalent to the condition that the long-time D is still sufficiently large to obtain an echo damping. Contrary to R_F which is a geometric quantity, T_d or D may be controlled via the friction factor ς. It may be decreased by increasing the temperature or, more effectively, by adding solvent as plasticizer. The Q^2-dependence of the scattering function $S(Q, t)$, Eq. (14) remains exponentially since $G_s(r, t)$ is Gaussian, only the time dependence of the distribution width $\langle r^2 \rangle$ alters.

The experiments were carried out with PEO [82,92] which has, as already mentioned, a high molecular mobility and is available in very high but, however, rather broadly distributed molecular weights. The nominal molecular weights are 2 Mio and 5 Mio, from light scattering $M_w = 1.2$ Mio and 4.0 Mio were determined. By addition of solvents (D_2O and C_6D_6) T_d is shifted into the time window of the experiment.

In Figs. 27 and 28, the results are shown. The 50% solutions were measured in the stray field (i.e. static gradient) experiment. In the data evaluation, T_2 was also fitted and obtained to 7.5 ms. The cross-over from segmental diffusion for $t < T_d$ to center of mass diffusion for $t > T_d$ is clearly detected, also the decrease of the temperature dependence for $t < T_d$. The experimental scatter of the data prevents a more exact evaluation of the time exponents better than ± 0.1, and the not very exactly known molecular weight and its distribution of the samples prevents the exact verification of the scaling laws of T_d and D on N, but it can be seen that the theory is at least qualitatively fulfilled.

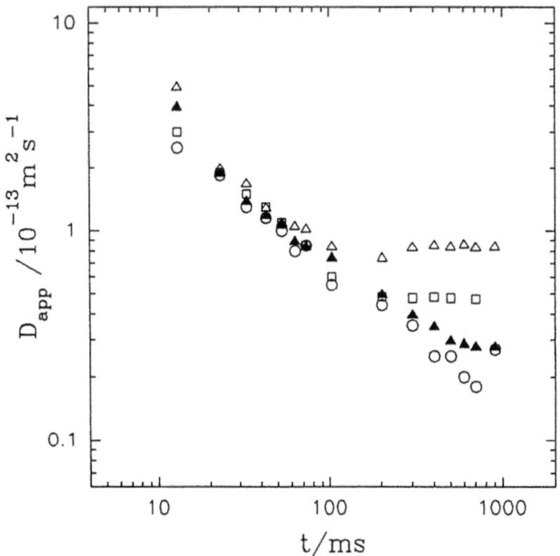

Fig. 28. Log-log plot of D_{app} on time t for PEO 5Mio 5% in D_2O at 303 K (○), 313 K (▲), 323 K (□) and 333 K (△)

3.6 Q-Space Imaging: Diffusion in a Confined Geometry

In Sect. 2.1 we saw that the long-time behaviour of $S(Q,t)$ reaches a constant value $A_0(Q)$ if $G_s(r,t)$ contains a time-independent part, called EISF in neutron scattering. This behaviour occurs if a non-vanishing correlation of the motion of the diffusing species exists due to spatial restrictions. Whereas the calculation of $S(Q,t)$ over the whole time and space range by means of the solution of the diffusion equation is difficult, the $A_0(Q)$ as the square of spatial Fourier-transform of the distribution $p(r)$, Eq. (11), is more easily obtained. Looking at the most simple case, the diffusion between two impenetrable walls with distance l, we have [21, 22]

$$A_0(Q) = (\sin(Ql/2)/(Ql/2))^2 \tag{61}$$

$\sin(Ql/2)/(Ql/2)$ are the Fourier components of a rectangular function of width l. The result of Eq. (61) is just the intensity distribution of diffracted light of wavevector Q behind a slit of width l.

A second important example is the diffusion within a sphere of radius R. The expression of $A_0(Q)$ is more complicated [22], for $QR \ll 1$ one gets

$$A_0(Q) = \exp(-Q^2R^2/5) \tag{62}$$

Whereas in the accessible space scale of neutron scattering many experiments allow us to observe an EISF, see Sect. 3.2, only a few such experiments in field gradient NMR are known because samples with diffusion barriers with a well-defined spatial geometry in the μm range are rare. We have to work with $Q_{max} l > 1$ and $2D_s t \gg l^2$. The second condition means that, in the diffusion time t, the molecule must have travelled a distance $\gg l$. With typical values of $D = 2 \times 10^{-9}\,\mathrm{m^2 s^{-1}}$ for low molecular liquids and $t = 0.1$ s we have $l < 20\,\mu\mathrm{m}$ and $Q \gg (20\,\mu\mathrm{m})^{-1} = 10^{-5}\,\mathrm{A}^{-1}$.

We will report on two examples where completely restricted diffusion was observed. The first is from biophysics and deals with the self diffusion of oil in rape seeds [93]. The oil in rape seeds is dispersed in small droplets whose diameters, however, have a distribution. In Fig. 29, $S(Q,t)$ for the oil self diffusion

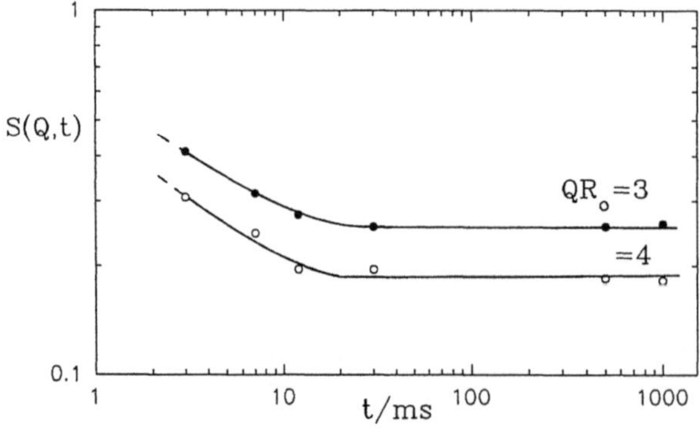

Fig. 29. $S(Q,t)$ as a function of t for oil self diffusion in rape seeds of variety "Sollux" at $T = 338$ K taken from Ref. [93]

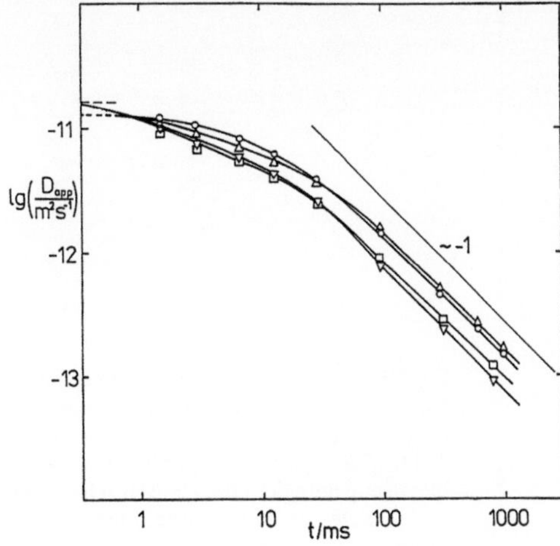

Fig. 30. Log D_{app} vs log t for two varieties of rape seeds and different moisture content. The *dashed lines* mark the self-diffusion coefficient of the pure oil. Taken from Ref. [93]

in rape seeds is depicted where a clear $A_0(Q)$ is reached for diffusion times exceeding 30 ms. The experiment gives an apparent self diffusion coefficient which decreases exactly with t^{-1}, see Fig. 30.

For the evaluation of the droplet radius a Gaussian mass distribution of the radii around R_0 was assumed

$$p(R)\mathrm{d}R \sim \exp(-(R-R_0)^2/\sigma^2) \tag{63}$$

$p(R)\mathrm{d}R$ is the mass contribution of spheres with radius between R and $R+\mathrm{d}R$, σ the distribution width of $p(R)$. With the distribution of Eq. (63) the dependence of $A_0(Q)$ on Q^2 was optimal fitted [93]. For R_0 a value of $\approx 0.8\,\mu$m was obtained, for $\sigma \approx 0.4\,\mu$m varying slightly for various varieties of the rape seeds. Since NMR relaxation measurements are common for determination of oil content of rape seeds, it should be easily possible to extent these spectrometers for field gradient NMR to obtain more precise data on the seeds.

The second example is again taken from polymer dynamics and shows the apparent diffusion coefficient in dependence on t for polybutadiene networks swollen in deuterated benzene at room temperature [94].

In this system, the polymer chains are fixed at chemically cross-link junctions and are therefore confined to a space in the region of their Flory radius (which is in the order of the junction distances).

Again, the apparent diffusion coefficient decreases with t^{-1}, $\langle z^2 \rangle$ is independent on t. We observe the spatially restricted fluctuations of the network chains between the fixed junction points. As expected, $\langle z^2 \rangle$ increases with increasing network chain length, and is found between 110 and 230 Å, but again we have difficulties with unknown distributions since the network chain lengths are broadly distributed and the form and width of the distribution is not exactly known. Therefore a highly desirable quantitative comparison with network theories [95] is not possible.

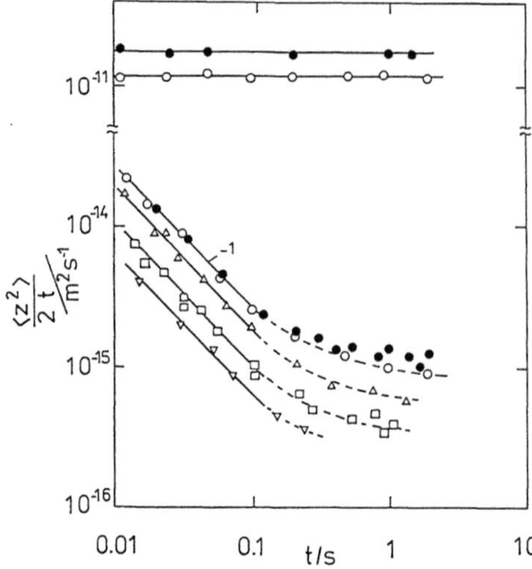

Fig. 31. $\langle z^2 \rangle/2t$ for segments of poly(butadiene) networks swollen in C_6D_6 in dependence on t at $T = 303$ K (*open symbols*) and 333 K (*filled symbols*). From bottom to top increasing network chain length. The *upper curve* shows D of the sol-content in the sample with highest swelling ratio. Taken from Ref. [94]

3.7 *Molecular Transport in Microporous Crystallites: A Propagator Representation*

For diffusion of adsorbates in zeolites and other microporous media, pulsed field gradient NMR experiments have provided a lot of fundamental results and insights into the diffusion mechanism [96]. In 1983, Kärger and Heink [97] at first explicitly used the description of the echo height $S(Q,t)$ in field gradient NMR as a Fourier transform of the averaged propagator, see Eqs. (7) and (9). They have calculated the averaged propagator $P(z,t)$ by the inverse Fourier transformation of the measured echo height of the PFG NMR experiment with respect to Q, see Eq. (9), and could visualize very instructively the intra- and inter-crystalline diffusion in zeolites by this propagator representation.

In zeolites, we have intracrystalline diffusion within the crystallites (D_{intra}) and long range diffusion if the diffusing molecules leave, during the diffusion time t, the crystallite in which they were situated at time $t = 0$. The long-range diffusion is determined by the product of $p_{\text{inter}}D_{\text{inter}}$ where p_{inter} is the relative number of molecules in the intercrystalline space and D_{inter} their diffusion coefficient. In the investigated system (ethane in NaX zeolites with different crystallite radii R) the activation energy of D_{intra} is about 4 kJ/mol, that of $p_{\text{inter}}D_{\text{inter}}$ about 25 kJ/mol and at $T \approx 173$ K D_{inter} and $p_{\text{inter}}D_{\text{inter}}$ are of the same order of magnitude.

In Fig. 32, the time dependence of the (averaged) propagator shows very evidently the following characteristic features of the diffusion in the bed of crystallites: at the lower temperature (153 K) D_{intra} is much larger than $p_{\text{inter}}D_{\text{inter}}$. Hence, for the smaller crystallites where the crystallite boundaries are reached within the diffusion time, we observe completely restricted diffusion within the

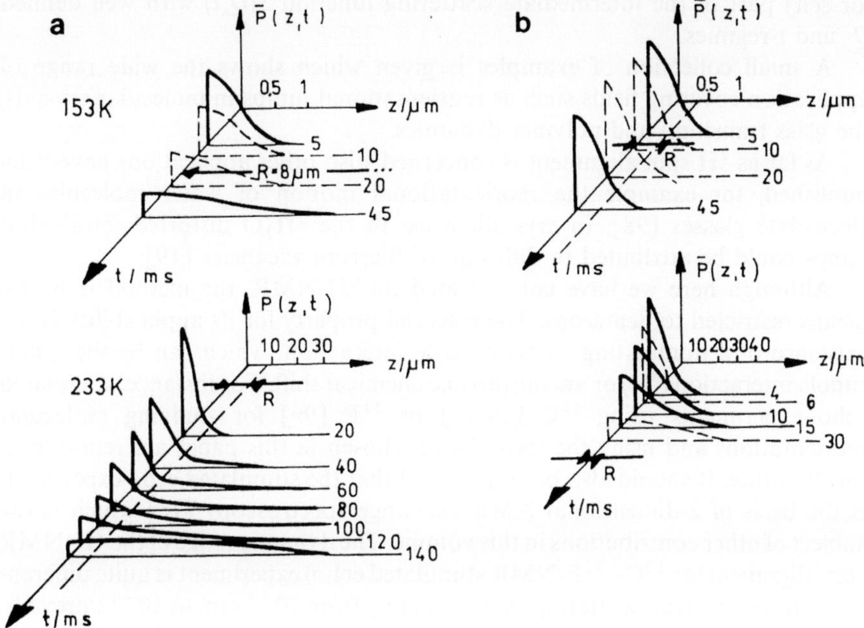

Fig. 32a, b. Propagator representation for the self diffusion of ethane in NaCaA zeolites. (**a**) 40 mg/g, $R \approx 8\ \mu m$, (**b**) 58 mg/g, $R \approx 0.4\ \mu m$. Only the part for $z > 0$ for the symmetric propagator is shown. Taken from Ref. [97]

crystallites. At the higher temperature (233 K) we meet the opposite situation. Now D_{intra} is smaller than $p_{\text{inter}} D_{\text{inter}}$. The propagator outside the crystallites broadens faster than within the crystallites.

4 Concluding Remarks

In this review, we have looked at some NMR stimulated echo experiments and have interpreted them in terms of the generalized incoherent scattering function. We have concentrated on situations where the stimulated echo represents a *single* particle phase correlation function, see Eq. (45). Two different types of experiments have been analyzed in detail:

a. ^{2}H-NMR spin alignment;
b. NMR in a magnetic field gradient.

The ^{2}H spin alignment contains information on geometry and timescale of single particle rotational dynamics which is essentially equivalent to the rotational part of the incoherent intermediate scattering function. The field gradient NMR stimulated echo experiment can be fully formulated in terms of the incoherent (or self) part of the intermediate scattering function $S(Q, t)$ with well defined Q- and t-regimes.

A small collection of examples is given which shows the wide range of application covering fields such as reorientational jumps in molecular crystals, the glass transition and polymer dynamics.

As far as ^{2}H spin alignment is concerned, also other applications have been published, for example the reorientational motion of water molecules in electrolyte glasses [98]. In crystalline ice Ih the ^2H_2O distorted tetrahedral jumps could be attributed to diffusion of Bjerrum vacancies [19].

Although here we have concentrated on ^{2}H-NMR, the method is by no means restricted to deuterons. The essential property for its applicability is the existence of a dominating single particle interaction, which can be the quadrupole interaction (^{2}H) or an anisotropic chemical shift. For instance, stimulated echo experiments using ^{13}C [29, 42] or ^{31}P [99] for studying molecular reorientations and using the formulation chosen in this paper are reported in the literature. It should also be emphasized that the stimulated echo experiment is the basis of 2-dimensional NMR exchange spectroscopy [16] which is the subject of other contributions in this volume. The dynamic range of the ^{2}H-NMR spin alignment (or ^{13}C-, ^{31}P-NMR stimulated echo) experiment is quite different from that of neutron scattering. A gap ranging from 10^{-8} s up to 10^{-4} s prevails. It would be extremely worthwhile to develop experiments capable of bridging this gap. At the magnetic resonance side one may think of stimulated ESR echoes in cases where the hyperfine interaction dominates the spectra [100]. At the scattering side there are attempts to increase the energy resolution of neutron scattering spectrometers [101, 102] or to use synchrotron radiation to produce intensive pulsed Mössbauer sources for Mössbauer-Rayleigh scattering experiments [103].

In field gradient NMR, up to now, most experiments reported in the literature are carried out at low Q detecting only hydrodynamic translational diffusion. The more interesting observations of restricted diffusion in structured systems revealing an EISF are at most seen only for larger Q. In PFG NMR the technical limitations are at a field gradient of about 50 T/m.

By using static magnetic field gradients the gradient NMR experiment is about to overcome the threshold of about 50 T/m. Using an anti-Helmholtz arrangement of split superconducting coils a gradient of up to 180 T/m is presently being installed at the universities of Mainz and Dortmund [104]. Such ultrahigh gradients can be of use only if extremely short rf pulses are used to enlarge the width of the excitation band as much as possible in order to maximize the thickness of the excited slice, i.e. to optimize the S/N ratio. Remind that a gradient of almost 200 T/m yields a maximum Q-value of more than 10^{-2} Å (Eq. (35) and Fig. 6) thus overlapping with the neutron scattering Q-range. Needless to say that for such Q-values there are a lot of perspectives concerning the study of dynamics in mesoscopic structures, confined geometries, soft matter etc.

To summarize: The interpretation of stimulated NMR echoes in terms of a generalized incoherent scattering function is not only a beautiful concept but also provides the framework for many experiments, especially in concert with neutron scattering.

Acknowledgements. We like to thank our coworkers in the Leipzig and Mainz groups, especially B. Geil, W. Heink, G. Hinze, J. Kärger and S. Wefing. For continuous support and many stimulating discussions we also thank D. Geschke, H. Pfeifer and H. Sillescu. One of us (F. F.) would also like to thank W. Petry (TU München) for the cooperation in the neutron scattering work.

5 References

1. Egelstaff PA (ed) (1971) Thermal neutron scattering, Academic, New York
2. Marshall W, Lovesey SW (1971) Theory of thermal neutron scattering, Clarendon Oxford
3. Lovesey SW (1984) Theory of neutron scattering from condensed matter, Clarendon, Oxford
4. Squires GL (1978) Introduction to the Theory of Thermal Neutron Scattering, Cambridge University-Press, Cambridge
5. Van Hove L (1954) Phys. Rev. 95: 245
6. Leadbetter AJ, Lechner RE (1979) In: Sherwood JN (ed) The plastically crystalline state, Wiley, New York, p 285; Lechner RE (1983) In: Berniere F, Catlow CRA (ed) Mass Transport in Solids, Plenum, New York, p 169–226
7. Sears VF (1984) Thermal-neutron scattering lengths and cross-sections for condensed matter research, Chalk River Nuclear Laboratories, Chalk River, Ontario, Canada
8. Bee M (1988) Quasielastic Neutron Scattering, Adam Hilger, Bristol, Philadelphia
9. Chudley GT, Elliott RJ (1961) Proc. Phys. Soc. 77: 353
10. Springer T (1972) Quasielastic Neutron Scattering for the Investigation of Diffusive Motions in Solids and Liquids, Springer Tracts in Modern Physics, Berlin
11. Dianoux AJ, Volino F (1977) Mol. Phys. 34: 1263 ; Volino F, Dianoux AJ, Heidelmann A (1979) J. Physique Lett. 40: L583
12. Dorner B (1972) Coherent inelastic neutron scattering in lattice dynamics, Tracts in Modern Physics 93: Springer, Heidelberg
13. Spiess HW (1983) Colloid & Polymer Sci. 261: 193; Sillescu H (1982) Pure & Applied Chem. 54: 619; Spiess HW (1985) Adv. Pol. Sci. 66: 23
14. Spiess HW, Sillescu H (1981) J. Magn. Res. 42: 381
15. Spiess HW (1980) J. Chem. Phys. 72: 6755
16. Schmidt C, Blümich B, Wefing S, Spiess HW (1986) Chem. Phys. Lett. 130: 84
17. Wefing S (1987) Ph.D. thesis, Mainz
18. Wefing S, Spiess HW (1988) J. Chem. Phys. 89: 1219
19. Fujara F, Wefing S, Kuhs WF (1988) J. Chem. Phys. 88: 6801
20. Tanner JE, Stejskal EO (1968) J. Chem. Phys. 49: 1768
21. Kärger J, Pfeifer H, Heink W (1989) Adv. Magn. Res. 12: 1
22. Callaghan PT (1991) Principles of Magnetic Resonance Microscopy, Clarendon, Oxford
23. Kimmich R, Unrath W, Schnur G, Rommel E (1991) J. Magn. Res. 91: 136
24. Fujara F, Geil B, Sillescu H, Fleischer G (1992) Z. Phys. B 88: 195
25. Fujara F, Wefing S, Spiess HW (1986) J. Chem. Phys. 84: 4579
26. Fujara F, Petry W, Schnauss W, Sillescu H (1988) J. Chem. Phys. 89: 1801
27. Solum MS, Zilm KW, Michl J, Grant DM, (1983) J. Chem. Phys. 87: 2940
28. Schmidt C (1987) Ph.D. thesis, Mainz
29. Rössler E (1986) Chem. Phys. Lett. 128: 330
30. Lausch M, Spiess HW (1983) J. Magn. Res. 54: 466
31. Andrew ER (1950) J. Chem. Phys. 18: 607; Allen PS, Cowking A (1967) J. Chem. Phys. 47: 4286
32. Wykoff RWG (1969) Crystal Structures, 2nd ed., vol. 6, part 1, Interscience, New York
33. Pschorn U, Rössler E, Sillescu H, Kaufmann S, Schäfer D, Spiess HW, (1991) Macromolecules 24: 398

34. Wefing S, Kaufmann S, Spiess HW (1988) J. Chem. Phys. 89: 1234
35. Andrew ER, Eades RG (1953) Proc. Roy. Soc. London Ser. A 218, 537
36. Haeberlen U, Maier G, Z Naturforsch, Teil A (1967) 22: 1236
37. Wemmer DE (1978) Ph.D. thesis, University of California, Berkeley; Mehring M (1983) Principles of High Resolution NMR in Solids, p. 59, Springer, Berlin
38. Cox EG, Cruickshank DWJ, Smith JA (1958) Proc. Roy. Soc. London Ser. A 247: 1
39. Oppenheim I, Schuler KE, Weiss GH (1977) Stochastic Processes in Chemical Physics, MIT Press, Cambridge (Mass.), USA
40. Van Steenwinkel R, Z Naturforsch A (1969) 24: 1526; O'Reilly D, Peterson EM (1972) J. Chem. Phys. 56: 5536; Noack F, Weithase M, von Schütz J (1975) Z. Naturforsch. A30: 1707, McGuigan S, Strange JH, Chezeau JM (1982) Mol. Phys. 47: 373
41. Fox R, Sherwood JN (1971) Trans. Faraday Soc. 67: 3364
42. Guillon T, Conradi MS (1985) Phys. Rev. B 32: 7076
43. Lutze U (1989) Ph.D. thesis, Mainz
44. Wong J, Angell CA (1976) Glass—Structure by Spectroscopy, Marcel Dekker, New York
45. Jäckle J (1986) Rep. Prog. Phys. 49: 171
46. Brawer SA (1983) Relaxation in viscous liquids and glasses, Am. Ceram. Soc., New York
47. Götze W, Sjögren L (1992) Rep. Prog. Phys. 55: 241
48. Diehl RM, Fujara F, Sillescu H (1990) Europhys. Lett. 13: 257
49. McDuffie GE, Litovitz jr. TA, (1962) J. Chem. Phys. 37: 1699
50. Dux H, Dorfmüller Th (1979) J. Chem. Phys. 40: 219
51. Wolfe M, Jonas J (1979) J. Chem. Phys. 71: 3252
52. Kintzinger JP, Zeidler MD (1972) Ber. Bunsenges. Phys. Chem. 77: 98
53. Kuhns PL, Conradi MS (1982) J. Chem. Phys. 77: 1771
54. Soltwisch M, Elwenspoek M, Quitmann D (1978) Mol. Phys. 34: 33 (1977) 35: 1221
55. Soltwisch M, Quitmann D (1979) J. Phys. C, 40: 666
56. Fujara F, Petry W, Diehl RM, Sillescu H (1991) Europhys Lett. 14: 563
57. Birge NO, Nagel SR (1985) Phys. Rev. Lett. 54: 2674
58. Champeney DC, Joarder RN, Dore JC (1986) Mol. Phys. 58: 337
59. Bengtzelius U, Götze W, Sjölander A (1984) J. Phys. C 17: 5915
60. Sjögren L, Götze W (1989) Springer Proceedings in Physics 37: 18
61. Götze W, in: Liquid, freezing and the glass transition, ed: Levesque D, Hansen JP, Zinn-Justin J (1991) pp 287–503, North-Holland, Amsterdam
62. Rössler E (1990) Phys. Rev. Lett. 65: 1595
63. Cuikermann M, Lane JW, Uhlmann DR (1973) J. Chem. Phys. 59: 3639; Laughlin WT, Uhlmann DR, (1972) J. Phys. Chem. 76: 2317; McLaughlin E, Ubbelohde AR, (1958) Trans. Faraday Soc. 54, 1804
64. Lohfink M, Sillescu H, 1st Tohwa University International Symposium, Fukuoka, Japan, 4–8 November 1991, Am. Inst. of Phys. Conf. Proc. No 256 (1992), p. 30, New York
65. McCall DW, Douglass DC, Falcone DR (1969) J. Chem. Phys. 50: 3839
66. Petry W, Bartsch E, Fujara F, Kiebel M, Sillescu H, Farago B (1991) Z. Phys. B 83: 175
67. Fujara F (1993) J. Molecular Structure 296: 285
68. Kiebel M, Bartsch E, Debus O, Fujara F, Petry W, Sillescu H (1992) Phys. Rev. B 45: 10301
69. Wuttke J, Kiebel M, Bartsch E, Fujara F, Petry W, Sillescu H (1993) Z. Phys. B 91: 357
70. de Gennes PG (1971) J. Chem. Phys. 75: 572
71. Edwards SF (1967) Proc. Phys. Soc. 92: 9
72. Stejskaj OE, Tanner JE (1965) J. Chem. Phys. 42: 288
73. Klein J (1978) Nature 271: 143
74. Coutandin J, Sillescu H, Voelkel R (1982) Makromol. Chem. Rapid Comm. 3: 649
75. de Gennes PG (1979) Scaling Concepts in Polymer Physics. Cornell University Press, Ithaca NY.
76. Doi M, Edwards SF (1986) The Theory of Polymer Dynamics. Clarendon, Oxford
77. Graessley WW (1982) Adv. Polymer Sci. 47: 67
78. Vasiljev GI, Skirda VD (1988) Vysokomol. Soed. A 30: 849
79. Appel M, Fleischer G (1993) Macromolecules in press
80. Sevreugin VA, Skirda VD, Maklakov AI (1986) Polymer 27: 290
81. von Meerwall E, Palunas P (1987) J. Polymer Sci. Polymer Phys. Ed. 25: 1439
82. Appel M (1992) Diploma thesis, Univ. Leipzig
83. Bachus R, Kimmich R (1983) Polymer 24: 964

84. Pearson DS, Ver Strate G, von Meerwall E, Schilling FC (1987) Macromolecules 20: 1133
85. Fleischer G (1987) Colloid Polymer Sci. 265: 89
86. von Meerwall ED, Grigsby J, Tomich D, van Antwerp R (1982) J. Polymer Sci. Polymer Phys. Ed. 20: 1037
87. Fleischer G (1984) Polymer Bull. 11: 75
88. Ngai KL, Rendell RW, Rajagopal AK, Teitler S (1985) Ann. NY Acad. Sci. 484: 150
89. Ref. 76, chap. 6
90. Richter D, Ewen B, Farago B, Wagner T (1989) Phys. Rev. Lett. 62: 2140
91. Butera R, Fetters LJ, Huang JS, Richter D, Pyckhout-Hintzen W, Zirkel A, Farago B, Ewen B (1991) Phys. Rev. Lett. 66: 2088
92. Fleischer G, Fujara F (1992) Macromolecules 25: 4210
93. Fleischer G, Skirda, VD, Werner A (1990) Europ. Biophys. J. 19: 25
94. Skirda VD, Doroginizkij MM, Sundukov VI, Maklakov AI, Fleischer G, Häusler KG, Straube E (1988) Makromol. Chem. Rapid Comm. 9: 603
95. Heinrich G, Straube E (1988) Adv. Polymer Sci. 85: 34
96. Kärger J, Ruthven DM (1992) Diffusion in Zeolites and Other Microporous Solids. Wiley, New York
97. Kärger J, Heink W (1983) J. Magn. Res. 51: 1
98. Hagemeyer A, Kanert O, Balzer-Jöllenbeck G (1989) Phys. Rev. B 39: 15
99. Rössler E (1992) Habilitationsschrift, Berlin
100. Patyal BR, Crepeau RH, Gamliel D, Freed JH (1990) Chem. Phys. Lett. 175: 453
101. Steyerl A, Drexel W, Malik SS, Gutsmiedl E (1988) Physica B 151: 36
102. Alefeld B, Badurek G, Rauch H (1981) Z. Phys. B 41, 231
103. Gerdau E, Rüffer R, Winkler H, Tolksdorf W, Klages CP, Hannon JP, Phys. Rev. Lett. 54, 835
104. Chang I, Fujara F, Geil B, Hinze G, Tölle A, Sillescu H (1994) J. Non-Cryst. Solids, in press

NMR Imaging of Solids

P. Blümler and B. Blümich*

Max-Planck-Institut für Polymerforschung, Ackermannweg 10, D-55128 Mainz, FRG
* Lehrstuhl für Makromolekulare Chemie, RWTH, Worringer Weg 1, D-52056 Aachen, FRG

Table of Contents

Methods and applications of NMR imaging of solids are reviewed. The prime concern of most imaging techniques is the achievement of high spatial resolution. Comparatively little work focuses on improvement and exploitation of NMR parameter contrast. Yet this seems to be the unique feature that makes NMR imaging superior to other imaging techniques, and provides a high potential for localization of previously unknown material heterogeneities.

Because outstanding applications of solid-state imaging are quite rare, this review is structured according to methodical aspects. In particular imaging methods which can be implemented with standard equipment are discriminated from those which require particular instrumentation. The applications most relevant for practical use of solid-state imaging are mainly in the first category. An attractive class of materials here is the one of elastomers.

1 Introduction

In the last few years, NMR imaging (MRI, *m*agnetic *r*esonance *i*maging) has proved to be extremely useful in medical diagnostics. The popularity of NMR as an imaging technique in medicine derives from the use of non-ionizing radiation as well as from the availability of contrast features which complement X-ray images and cut down the use of contrast agents. In particular, soft tissues can be discriminated. The fact that the method is noninvasive is a prerequisite for this application. The noninvasiveness of the technique and the contrast features are illustrated in Fig. 1 by sagittal images of the heads of the authors. White and gray brain tissue can be clearly differentiated.

NMR imaging is a rapidly expanding field of research, and has been reviewed on several occasions. In consideration that readers may have different interests in imaging methods, applications, and related problems, reviews and monographs can be characterized by topics as follows:

- Solid-state NMR spectroscopy [1–3]
- Principles of NMR imaging [4–9]
- Imaging of solids [8, 10–12]
- Imaging techniques and fundamentals [4–9, 13–18]
- Medical applications [4, 14–16, 18–23]
- Imaging in materials science [11, 13, 24–27]
- Imaging of biological samples [8, 11, 13]
- Imaging of diffusion and flow [8, 11, 28–30]

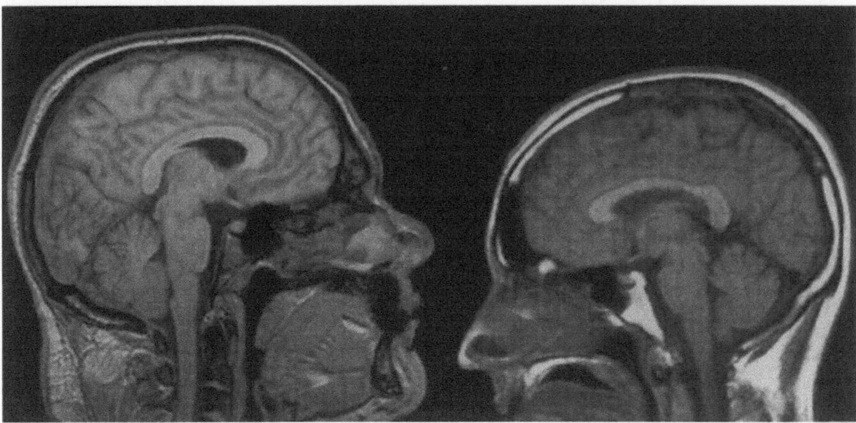

Fig. 1. Application of NMR imaging in medicine: Medio-sagittal slices through the heads of P. Blümler (left) and B. Blümich (right). Radiologists confirm that the features displayed in these images are in the normal range of anatomical deviations. Both images appear slightly different in their tissue contrast (T_2 weightened images) due to different experimental setups and slice thickness. Measured with a Philips GYROSCAN S15 ACS II. (Reproduced with permission from Thelen and Schauß, Universitätskliniken Mainz)

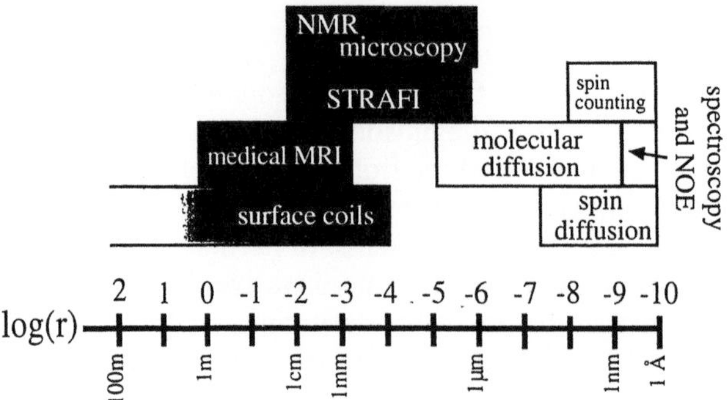

Fig. 2. Diagram outlining the range of spatial information available from different NMR experiments, ranging over ten orders of magnitude [10]. The methods using direct space encoding by gradients are shaded, but all other spatial information can be included for creation of image contrast. The model-dependent spatial information obtained by spin counting [49, 50], spin [51, 52] and molecular diffusion [8, 53–56] is available as an average over local geometries. The nuclear Overhauser effect (NOE) is yet another method of interrogating nuclear neighborhoods. Even the chemical shift contains spatial information from the local electron density

- EPR imaging [31–36]
- NQR imaging [37–40]
- Imaging hardware [8, 14–16, 31, 41, 42]
- Typical imaging artifacts [43], especially susceptibility [8]
- Digital image processing [44–48]

NMR provides various ways to investigate spatial structures. Some NMR techniques suitable for this purpose are identified in Fig. 2 [10] together with the appropriate spatial resolution scale. At one end the internuclear distance scale can be investigated by probing the dipole–dipole interaction with techniques like spin counting, NOE, and spin diffusion. In the intermediate range on the logarithmic scale, self diffusion can be probed. These techniques provide the spatial information only in an indirect way. The direct space encoding techniques can be used above one micrometer (shaded). These imaging experiments are the subject of this review. Nevertheless, the spatial information of the other methods can be incorporated as image contrast by use of suitable pulse sequences. Such images display a distribution of a microscopic parameter on a macroscopic scale.

1.1 Spatial Resolution

Every NMR spectroscopist is familiar with the fact that inhomogeneous magnetic fields cause line broadening and degrade the spectral resolution. Therefore, the homogeneity of the field is optimized by shimming the magnet [57]. Shimming is achieved by applying additional magnetic fields which are usually produced by currents in coils. A nearly reversed approach has to be used for imaging. A linear change of magnetic field with space is achieved by application of a constant magnetic field gradient. Because the Larmor frequency is proportional to the applied magnetic field, this provides a linear dependence of the precession frequency on the space coordinate. The NMR spectrum acquired under such conditions corresponds to a projection of the sample as a function of the space coordinate in the direction of the gradient. Both magnetic fields (static $\boldsymbol{B}_0$ and radio frequency excitation field $\boldsymbol{B}_1$) involved in NMR experiments can be made spatially dependent for this purpose.

The traditional realization of NMR imaging modifies the homogeneous, static magnetic field $\boldsymbol{B}_0$ by superposition of field gradients which can be switched on and off during a pulse sequence. An NMR signal detected in the presence of such a field gradient shows a spread in the NMR spectrum, depending on the strength and spatial variation of this additional field. The sample shape can only be discerned, however, if the natural distribution of Larmor frequencies is smaller than that caused by the field gradient. In general, this is no problem for liquids, because the line width is narrowed by fast molecular motion to a few Hertz or less. For solids, however, the molecular motion is slowed down, resulting in line widths three to five orders of magnitude broader than in liquids. Thus, the spatial resolution for solids is worse by the same factor, if observed by traditional liquid-state imaging experiments at the same gradient strength. This problem can be overcome by two fundamentally different methodical approaches. One is to strengthen the applied gradients and the other is narrowing of the NMR line widths.

The resolution limit is one reason for the fact that solid-state imaging is far less familiar than liquid-state imaging, even within the magnetic resonance community. Nevertheless, the first publications of NMR imaging in 1973 were addressed to both, liquid [58–60] and solid materials applications [61, 62]. Other reasons may be the demanding experimental procedures and the different importance of medical and materials research to the needs of human society. The noninvasive character of the method is less important for imaging of non-living objects, where slice selection can be performed by cutting the sample, unless the material heterogeneities to be investigated depend on such properties as temperature or applied strain. In fact, many solid samples need to be cut for imaging, because the *r*adio *f*requency (rf) power needed for excitation scales with the fifth power of the dimension of concluding samples. For solids, pulse durations as short as 1 μs are needed, which can be achieved with current technology at power levels of the order of 1 kW. This restricts the sample size to diameters of less than 10 mm unless surface techniques are used.

NMR microscopy is devoted to high spatial resolution, and the term is applied whenever structures smaller than or equivalent to the optical resolution of the human eye (some ten micrometers) can be resolved by NMR imaging [8, 13, 63, 64]. Due to the general insensitivity of NMR, however, the spatial resolution of NMR microscopy is not high enough to compare favorably with other imaging techniques like light, electron, or X-ray microscopy [14, 27, 65], or even scanning tunnel and atomic force microscopy. On the other hand, a combination of atomic force microscopy and magnetic resonance has recently been invented by Sidles et al. [66–68]. Nevertheless, NMR imaging has considerable advantage to offer over other methods [69]. First, applications of the technique are not restricted to surfaces, because NMR imaging is a three-dimensional method. A second and more important advantage is the image contrast, which is not only different from that of other techniques, but also adjustable for many purposes and variable within wide ranges. The relaxation times of NMR signals [70] are only one way to discriminate different materials by their differences in molecular mobility [71–75]. Others are diffusion and velocity coefficients of liquids in solids, resonance frequencies, spectral signal intensities, and line shapes. Variations in molecular dynamics, molecular order, and chemical composition can be detected in this way with spatial resolution of a few micrometers, which is sufficient for most technical purposes.

1.2 Solid-State Imaging

Despite the resolution and sensitivity problems in imaging of solids, this class of materials has some important advantages over biological samples. First, the state of the sample is usually time independent, allowing over-night experiments. Molecular diffusion, which often limits the spatial resolution in liquids [8, 13], is insignificant in most solids. Changes in temperature, pressure, strain, as well as mechanical spinning can be applied to the sample. Furthermore, the investigated material can be modified by tracer isotopes to obtain additional information, limited only by chemical and financial restrictions.

Taken together, these considerations show that it is rewarding to apply NMR imaging techniques to map NMR nuclei in rigid solids. Sufficient spatial resolution may be obtained for many questions of interest, provided that suitable methodical and instrumental modifications are followed. Many of the solid-state imaging techniques have been demonstrated on phantoms. Only a few applications are reported from industry and materials research, due to the fact that the requirements for probes, gradient drivers, and the rf pulse performance are often quite demanding for spatial resolution better than one millimeter.

On the other hand, many investigations of liquids in solids have been performed with standard equipment. Examples are the flow of liquids in solids [76–81] and chromatography columns [82, 83], diffusion coefficients of solvents in polymer matrices [84–96] and swollen polymers [91, 92, 94, 97–101], and gas distribution in polymers [102, 103] and in graphite [104]. Samples involving liquids in solids offer the possibility of studying both the liquid in the solid as well as the structure of the solid matrix as it has been modified by the liquid. This idea has been used to investigate the liquid uptake by polymers and polymer composites [94, 102, 105–113] and plants [114–118], as well as the distribution of oil and water in rocks and porous media [119–130], and of liquids (binder) in ceramics [94, 131–135]. Investigations of the time and space dependence of curing processes in polymerization [136–140] and adhesive bond formation [141–144] is another promising field of research.

These and other experiments demonstrate the enormous profit that can be gained from the use of NMR imaging for the characterization of materials. A rapid growth in applications of the imaging technique can be anticipated as long as methodical and instrumental developments continue to expand as rapidly as they do at present.

1.3 About this Review

In the following sections, the principles of liquid-state imaging are reviewed. Those methods are then shown to be applicable to imaging of solid materials which per se have narrow lines. Thus, in this case, few or no changes have to be made to the standard imaging techniques. Methods based on the use of strong gradients for improved spatial resolution in rigid solids with wide NMR lines are reviewed next, followed by techniques which exploit the complementary approach by use of line-narrowing. Finally, selection methods are discussed, which are helpful in tailoring contrast and information contents to particular questions of interest. They involve the selection of spectroscopic information on one side and the selection of the sensitive volume by selective excitation and surface coils on the other side.

The presentation primarily focuses on different methods, and applications are shown for purpose of illustrating features of the method. Unless another nucleus is explicitly mentioned, the discussion refers to ^{1}H, the most sensitive stable and the most abundant NMR nucleus in organic matter. Various references are cited, particularly in Sect. 6, which deal with imaging techniques applied to liquids. However, many of these concepts can be applied also to solid materials in combination with suitable line-narrowing techniques.

2 Imaging Principles

The principles of NMR imaging are outlined. It is shown, how spatial resolution can be obtained. Two- and three-dimensional imaging schemes are reviewed and compared to diffraction and scattering techniques. The limits imposed upon spatial resolution by the sensitivity of the technique are discussed, and the important issue of contrast parameters and their applications is addressed.

2.1 Frequency Encoding

2.1.1 The NMR Pulse Response in a Magnetic-Field Gradient

Neglecting all other relaxation but the transverse relaxation time (T_2), the NMR time-domain response $s(t)$ to a single pulse is proportional to

$$s(t) \propto \boldsymbol{M}_0 \exp[-(1/T_2 - \mathrm{i}\omega_0)t]. \tag{1}$$

Equation (1) applies in a homogeneous magnetic field, where $\boldsymbol{M}_0$ is the thermodynamic equilibrium magnetization of the entire sample and ω_0 the Larmor frequency. Thus, $\boldsymbol{M}_0$ can be written as the volume integral over the space-dependent spin density $p(\boldsymbol{r})$,

$$\boldsymbol{M}_0 = \int p(\boldsymbol{r})\mathrm{d}\boldsymbol{r}, \tag{2}$$

where $\boldsymbol{r}$ is the space vector with components x, y, and z.

A distribution of the spin density can be resolved in space, if the precession frequency is made dependent on space by introduction of an additional field G to the magnetic field. G is a second rank tensor with nine components which are defined as the spatial derivatives of the magnetic field,

$$\mathsf{G} = \begin{bmatrix} \partial B_x/\partial x & \partial B_y/\partial x & \partial B_z/\partial x \\ \partial B_x/\partial y & \partial B_y/\partial y & \partial B_z/\partial y \\ \partial B_x/\partial z & \partial B_y/\partial z & \partial B_z/\partial z \end{bmatrix}. \tag{3}$$

If the derivatives are independent of space (constant components), the magnetic field varies linearly with space. The components of the gradient tensor are related by Maxwell's field equations [145].

The linearly space-dependent part of the magnetic field is called the field gradient. It is generated by an extra set of coils and is added to the homogeneous magnetic fields given by the static field $\boldsymbol{B}_0$ and the time-dependent rf field $\boldsymbol{B}_1$ perpendicular to $\boldsymbol{B}_0$,

$$\boldsymbol{B}(\boldsymbol{r}) = \boldsymbol{B}_0 + \boldsymbol{B}_1 + \mathsf{G}\cdot\boldsymbol{r}. \tag{4}$$

The gradient components are often referred to as the gradients per se. They can be applied either quasi-statically to modify $\boldsymbol{B}_0$ or oscillating with radio frequency to modify $\boldsymbol{B}_1$. The first case is standard in most imaging experiments. The gradients are said to be applied in the laboratory frame. Here the maximum value of the gradient field $\mathsf{G}\cdot\boldsymbol{r}$ typically is less than 1% of the strength of the homogeneous magnetic field $\boldsymbol{B}_0$. However, in some cases the gradients are oscillating at the NMR frequency to provide a component, which rotates about the static field $\boldsymbol{B}_0$. Then the gradients are said to be applied in the rotating frame. Maximum values of the gradient field in this case are of the order of the strength of $\boldsymbol{B}_1$. This situation is excluded for most part in the following.

Neglecting $\boldsymbol{B}_1$ in Eq. (4), the NMR frequency $\omega_0 = -\gamma \boldsymbol{B}_0$ becomes dependent on space,

$$\omega_0(\boldsymbol{r}) = -\gamma \boldsymbol{B}(\boldsymbol{r}) = -\gamma(\boldsymbol{B}_0 + \mathsf{G}\cdot\boldsymbol{r}). \tag{5}$$

The maximum strength of the laboratory gradient field is small enough to be treated as a perturbation of $\boldsymbol{B}_0$, and the gradients are applied in a coordinate frame whose symmetry axis is parallel to the z-axis defined by the direction of $\boldsymbol{B}_0$. In this case, only three components of the gradient tensor determine the NMR frequency to first order. The other components, though non-zero by Maxwell's equations, can be neglected. The three relevant terms of the gradient tensor can be concatenated to form the gradient vector $\boldsymbol{G}$,

$$[\partial B_z/\partial x, \partial B_z/\partial y, \partial B_z/\partial z] \equiv [G_x, G_y, G_z] \equiv \boldsymbol{G}. \tag{6}$$

This practical simplification is used in most of the magnetic resonance imaging literature. For parallel z-axis, Eq. (5) is well approximated by

$$\omega_0(\boldsymbol{r}) \cong -\gamma(\boldsymbol{B}_0 + G_x x + G_y y + G_z z) = -\gamma(\boldsymbol{B}_0 + \boldsymbol{G}\cdot\boldsymbol{r}). \tag{7}$$

However, if the *g*radient *c*oordinate *f*rame (GCF) is tilted in the *l*aboratory *c*oordinate *f*rame (LCF), then the other gradient elements in Eq. (3) may no longer be neglected. A situation where this is the case is *m*agic-*a*ngle sample *s*pinning (MAS) NMR imaging. Here the z-axis of the LCF is tilted by the magic angle of 54.7° with respect to the z-axis of the GCF [145].

Combining Eqs. (1), (2) and (7), the NMR time signal in the presence of a time-independent field gradient is given as

$$s(t, \boldsymbol{r}) \propto \int \rho(\boldsymbol{r}) \exp[-(1/T_2 - \mathrm{i}\omega_0 + \mathrm{i}\gamma \boldsymbol{G}\cdot\boldsymbol{r})t]\,\mathrm{d}\boldsymbol{r}. \tag{8}$$

In practice, spectral dispersion has to be included, for example by the chemical shift σ. Thus, Eq. (7) is modified according to

$$\omega_0(\boldsymbol{r}) \cong -\gamma[\boldsymbol{B}_0(1-\sigma) + \boldsymbol{G}\cdot\boldsymbol{r}] \equiv \omega_S - \gamma \boldsymbol{G}\cdot\boldsymbol{r}, \tag{9}$$

so that a second integration over ω_S has to be included in Eq. (8) to account for the distribution of magnetization components with different resonance frequencies ω_S in the NMR spectrum. Furthermore, each other NMR parameter has to be treated as space dependent. In our simplified description we only have to add a third integration over T_2

$$s(t, \boldsymbol{r}) \propto \int\int\int \rho(\boldsymbol{r}, \omega_S, T_2) \exp[-(1/T_2 - \mathrm{i}\omega_S + \mathrm{i}\gamma \boldsymbol{G}\cdot\boldsymbol{r})t]\,\mathrm{d}T_2\,\mathrm{d}\omega_S\,\mathrm{d}\boldsymbol{r}. \tag{10}$$

2.1.2 Spatial Resolution

If the NMR spectrum $B(\omega)$ is the same for each volume element of the sample, that is, if $\rho(\boldsymbol{r}, \omega_S, T_2) = \rho(\boldsymbol{r})$, Eq. (10) simplifies,

$$s(t, \boldsymbol{r}) \propto \int \rho(\boldsymbol{r}) \exp[-(1/T_2 - \mathrm{i}\omega_S + \mathrm{i}\gamma \boldsymbol{G}\cdot\boldsymbol{r})t]\,\mathrm{d}\boldsymbol{r}. \tag{11}$$

The spatial distribution of these parameters, that is the image, can be obtained by Fourier transformation of the *f*ree *i*nduction *d*ecay (FID) (11),

$$S(\omega, \boldsymbol{r}) \propto \int \rho(\boldsymbol{r})[1/T_2 - \mathrm{i}(\omega_S - \omega - \gamma \boldsymbol{G}\cdot\boldsymbol{r})]^{-1} \mathrm{d}\boldsymbol{r}, = \int \rho(\boldsymbol{r}) B(\omega - \gamma \boldsymbol{G}\cdot\boldsymbol{r}) \mathrm{d}\boldsymbol{r}. \tag{12}$$

Thus, the NMR spectrum $S(\omega, r)$ in the presence of a field gradient is a projection of a spatial distribution function $\rho(r)$ convoluted with the NMR spectrum $B(\omega)$ in the absence of a field gradient. In general, $B(\omega)$ is called the point-spread function [17, 146, 147]. Due to that convolution it is clear that the spatial resolution $1/\Delta r$ achievable by direct frequency encoding is limited by the width of the NMR spectrum, or, in case of one line only, by the line width $\Delta\omega = 2/T_2$. This is illustrated schematically in Fig. 3.

The minimum distance Δr, which can be resolved, is defined by the line width $\Delta\omega$ (or T_2) and the gradient strength G

$$\Delta r = \frac{\Delta\omega}{\gamma G} = \frac{2}{T_2 \gamma G} \tag{13}$$

It is obvious that if the line width increases, the resolution decreases accordingly, unless the gradient strength is increased. This explains the difficulties of NMR imaging of solids, where line widths are broader than in liquids often by a factor of 10^5 and more, because the dipole–dipole and quadrupole couplings are no longer averaged out as a consequence of reduced mobility.

The simplest solid-state imaging experiments require detection of an FID signal in the presence of constant gradients. Because the already fast signal decay is accelerated by the frequency dispersion from the gradients, echoes are used to overcome receiver-dead time and gradient-switching times. Assuming the point-spread function (spectrum) $B(\omega)$ as a delta-function, the Fourier

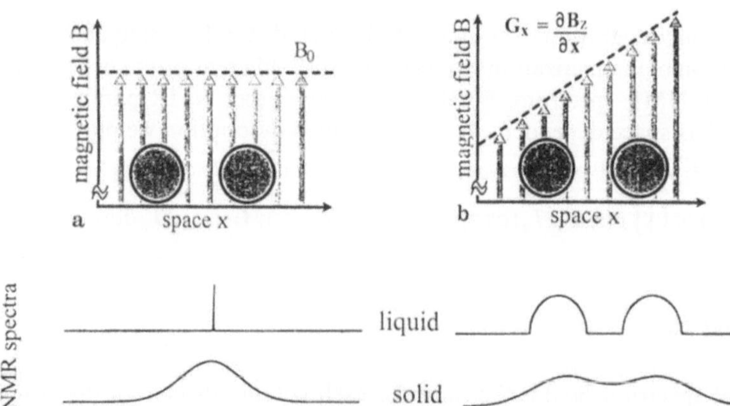

Fig. 3a, b. Effect of field gradients on the NMR signal for solids and liquids: **a** Sample (symbolized by two dark circles) in a homogeneous field, **b** Applied gradient. *Top*: Schematic drawing of the field dependence across the sample. *Bottom*: NMR spectra for liquid and solid sample material

transform of the echo decay provides a projection of the object as a function of the space coordinate in direction of the gradient. Such a projection is sometimes called a one-dimensional image. From a set of projections acquired under incremented angles of the magnetic field gradient an image can be constructed [59, 60, 148] by the back projection method (cf. Sect. 2.3).

2.2 Phase Encoding

Gradient amplitude and time dependence are adjustable during imaging experiments under control of the spectrometer. Therefore, space encoding can also be achieved in an evolution period prior to data acquisition. By analogy to two-dimensional NMR spectroscopy, the evolution of the spins under the influence of the gradient is reflected in a phase shift $\Delta\varphi$ of the signal acquired during the detection period. In this case, Eq. (10) can be rewritten as

$$s(t_1, t_2, \boldsymbol{r}) \propto \exp[\Delta\varphi(t_1)] \iiint \rho(\boldsymbol{r}) \exp[-(1/T_2 - \mathrm{i}\omega_\mathrm{S} + \mathrm{i}\gamma\boldsymbol{G}\cdot\boldsymbol{r})t_2] \mathrm{d}T_2 \mathrm{d}\omega_\mathrm{S} \mathrm{d}\boldsymbol{r}, \tag{14}$$

where

$$\Delta\varphi(t) = \gamma\boldsymbol{G}\cdot\boldsymbol{r}\Delta t, \quad \text{or} \quad \Delta\varphi(t) = \gamma\Delta\boldsymbol{G}\cdot\boldsymbol{r}t. \tag{15a/b}$$

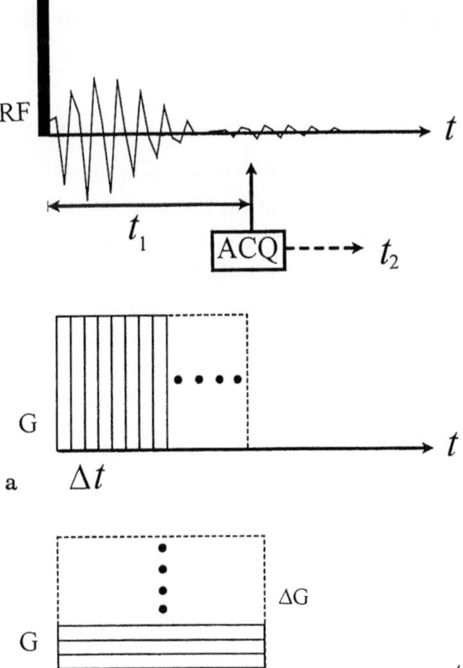

Fig. 4a,b. Principle of phase encoding: *Top*: The NMR signal is sampled after a fixed evolution time t_1. The acquisition (ACQ) can be realized point-by-point (*solid arrow*) or spread over the next time dimension (*dotted arrow*). Spatial encoding of the signal can be generated by **a** increasing the duration of the gradient with a fixed strength by Δt, or **b** by increasing the gradient strength by ΔG during t_1

Phase modulation can be realized in two ways, a) by incrementing the time in steps of Δt until the end of a fixed evolution time t_1 with a fixed gradient strength G, and b) by incrementing the gradient strength in steps ΔG at constant time t_1. Both experiments are schematically shown in Fig. 4. Compared with frequency space encoding, these experiments have the disadvantage that for the measurement of a spatial projection with n points n experiments are necessary, because a single point or FID has to be sampled after each time increment or gradient step. Then, a simple Fourier transformation along the phase encoded dimension uncovers the image. This form of spatial encoding is very time consuming, even more so because the signal decays during t_1.

In practice, scheme a) is used in combination with the continuous time axis of direct detection ($t_1 = n\Delta t$ in Eq. (15a)) and, in this case, turns into frequency encoding [149]. Scheme b) provides the advantage of constant phase evolution resulting from spin interactions other than the applied gradient by keeping t_1 constant. Thus, the phase change depends entirely upon the applied gradient strength. This technique is called *spin-warp* imaging [150] and is routinely used for encoding of indirectly detected space dimensions in Fourier imaging.

For phase encoding by the spin-warp technique, the minimum resolvable distance Δr is no longer determined by the line width, but by the maximum gradient amplitude G_{max},

$$\Delta r = \frac{2\pi}{\gamma G_{max} t}. \tag{16}$$

Because the spatial resolution is independent of the line width, spin-warp phase encoding is important for NMR imaging of solids (cf. Sect. 4.2).

2.3 k-Space

The concept of phase encoding of the space information directly leads to the description of NMR imaging methods in Fourier space, or $\boldsymbol{k}$-space [5, 6, 8, 151–153]. It should be emphasized the $\boldsymbol{k}$-space is nothing but a different interpretation of the time signal. Usually time is recognized as the Fourier conjugate to frequency (or inverse time). However, we are more interested in space than frequency (cf. Sect. 2.1), time is to be replaced by the Fourier conjugate of space. Consequently, the detected time-dependent signal is described as a trajectory through reciprocal space ($\boldsymbol{k}$-space). When neglecting relaxation and frequency dispersion, that is, assuming $B(\omega)$ as a delta-function in Eq. (12), the FID in the presence of a time-dependent magnetic field gradient can be written as,

$$s(\boldsymbol{k}) = \int \rho(\boldsymbol{r}) \exp[\mathrm{i}\boldsymbol{r}\cdot\boldsymbol{k}(t)]\,\mathrm{d}\boldsymbol{r}, \tag{17}$$

where

$$\boldsymbol{k}(t) = -\gamma \int_0^t \boldsymbol{G}(t')\,\mathrm{d}t'. \tag{18}$$

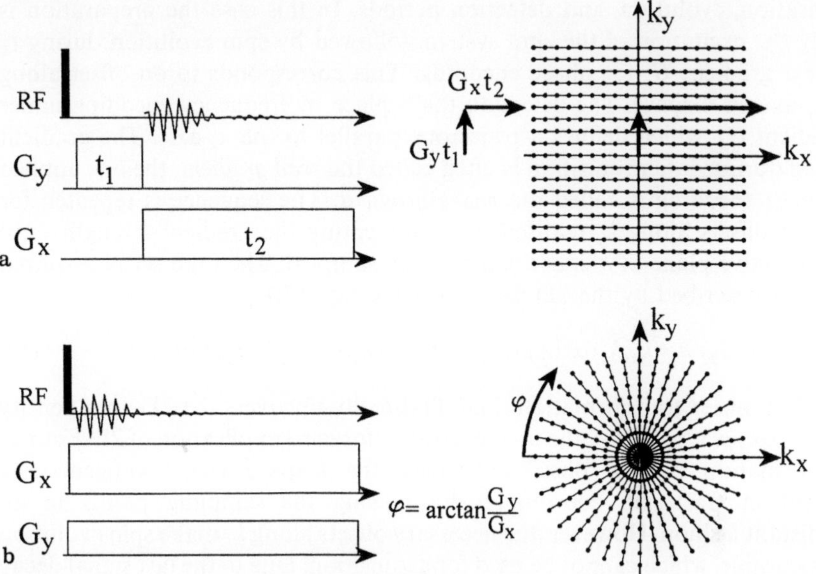

Fig. 5a, b. Interpretation of different two-dimensional imaging techniques by $\boldsymbol{k}$-space trajectories: **a** 2DFI method **b** BP method. The pulse sequences are shown on the *left* and the corresponding trajectories in $\boldsymbol{k}$-space on the *right*. The *solid lines* mark sampling of an FID and the *gray lines* sampling of echoes. See text for details

The spin density $\rho(\boldsymbol{r})$ is then simply obtained by Fourier transformation of $s(\boldsymbol{k})$. As the phase of the precessing transverse magnetization changes with time t under the influence of a gradient $\boldsymbol{G}$, the magnitude of the $\boldsymbol{k}$ vector changes accordingly. For constant gradient direction, the direction of $\boldsymbol{k}$ is parallel to $\boldsymbol{G}$, and the sign of $\boldsymbol{k}$ depends on the time dependence of $\boldsymbol{G}$.

To sample the image information, the entire $\boldsymbol{k}$-space must be covered in the course of the experiment. Different 2D- and 3D-imaging experiments can readily be discriminated by the shape of the paths along which 2D and 3D $\boldsymbol{k}$-space is scanned. Often the 3D spin density is reduced to two dimensions by slice selection prior to space encoding (cf. Sect. 6.2) [154–160] or a projection is acquired over the third space dimension. Therefore, the discussion is restricted to two-dimensional imaging techniques.

2.3.1 Fourier Imaging

A typical *two-d*imensional *F*ourier *i*maging (2DFI) scheme is shown in Fig. 5a. It is a combination of spin-warp phase encoding (here in the y direction) and frequency encoding (here in x direction), and only one of many possible 2DFI approaches. Like every 2D-NMR sequence [161] it can be subdivided into preparation, evolution, and detection periods. In this case the preparation is simply the excitation of the spin system, followed by spin evolution during t_1 under a gradient G_y for phase encoding. This corresponds to an offset along the k_y-axis in $\boldsymbol{k}$-space. The detection takes place by frequency encoding under a gradient G_x, which defines a trajectory parallel to the k_x-axis. The gradient applied during data acquisition is often called the *read gradient*, the one applied for phase encoding is called the *phase gradient*. The sequence is repeated for different offsets along k_y realized by incrementing the gradient strength until the respective plane of $\boldsymbol{k}$-space is uniformly sampled. Then the set of acquired signals is described by the 2D data array (cf. Eq. (17))

$$s(t_1, t_2) \equiv s(k_x, k_y) \propto \iint \rho(x, y) \exp\{\mathrm{i}[xk_x(t_2) + yk_y(t_1)]\} \mathrm{d}x\mathrm{d}y, \tag{19}$$

and 2D-Fourier transformation (2DFT) directly uncovers the 2D spin density distribution $\rho(x, y)$ that is, the image. Straightforward application of 2D Fourier transformation is now possible because the $\boldsymbol{k}$-space was homogeneously scanned in Cartesian coordinates by placing the sampling points in an equidistant fashion. However, the necessary offsets along k_y make spin evolution indispensable, which cannot be used for acquisition. Due to the fast signal decay for solids, this is a severe disadvantage for the applicability of the method, because only little signal intensity can be detected after the phase encoding time t_1. The resultant loss in sensitivity becomes an additional handicap, because gradient switching times often have to be accommodated. However, the problem can be alleviated by refocusing the signal with an echo sequence. This additionally allows the sampling of negative $\boldsymbol{k}$ coordinates (gray lines in Fig. 5) in the frequency encoded dimension due to the time inversion by the echo (cf. Eq. (18)) [162, 163].

If the imaging process is to be extended to three dimensions, the technique can be modified by varying two perpendicular gradients independently during the evolution time t_1, and the gradient direction angle needs to be stepped in three dimensions. However, the acquisition of such a 3D matrix is extremely time consuming. Because in many cases only a few planes are of interest in the third space dimension, experimental time can be saved by tomographic excitation of the desired slices (cf. Sect. 6.2) or by physically slicing the sample.

2.3.2 Back Projection Imaging

Another scheme of $\boldsymbol{k}$-space scanning is used for image reconstruction from projections (*b*ack *p*rojection, BP) [148, 164]. The idea is completely analogous to image reconstruction in computed X-ray tomography [65] and was introduced very early to NMR imaging [58–60]. A set of 1D projections is measured via frequency encoding for different "views" of the sample by incrementing the angle of the gradient relative to some fixed direction in the sample. This can be realized either by sinusoidal modulation of the strength of two perpendicular gradients or by sample rotation under a gradient with fixed direction. Therefore, the angle variation should cover a range from 0° to 180°.

The principle imaging scheme and the corresponding $\boldsymbol{k}$-space trajectories are shown in Fig. 5b. Because $\boldsymbol{k}$-space is scanned in cylindrical coordinates, a 2D Fourier transformation cannot immediately be applied to recover the image. The data points have to be previously interpolated onto a Cartesian $\boldsymbol{k}$-space coordinate frame. From Fig. 5b it is obvious, that $\boldsymbol{k}$-space is sampled more densely close to its origin. A denser accumulation of information at small $\boldsymbol{k}$ values corresponds to a better reconstruction of coarse image features than of fine details. To avoid overemphasing the correlated image structures, the projections have to be filtered during the course of reconstruction. For this reason, the proper algorithm for reconstruction is referred to as *filtered backy projection* (BP) [148]. This filtering process can in fact be interpreted in terms of an interpolation of the experimental data onto a different, more rectangular grid. By rule of thumb, the necessary number of projections can be estimated at the square root of *vo*lume *cel*ls (voxel) in the 2D image sufficient to uncover the desired resolution per voxel. Compared with the 2DFI method, the BP method offers a sensitivity advantage, because neither phase encoding nor gradient switching is required in the presence of transverse magnetization. Therefore, the BP method is often applied for imaging of solids. To increase *s*ignal-to-*n*oise (S/N) echoes are preferred here as well, which cut down receiver dead time problems.

2.3.3 Other Imaging Techniques

Extremely fast sequences for imaging have been developed for liquids, by which a 3D image can be measured in less than 100 milliseconds [5, 7, 165–168]. These techniques make use of excitation with single, small flip angle pulses in combination with gradient echoes. A gradient echo is generated by returning to the origin in $\boldsymbol{k}$-space. The simplest way to achieve this is by sign inversion of the gradient amplitude after evolution for a time τ. A gradient echo is then observed at time 2τ. This concept can be used also for solid-state imaging where gradient echoes are generated by rapidly oscillating gradients [169–172]. In combination with line-narrowing techniques (cf. Sect. 5), this is a very promising approach for rapid NMR imaging of solids. Far more schemes with different trajectories in $\boldsymbol{k}$-space have been investigated. [5, 6, 173–178], but so far none has been applied to imaging of solids.

If frequency encoding is omitted in the 2DFI technique, and only phase encoding is used, the directly detected NMR signal carries no space but only spectroscopic information. Methods of this kind are referred to as *spectroscopic imaging* [179]. Spectroscopic resolution can also be obtained for the back-projection technique if the gradient strength is varied in addition to the gradient orientation [180–183].

2.3.4 NMR Imaging in Terms of Scattering

In contrast to scattering and diffraction experiments, the spatial resolution does not depend on the wavelength of the irradiation, which is of the order of one meter in NMR. Rather it is given by a fictitious wavelength in $\boldsymbol{k}$-space [4–6, 61, 62],

$$\lambda = \frac{2\pi}{|\boldsymbol{k}|}. \tag{20}$$

Scattering and NMR imaging techniques, however, are similar in terms of their common description in $\boldsymbol{k}$-space [61, 62]. This is illustrated in Fig. 6 where a

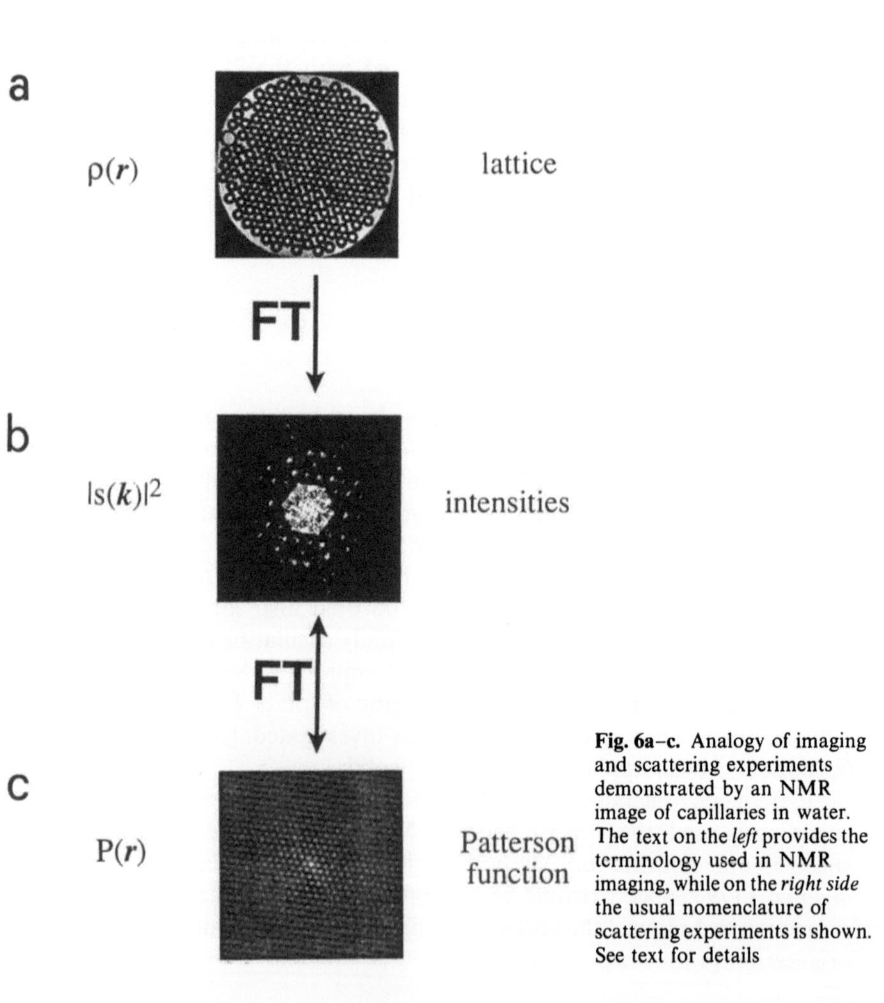

Fig. 6a–c. Analogy of imaging and scattering experiments demonstrated by an NMR image of capillaries in water. The text on the *left* provides the terminology used in NMR imaging, while on the *right side* the usual nomenclature of scattering experiments is shown. See text for details

lattice for diffraction is given by an NMR image of a relatively regular arrangement of capillaries in water. Unlike scattering experiments, the NMR signal can be detected in a phase sensitive fashion. Therefore, the Fourier transform of the signal acquired in $\boldsymbol{k}$-space reconstructs the lattice in real space. By X-ray and neutron scattering, only signal intensities can be detected, not the signal phase. Thus, the NMR signal corresponding to a diffraction pattern is the magnitude square of the $\boldsymbol{k}$-space signal, as illustrated in Fig. 6b. The Fourier transform of the diffraction pattern is the *Patterson function* (Fig. 6c), which is equivalent to the auto-correlation function of the image (Fig. 6a). In contrast to a real image, the auto-correlation function yields information on displacements $\Delta \boldsymbol{r}$ only. The Fourier-conjugate of the displacement space is often called $\boldsymbol{q}$-space [8]. It can be probed by NMR with spatial resolution in the range of 1 nm to 10 μm by particle diffusion experiments in magnetic field gradients (cf. Fig. 2) [53, 54, 184]. The relationship between the image and the Patterson function is of use not only for illustration of the close relationship between imaging and scattering but it can also be exploited, for instance, for computer assisted analysis of image structures [185].

2.4 Spatial Resolution and Signal-to-Noise Limits

It may be concluded that the spatial resolution for NMR imaging should be superior to those of other imaging techniques; this is because for phase encoded signals, it depends on experimental parameters only and there seems to be no other limit to adjusting their range of values as necessary for the resolution desired. This indeed is true, but only as long as sensitivity and instrumental imperfections are neglected.

2.4.1 The Phase-Encoded Dimension

Compared to other spectroscopic methods, NMR is rather insensitive, because Boltzmann population differences are small due to the weak Zeeman splittings of the energy levels at high temperature. In an optimistic estimate, the experiment is sensitive to some 10^{15} ^{1}H spins in water at room temperature, which corresponds to a detectable volume of approximately 15 μm^3. Smaller volumes can be considered as part of this minimum detectable voxel, and as such are measurable only by their influence on the average signal of the minimum voxel. In theory, the signal accumulation could be prolonged ad infinitum with coherent signal addition for signal-to-noise (S/N) improvement. In practice instrumental drifts of the signal phase, for instance by changes of $\boldsymbol{B}_0$, would destroy the spatial resolution [13]. Therefore, noise has less of an influence on the limits of spatial resolution, but it defines the total time of the experiment necessary to acquire an image with suitable S/N (see below). Other limitations arise from diffusion of molecules from one voxel into another during the delay between excitation and detection. This is a serious problem for high resolution NMR imaging of liquids in unconfined environments [8, 186, 187] but not for the solid state.

2.4.2 *The Frequency-Encoded Dimension*

The spatial resolution limit for the directly detected, frequency-encoded dimension is different from the before-mentioned case. Even if the spectral line width is suitably narrow, chemical shift dispersion and magnetic field inhomogeneities from susceptibility differences inside the heterogeneous sample cause blurring of the image structures [8, 188]. Because an image recorded in this fashion can be described by a convolution of the spin density with the NMR spectrum (cf. Eq. (12)), resonance lines at different chemical shifts give rise to a superposition of shifted spin density images. If the chemical shifts are the same over the whole sample, this blurring can be removed by deconvolution. If not, phase encoding in all directions or special pulse sequences (e.g. the CPMG technique, cf. Sect. 5.1) can be applied to remove such an artifact.

The effect of susceptibility artifacts is similar. If the susceptibility varies inside the sample, the resonance frequencies are shifted by additional fields. Thus, the corresponding image structures appear at locations different from where they actually are. This causes local superpositions of signals in the image, which appear like shadows around those positions (cf. Fig. 8). Susceptibility artifacts can be useful to detect small inhomogeneities, which are amplified by this effect and would be difficult to observe otherwise. They can be distinguished from other distortions by reorientation of the sample geometry relative to the frequency encoding gradient. However, strong changes of susceptibility can be understood as an additional gradient causing a broadening of the line width and reducing the spatial resolution. They can be removed again by suitable pulse sequences or total phase encoding. Because both interactions, chemical shift and susceptibility, scale with the $\boldsymbol{B}_0$ field, they can be reduced for frequency encoding by low $\boldsymbol{B}_0$ field strengths and gradients strong enough to gather all distortions within a single voxel.

2.4.3 *Spatial Resolution and Sensitivity*

All of the sensitivity problems discussed previously represent general limits for NMR imaging. However, the fast decay of the signal due to short T_2 for solids limits the time during which the spin system can be observed and manipulated by gradients. This restricts the sensitivity for phase encoding with long evolution times. Furthermore, the corresponding wide lines are problematic for frequency encoded imaging with high spatial resolution. This interwining relationship between sensitivity and spatial resolution will be discussed in the following with reference to the influence of gradient strength and line width on the S/N or the total acquisition time t_{tot} [10].

For a given experimental sensitivity, the bandwidth of a voxel is chosen large enough so all distortions discussed above are contained inside. For an additional increase of the gradient strength, the total spectral width of the experiment has to be increased by the same factor as the gradient strength. The bandwidth and, thus the noise per voxel, is unchanged by this procedure, but the signal is spread amongst a larger number of voxels. Thus, the noise per voxel remains constant while the signal is reduced in proportion to the gradient strength. The signal-to-noise ratio is therefore inversely proportional to the gradient strength. To obtain the same S/N as before, the number of signal acquisitions has to be increased. Because the S/N improves in proportion to the square root of the number of signals averaged, the total recording time for the new image is proportional to the square of the gradient strength,

$$t_{tot} \propto G^2. \tag{21}$$

In other words, to acquire an image with twice the spatial resolution and the same S/N as the previous one, four times more sampling time must be spent. Similar considerations apply when the effect of broader line widths on the experimental time is discussed. If the same spatial resolution and S/N for a sample with double the intrinsic line width $\Delta\omega$ is desired, the voxel bandwidth has to be doubled to contain all distortions inside. Because the noise per voxel is proportional to the square root of its bandwidth, the total acquisition time is directly proportional to the line width,

$$t_{tot} \propto \Delta\omega. \tag{22}$$

This imposes a stringent limit on NMR imaging of solids, because their line widths are several orders of magnitude broader than for liquids.

Equations (21) and (22) have to be taken into account for each additional dimension of an image. A doubling of the gradient strength for all three dimensions must be paid for with an experiment lasting 64 times as long. Therefore, the optimal gradient strength for an imaging experiment is the smallest providing the necessary resolution. Apart from these general limits, sensitivity and resolution are different for the individual imaging experiments. More detailed reviews about this topic can be found elsewhere [4, 8, 63, 64, 186, 187, 189].

2.5 Contrast Parameters

As mentioned in the introduction, the contrast parameters are the primary asset for justifying use of NMR in materials imaging. Other imaging methods are capable of providing better spatial resolution, but their contrast is determined by just one or a few parameters. The contrast ΔI is defined as the relative difference in image intensities $I(\boldsymbol{r})$ of neighboring voxels i and j,

$$\Delta I = \frac{I(\boldsymbol{r}_i) - I(\boldsymbol{r}_j)}{|I_{max}|}, \tag{23}$$

where I_{max} is the maximum of $I(\boldsymbol{r})$. In principle, all molecular and macroscopic properties measurable by NMR can be conceived to influence the image contrast. Some examples are listed in Table 1.

The fundamental contrast parameter is the spin density of a given NMR nucleus. In medical applications, the NMR image is usually determined by this quantity in combination with relaxation times. More recently, images showing signal intensities at particular resonance frequencies have been processed from a spectroscopic image, because they depict the spatial distribution of particular chemical species [190]. In solids, spectroscopic images of wideline spectra can be highly informative, because the line shapes carry information about molecular order and slow molecular dynamics (10^{-5} s to 10 s) [191, 192] (cf. Fig. 16). Relaxation times like T_1, T_2, $T_{1\rho}$, and T_D, which denotes relaxation in the local fields of dipole–dipole interaction, provide another access to molecular dynamics, encompassing a wide range of correlation times τ_c (10^{-2} s $> \tau_c > 10^{-12}$ s). Many of these NMR parameters can be correlated by simple recalibration with macroscopic material properties or external parameters like temperature, viscosity, particle velocity, reaction rates, stress, strain, etc. In such a way, a variety of chemical and physical properties of the material can be mapped.

Table 1. Contrast parameters

Spin density	numbers of particular NMR nuclei
Relaxation times	
spectral densities of molecular motions at once and twice the Larmor frequency	T_1
spectral densities of molecular motions at zero, once, and twice the Larmor frequency	T_2
spectral densities at the precession frequency in the rf field	$T_{1\rho}$
relaxation in local dipolar fields	T_D
Chemical shift, NMR line shape	
chemical species	isotropic chemical shift
susceptibility	additional frequency shift
slow molecular motion (time scale and geometry)	shape of the wide line spectrum
molecular order	shape of the wide line spectrum
Coupling strength	
homonuclear dipolar couplings	lineshape
heteronuclear dipolar couplings	CP efficiency
microstructure	spin diffusion
J coupling	multiple-quantum coherences
Flow and diffusion [8]	amplitude and phase shifts of echoes in field gradients
External parameters	
temperature [193–195]	from chemical shifts and relaxation times
reaction kinetics	from chemical shifts and relaxation times
viscosity [196]	from relaxation times
stress-strain [197–200]	from relaxation times
contrast agents	noble gases (e.g. ^{129}Xe), O_2, liquid ingress

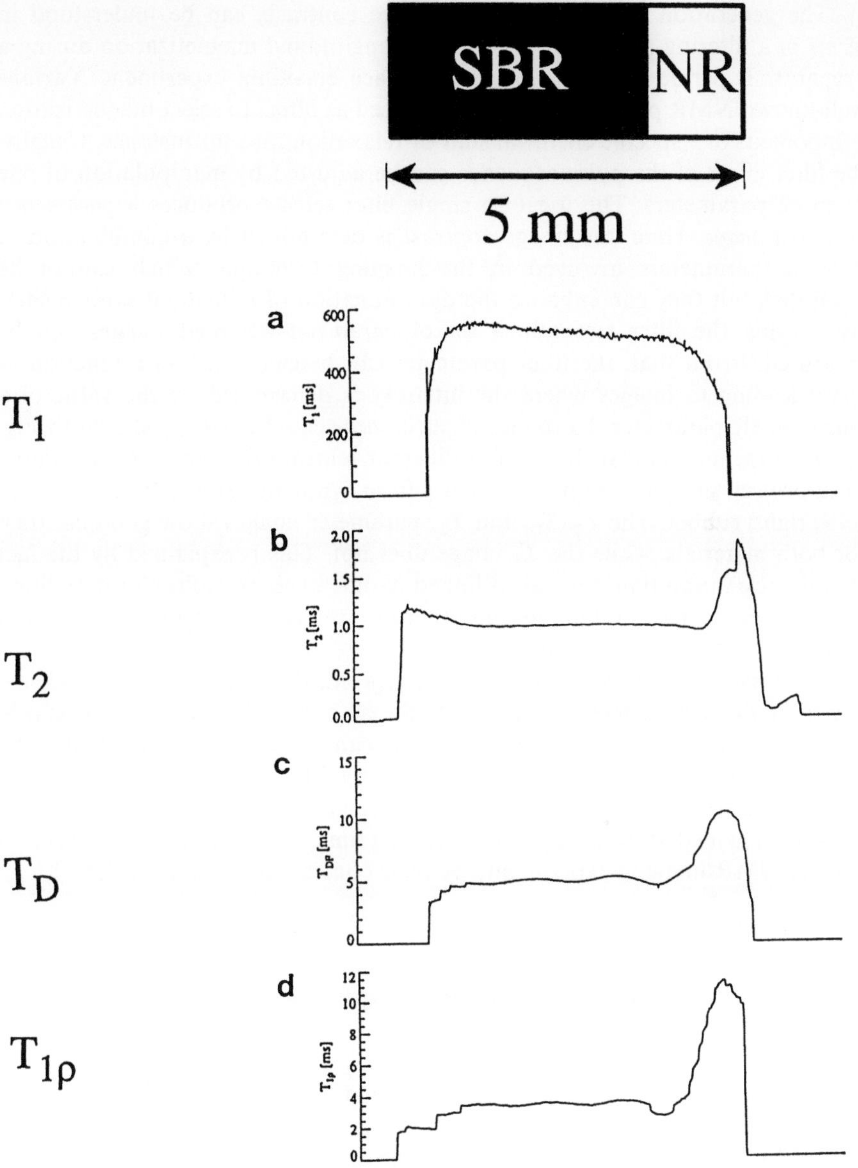

Fig. 7a–d. 1D parameter images of a tire sample showing two rubber layers of the tire: a layer of synthetic rubber (SBR, *left*) forming the tread and the first part of a layer of natural rubber (NR, *right*). The parameter images were calculated from sets of images measured with different adjustments of the T_1 filter (**a**), T_2 filter (**b**), dipolar filter (**c**), and $T_{1\rho}$ filter (**d**)

The generation of many types of image contrasts can be understood in terms of a filtering process applied to the longitudinal magnetization during a preparation period prior to the actual space encoding experiment. Various well-known NMR pulse sequences can be used as filters to select magnetization components of a specific chemical shift or relaxation rate, for instance. Usually, the filter effect of the pulse sequence can be adjusted by manipulation of one or more parameters. The use of a single filter setting produces a *parameter-weighted image*. Here, the image contrast is determined by a combination of different parameters involved in the imaging technique which cannot be separated, but they can enhance the discrimination of individual sample sites. By varying the filter strength, a set of parameter-weighted images can be measured. From that, the filter parameter can be calculated as a function of space leading to images where the intensity is determined by the value of a single NMR parameter. Examples of such *parameter images* are shown in Fig. 7, where the spatial distributions of different relaxation times were calculated for a projection of a section of a car tire from synthetic (SBR, left) and natural (NR, right) rubber. The T_2, T_D, and $T_{1\rho}$ parameter images show good contrast for both materials, while the T_1 image does not. This is explained by the fact that the relaxation time contrast is linked to differences in molecular mobilities, and the time scales of these motions enter into the relaxation times in different manners via the spectral densities.

Contrast parameters directly related to specific delays in an imaging scheme (e.g. T_1 to the recycle delay TR and T_2 to the echo time TE) have not necessarily be introducted by a preparation time but can be achieved even within the imaging sequence. In general, S/N is reduced by filtering, because only a subset of magnetization components gives rise to image formation. Thus, enhanced contrast is gained at the expense of measuring time. Therefore, it can be useful to classify NMR imaging experiments by their contrast-to-noise ratio [201].

3 Imaging of Solids with Standard Equipment

Standard equipment can be used for imaging of solid samples that have narrow intrinsic line widths to give satisfactory spatial resolution. For such materials, the NMR imaging methods discussed in the previous chapter can be employed without major modification. Suitable materials are plastic solids with high molecular mobility, e.g. rubbers or polymers at elevated temperatures, and highly oriented rigid solids like fibers and crystals. If isotopes are investigated al low concentration, the experimental procedures are also simplified, because then the homonuclear dipole–dipole couplings can often be neglected, and the heteronuclear couplings can be decoupled from abundant nuclei.

So far the simplicity of the experiment is a key to acceptance as a routine method, and applications of significance to industry rely on standard Fourier and back-projection techniques. The information available in such images can be extracted not only by visual inspection but also by suitable data processing. To illustrate this point, numerical techniques are reviewed that exploit a priori spectral or spatial knowledge about the sample and lead to improved signal-to-noise ratios and to better spatial resolutions.

3.1 Imaging of Elastomers

Elastomers are objects most suitable for NMR imaging. They are rich in protons, the most sensitive NMR nucleus, and their elastic properties require high molecular mobility in combination with form stability. In a simplistic fashion, they can be described as cross-linked liquids. Their macroscopic properties can be varied from those of viscous liquids to rigid solids by changing cross-link density and filler concentration. Their glass transition temperature T_g is well below room temperature, so their NMR spectra are characterized by relatively narrow line widths (100 Hz–3 kHz are typical) corresponding to large values of T_2.

Most NMR imaging investigations have focused on detection and characterization of the homogeneity of the polymer matrix, the cross-link density, and the distribution of filler particles for different sample histories and applications [26, 74, 75, 139, 140, 122, 202–211]. Because the magnetic susceptibility may change considerably at the interface between polymer matrix and filler particles, the image contrast can often be enhanced by the use of gradient echo techniques.

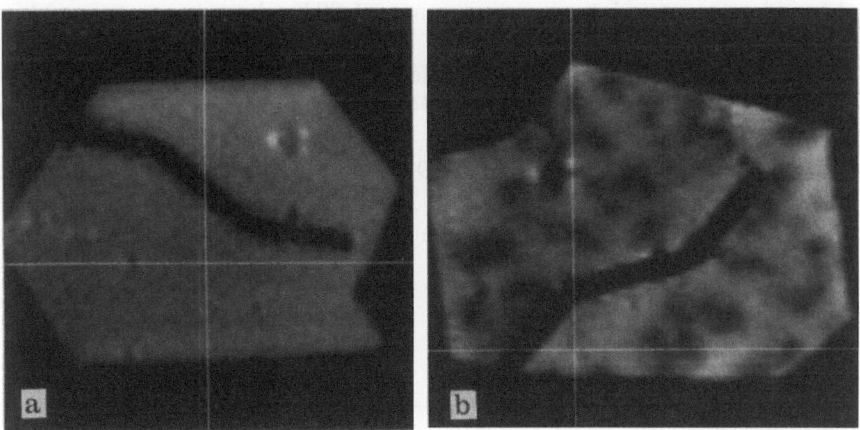

Fig. 8a, b. a Typical NMR spin-echo image of a tire tread with good dispersion of carbon black and a spatial resolution of $150 \times 150 \times 350\,\mu m^3$. Total measuring time 2 h. **b** NMR spin-echo image of a tire tread with poor carbon black dispersion. The "arrow-head" figures at the *top right* in **a** and *left, close to center* in **b** indicate strong susceptibility changes due to conductive coke particles. $TE = 1.3$ ms, $TR = 1$ s for both images. (Reproduced with permission from Komoroski et al. [204–206])

In spin-echo images, the susceptibility changes cause typical artifacts (cf. Sect. 2.4). This is illustrated in Fig. 8 with images of tire treads for good (a) and poor (b) carbon black dispersion [204–206].

Because most rubbers contain unsaturated olefinic groups, they react easily at higher temperatures with oxygen. The associated aging and deterioration process can be studied in situ and characterized in terms of oxygen diffusion and reaction kinetics [71–75, 197–199, 212]. For the detection of the associated material change, image contrast based on slow segmental mobility is most sensitive and can be enhanced by use of magnetization filters (cf. Fig. 7).

Another application of NMR imaging to elastomers is the localization of stress distributions. Here, the noninvasive character of the method as well as the possibility to generate images within non-transparent samples in three dimensions are unique assets. Mechanical stress influences the segmental mobility in the slow motion regime. Thus local stress and strain in elastomers can be investigated by T_2 and $T_{1\rho}$ images, which can be recalibrated to stress and strain images [197–200].

3.2 Imaging at Elevated Temperature

Although some thermoplastic polymers can be imaged with good resolution using standard equipment and sufficiently strong gradients [110, 154], signal-to-noise ratio and spatial resolution can be enhanced (cf. Sect. 2.4) by elevating the sample temperature [194]. In this way, the molecular mobility is increased resulting in further averaging of the dipole–dipole interactions, and thus in longer values of T_2 and narrower lines. Significant changes in T_2 are already observed in a narrow temperature range of 30 K above room temperature (Table 2).

Sample heating is particularly efficient in amorphous samples above their glass transition temperatures [84, 136, 194]. Although the non-invasive character of NMR imaging is somewhat sacrificed, standard equipment and large samples can be used. An example is given in Fig. 9 with images of a polypropylene sample at two different temperatures [194]. At 300 K the image signal is barely above the noise level, while at 300 K even the three 3 mm diameter holes are clearly

Table 2. Temperature dependence of T_2 [ms]

Sample	$B_0[T]$	$T_g[K]$	T_2[ca. RT]	T_2[330 K]	T_2[370 K]
epoxy resin [196]	4.7	—	0.16 [293 K]	4.0	25.9
vulcanized rubber [194]	4.7	< RT	1.1 [300 K]	1.57	1.69
*poly(vinyl acetate) [194]	4.7	308	0.05 [310 K]	0.12	182
*polypropylene [194]	2.0	255	0.37 [300 K]	0.61	—
**water in nylon 6.6 [194]	4.7	—	1.54 [310 K]	4.31	10.3

*long relaxation component.

**strong association of water with polymer matrix at room temperature

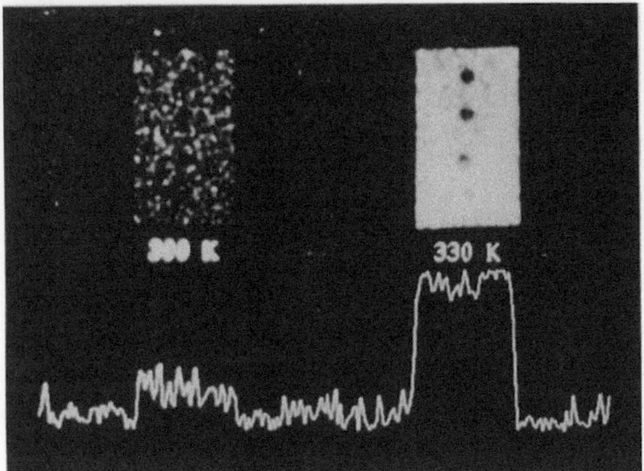

Fig. 9. Images of a polypropylene sample (12 mm × 9 mm × 5 mm) measured at 300 K (*left*) and 330 K (*right*). Three holes with a diameter of 3 mm were drilled into the sample with varying depths for testing the spatial resolution. With $TE = 2.1$ ms and $TR = 0.5$ s the total measurement time was 2.5 min. (Reproduced with permission from Hall et al. [194])

discerned. This approach can be employed also for in situ imaging of polymers during standard industrial processing like polymerization [25, 137–140], injection molding, and curing [196].

3.3 *Highly Ordered Materials*

Materials with high molecular orientation, for instance crystals, oriented liquid crystalline polymers [110, 193, 213–215], fibers, and polymers aligned by mechanical strain, injection-molding, or extrusion [216, 217] exhibit narrow NMR lines, because in these cases the angular dependence of the NMR frequency in the solid state (cf. Eq. (24)) is restricted to a rather limited range of orientation angles [1, 218–220]. This effect was utilized in ^{13}C NMR imaging to characterize the surface-core layer structure and the molecular alignment in a drawn, injection-molded tensile bar of syndiotactic polypropylene (Fig. 13) [216, 217].

The line widths of quadrupolar nuclei range from a few kHz to several MHz, depending on the quadrupole coupling constant, which is determined by the strength of the electric field gradient at the site of the nucleus, the quadrupolar moment and spin of the nucleus, and by the molecular dynamics. Furthermore, a possible asymmetry of the electric field gradient influences the line width. The ^{23}Na ($I = 3/2$) resonance in Na-β-alumina is narrowed to 1.1 kHz due to two-dimensional ion diffusion, rather than high orientation. Imaging was used to determine the interdiffusion coefficient for the displacement of sodium by potassium, and the strong dependence of the line width on the electric field gradient was used to characterize hair cracks [213]. In NaCl, the cubic symmetry

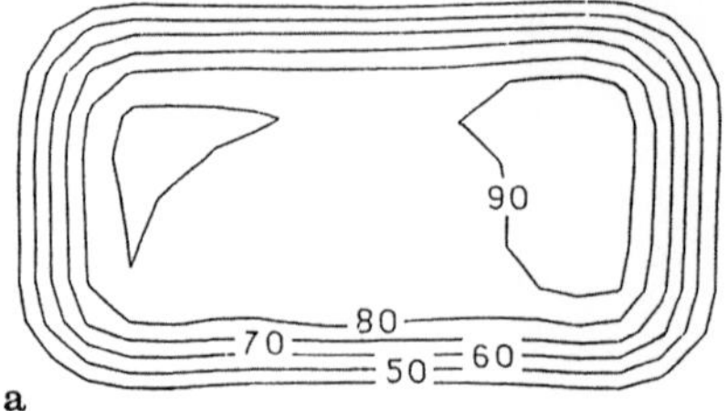

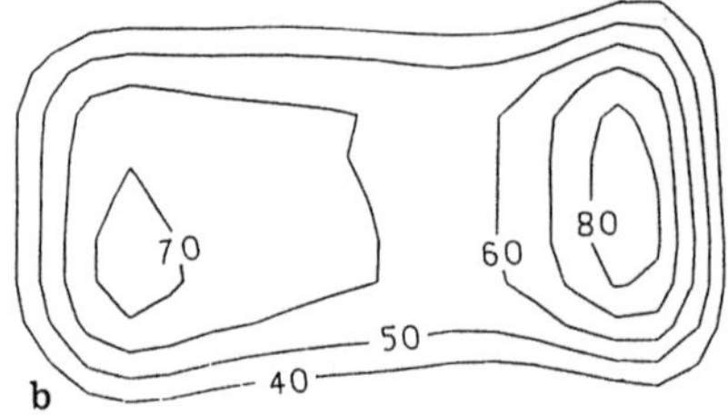

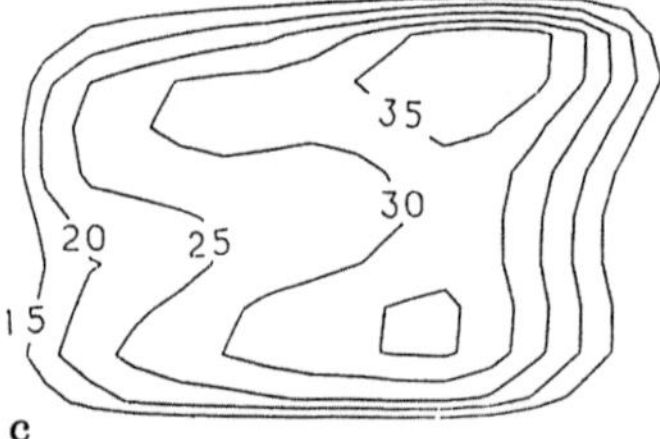

Fig. 10a–c. ^{23}Na images depicting the spatial distribution of NMR intensity in a NaCl sample 10 × 5.4 × 2 mm): **a** before, **b** after extensive damage of the crystal by mechanical impact. The rightmost quarter was outside the strike region. **c** The difference between **a** and **b** in %. The gradient strength was 150 mT/m resulting in a spatial resolution of about 1 mm. (Reproduced with permission from Suits et al. [214])

quenches the quadrupole coupling, and a line width of 2 to 3 kHz is observed. However, deviations from cubic symmetry caused by defects or stress lead to considerable line broadening, which can be exploited for defect analysis as illustrated in Fig. 10 [214, 215]. These images demonstrate, how small deviations from cubic symmetry can be detected on a macroscopic scale.

The same principle was applied to image temperature profiles [193]. In KBr crystals, the cubic symmetry reduces the line widths of ^{81}Br ($I = 3/2$) to about 1 kHz. Thus, standard imaging techniques can be applied yielding acceptable spatial resolution (here about 1 mm for a gradient of 80 mT/m). For this sample, T_1 is dominated by the quadrupole interaction of the nucleus and the surrounding electrons of the crystal lattice. The strong temperature dependence of T_1 relaxation can be used to generate image contrast by application of a saturation recovery sequence prior to spatial encoding of the acquired signal. The difference of two images, one measured at uniform temperature and the other in a temperature gradient, then maps the local temperature differences.

3.4 Imaging of Diluted Isotopes

As already discussed, the homonuclear coupling of abundant nuclei (e.g. ^{1}H or ^{19}F) is a severe problem for NMR imaging resulting in a decrease of spatial resolution and S/N. However, if the observed isotope is diluted due to low chemical concentration or natural abundance, the homonuclear dipolar couplings can often be neglected, because the coupling strength scales with $1/r^3$ (r = distance between two coupling isotopes). Typical spin $I = \frac{1}{2}$ nuclei, which can show such behavior because of their low natural abundance, are ^{13}C, ^{15}N, and ^{29}Si, while ^{31}P, for example, can show it for reason of the chemical structure.

Although the distance between the individual spins is large, heteronuclear couplings with surrounding abundant nuclei (mainly ^{1}H) may cause line broadening. In comparison with homonuclear couplings (cf. Chapter 5) heteronuclear couplings can readily be removed [221, 222] by CW (*c*ontinuous *w*ave) irradiation at the proton frequency during sampling of the signal from the diluted nuclei. Furthermore, the sensitivity can be enhanced by *c*ross-*p*olarization (CP) of magnetization [223, 224] from neighboring abundant nuclei. Cross-polarization itself can be used as an additional possibility for contrast enhancement [225].

The first experiments on ^{13}C phantoms were done by Szeverenyi and Maciel [226]. The experiment used a double-resonance probe with a switchable gradient system. After the gradient had stabilized, a conventional CP scheme was used to enhance ^{13}C magnetization. The gradients were slowly turned off after storing the magnetization along the *z*-axis by an additional 90° pulse. Following a further 90° pulse, the FID was acquired whilst the protons were decoupled. By changing the gradient strength or the length of evolution time, a spectroscopic image was obtained. For the measurement of frequency encoded projections, distortions by the chemical shift, susceptibility, and insufficient heteronuclear decoupling was averaged out by use of the *C*arr-*P*urcell-*M*eiboom-*G*ill (CPMG) sequence.

The most suitable nucleus for an NMR imaging experiment depends on the material investigated. For polymers, ^{13}C [216, 217, 225–228] (cf. Fig. 13) is an excellent isotope showing more chemical information than ^{1}H; for biomedical investigations, ^{13}P [133, 229] and ^{15}N [230] are useful while ^{29}Si is promising for studies on glasses and ceramics as well as quadrupolar nuclei like ^{11}B [133, 231, 232] or ^{27}Al (cf. Sect. 4.3).

3.5 Line-Narrowing and S/N Improvement by a priori Knowledge

A priori knowledge can be used for the improvement of spatial resolution and signal-to-noise ratio: a) by removing image blurring from different chemical shifts and broad lines, and b) by accounting for known spatial features in image processing. Although these techniques are very powerful for removing an overall blurring of an image, they fail to reveal a fine structure which varies in different portions of the sample.

a) As in most NMR spectroscopy experiments, a significant amount of a priori knowledge about the sample is available in NMR imaging. Such information may comprise, for instance, the NMR spectrum's shape including chemical shifts, line widths, and spectral amplitudes. All or part of this information can be used to improve the signal-to-noise ratio or the spatial resolution by applying various image processing routines like filters and prediction or iteration algorithms [233]. Spectral estimation and analysis methods such as maximum entropy, maximum likelihood [234–243], and Bayesian analysis [244–246], which were originally proposed for other experiments, are well known in NMR spectroscopy and can be used in problems related to imaging as well. Frequency encoded projections, for example, can be described as a convolution of the line shape function with a point-spread function. Both can be separated by inversion of Eq. (12). However, inversion by deconvolution is often ill-defined due to the numerical overestimation of very small values, which results in increased noise. This can be avoided by filtering, prediction, and estimation of the inverse [44, 45, 247–251] and by analytical inversion of simulated functions (sum of Lorentz and Gauss functions) [213]. When applied to NMR imaging, blurring by instrumental imperfections can be removed [252, 253] as well as chemical shift artifacts [123, 239, 254–256] and distortions from wide-lines in solids [213, 222].

b) A priori knowledge about known spatial features may be available from images obtained by other imaging techniques. Its usefulness was demonstrated for sensitivity enhancement in imaging of objects with cylindrical symmetry [96, 257–259], edge enhancement [260], object reconstruction from a set of limited projections in back-projection experiments by knowledge of the shape [261–263], and a priori knowledge of the arrangement of substructures [264].

4 Imaging of Solids with Wide Line Methods

The methods reviewed in this section rely on efficient gradient amplification for improved spatial resolution (cf. Eq. (13)), or they are designed to achieve spatial resolution independent of the line width. Methods that employ line-narrowing techniques are discussed separately in the Sect. 5.

The straightforward way to increase spatial resolution is to increase the gradient strength. To this end, the stray field of superconducting magnets can be used. *Stray-field imaging* (STRAFI) is similar to early sensitive-point methods like FONAR (*field focused nuclear magnetic resonance*) [265, 266] and techniques with CW excitation in static *field* gradients in NMR [59, 60] and EPR imaging [33, 250, 251, 267]. Constant-time imaging is a simple and robust method based on indirect space encoding by the spin-warp technique (cf. Sect. 4.2) [268]. Multiple-quantum imaging exploits an amplification of the applied field gradient by the faster phase evolution of multiple-quantum coherences [269].

4.1 Stray-Field Imaging

Stray-field imaging uses strong, static gradients which are readily available in the stray fields of superconducting NMR magnets [270, 271]. If the probe is suitably displaced from the region of homogeneous field in the magnet center, a region of constant field (typically 40% of the original $\boldsymbol{B}_0$ field strength) and constant gradients (40 T/m for 4.7 T and 80 T/m for 9.4 T magnets) can be found [271]. In such a strong field gradient, a conventional rectangular rf pulse becomes a selective pulse which excites a thin slice in the sample (e.g. 35 to 70 μm for a pulse length of 8 μs and a line width of 20 kHz). The individual points of a projection can be measured stepwise by mechanical shifting of the sample through this sensitive plane. For a given line width and gradient strength the spatial resolution is determined by the bandwidth of the rf pulse and by the precision of the mechanical sample positioning device. 2D and 3D images can be constructed by additional rotation of the sample. The quality of the STRAFI method is illustrated in Fig. 11 by a ^{1}H image of a phantom made of Plexiglass blocks which were tied together by adhesive tape. The FID decays extremely rapid in such strong field gradients. Therefore, solid-echo trains are used for sensitivity enhancement and to overcome receiver dead time. Information about the spin density is obtained from the sum of all echoes; relaxation contrast is calculated from the decay of the echo amplitudes.

The method is characterized by the following features: 1) The spatial resolution is virtually independent of the intrinsic line width as well as magnetic field distortions by changes of susceptibility [272]. Therefore, rigid materials including metal composites can be imaged [271]. 2) The method cannot yet be combined with spectroscopic resolution to access further

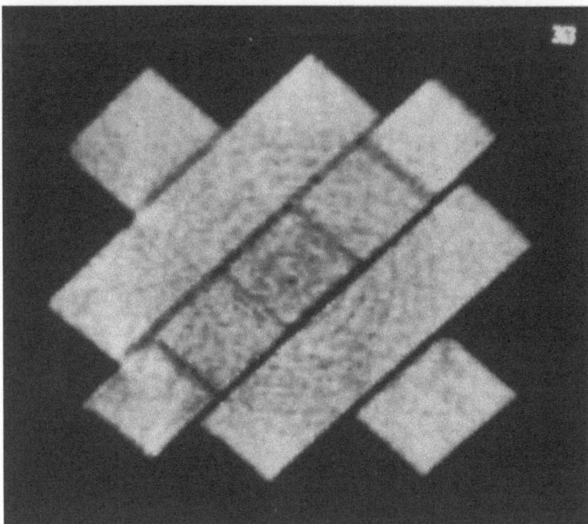

Fig. 11. Stray field ^{1}H image of a phantom composed of several pieces of plexiglass with gaps of 100 μm to 300 μm. (Reproduced with permission from K. Zick [271])

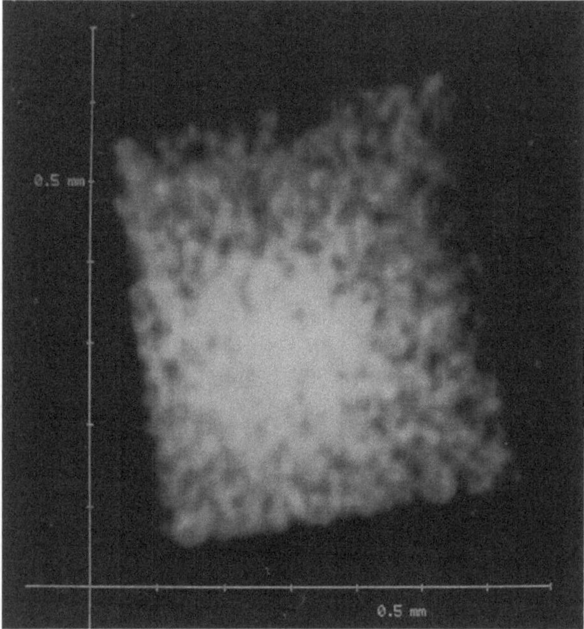

Fig. 12. Frequency encoded stray field image of a PMMA sample ($0.5 \times 0.6 \times 0.8\,mm^3$). The rf pulse duration was $0.4\,\mu s$ and the in-plane resolution is $10\,\mu m$. (Reproduced with permission from K. Zick [273])

parameter contrast. 3) The fixed gradient direction requires repositioning of the sample. Not the repetition time TR but mechanical complexity of the corresponding probe limits the speed of measurement, because individual planes of the sample are excited separately so that there is no need for a recycle delay.

The same principle can be used for frequency encoded imaging [273]. Then, the excitation pulse must be non-selective with a very large bandwidth. Figure 12 depicts an image of a PMMA sample obtained in this way. However, applications of frequency encoded stray-field imaging are restricted to rather small samples and are limited by the maximum bandwidth of transmitter and receiver, as well as the correspondingly broad spectrum of thermal noise acquired with the data.

4.2 Constant-Time Phase Encoded Imaging

The principle of constant-time phase encoding has been already introduced in Sect. 2.2 by the spin-warp technique. The basic experimental scheme is depicted in Fig. 4b. For an n-dimensional image, n gradients ($n = 1, 2$ or 3) of independently varied strengths are applied simultaneously during the evolution time of fixed duration t_1. A single data point is acquired at time t_1 for each gradient combination [274]. Depending on the switching times of the gradients and on the length of t_1, the gradients can be possibly turned off before data acquisition. As long as the gradient amplitudes are varied linearly, this has no effect on the space encoding. Because the linear (Zeeman and chemical shift interactions) and bilinear interactions (dipole–dipole and quadrupole) commute to first order at high field approximation, they can be discriminated from the influence of a field gradient when the signal is observed after a constant time, during which only the strength of the field gradient is changed. Then signal amplitude and phase are only modulated by the changing gradient strength whereas the amplitude and phase evolution due to other interaction remains constant (cf. Eq. (14) and (15)). Thus, the spatial resolution is no longer a function of the line widths [268, 274, 275], but is determined by the maximum gradient strength (cf. Eq. (16)).

Nevertheless, the line width determines the signal decay in the absence of a gradient and thus the signal-to-noise ratio. The wider the line, the lower the signal amplitude at the sampling time t_1. In the presence of a field gradient the signal decays even faster. This and the fact that every point in $\boldsymbol{k}$-space is measured separately after a recycle delay, which can be quite long for solids, make this method very time consuming [10, 275]. The flip angle of the excitation pulse, however, can be optimized for maximum S/N. Furthermore, the evolution time can be prolonged if the magnetization is refocused by solid echoes [276], Jeener-Broekaert and alignment echoes [197–199, 277, 278] for systems with spin $I = 1$, magic-echoes (cf. Sect. 5.2), and MAS echoes [279, 280]. Then ultra-fast gradient switching is less stringent, and the signal can be sampled in the absence of the gradient. In this way the sensitivity can be improved

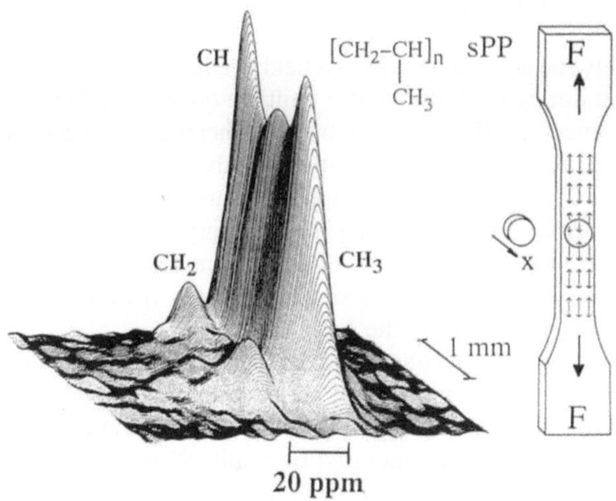

Fig. 13. ^{13}C-spectroscopic image of an injection-molded, drawn tensile bar of syndiotactic polypropylene (geometry *on the right*). A small piece was cut out from the center and oriented with the drawing direction parallel to $\boldsymbol{B}_0$. On the spectroscopic axis methylene and methine resonances can clearly be distinguished. From their line shapes the orientational distributions were calculated for the surface and core parts of the sample. Different material densities can be recognized for both regions on the spatial axis. (from Günther et al. [216, 217])

substantially and the gradient-free detection time can be used for acquisition of the spectroscopic wideline response.

Figure 13 demonstrates the use of this technique for ^{13}C spectroscopic imaging of a tensile bar of syndiotactic polypropylene after drawing to 500%. After cross-polarization, high-power proton decoupling was used to observe the ^{13}C chemical shift. During the constant evolution time, the ^{13}C magnetization was refocused by a Hahn echo to eliminate phase evolution due to chemical shift and magnetic field inhomogeneity; this ensures a phase-sensitive representation of the spectroscopic image. Before turning the gradients off, the transverse components of the ^{13}C magnetization were stored along the z-axis in two separate experiments (cf. Sect. 5.1) and subsequently restored to transverse magnetization for acquisition of the spectroscopic response in the absence of a gradient.

The spectroscopic dimension reveals the CH and CH_2 signals. Due to high mobility of the CH_3-groups, the corresponding signal cannot be enhanced by cross-polarization. A signal variation is observed along the space axis indicating higher material density for the surfaces than for the core of the injection-molded tensile bar. Analysis of the line shapes provides information about the orientational-distribution function. The variance of the orientational distribution function is larger for the core layer than for the surface layers. Constant evolution time imaging has also been applied to imaging of quadrupolar nuclei (like ^{11}B) in biological investigations [232] and ^{27}Al in ceramics [281].

4.3 Multiple-Quantum Imaging

An elegant way to amplify the effect of an applied field gradient is the encoding of the spatial information during evolution under multiple-quantum coherences. Multiple-quantum coherences are characterized by the coherence order p which is defined by the difference of the magnetic quantum numbers of the respective eigenstates. The oscillation frequencies of multiple-quantum coherences are p times larger than the Larmor frequency ω_0 [282] causing a p times faster evolution of the spin coherence in the presence of a gradient. This is analogous to an amplification of the gradient strength by the factor p resulting in an improvement of the spatial resolution by the same factor.

The theory of multiple-quantum coherences is reviewed elsewhere [161, 283–287]. Such coherences occur in systems of coupled spins. In solids, they can arise for quadrupolar spins and for $I=\frac{1}{2}$ spins that are coupled by dipole-dipole interactions. They can be generated by a suitable pulse sequence. Single-quantum coherences ($|p|=1$) correspond to dipolar radiation and can be observed directly. Multiple-quantum coherences ($|p| \neq 1$) can be observed only indirectly. Therefore, all multiple-quantum experiments necessarily have at least two dimensions.

Multiple-quantum imaging was first investigated for dipolarly coupled protons in adamantane (Fig. 14) [269]. Apart from nanoscopic-phase bound-

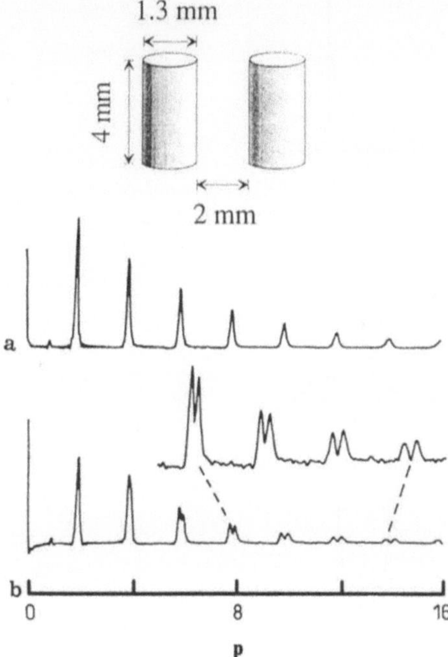

Fig. 14a, b. ^1H multiple-quantum spectra of an adamantane phantom (*upper inset*) without (**a**) and with (**b**) a static gradient (48 mT/m). The evolution time and phase of the preparation pulse sequence were incremented in 32 steps of 0.1 μs and $2\pi/32$, respectively. By use of a symmetric pulse sequence only even multiple-quantum coherences were detected. For clarity, orders 8 through 14 are presented in the *lower inset* on an expanded scale. Although the experiment was conducted in a two-dimensional fashion, the stacks of the 2D-matrix are plotted consecutively to demonstrate the increase in spatial resolution of the two cylinders with higher mulitple-quantum coherences. Therefore, the plot axis shows the spatial dimension of the experiment for each coherence order *p*. (Reproduced with permission from Garroway et al. [1269])

aries, the multiple-quantum preparation period determines the order to which the coherences are allowed to build up ($p = 20$ for a preparation period of 396 μs in Fig. 14). Because the phase of different coherence orders is proportional to $p\omega_0 t$, the contribution of a given coherence order can be separated by manipulation of phases [161]. Prior to detection, the multiple-quantum signal is converted to observable single-quantum coherence. The space information is detected indirectly in the fashion of constant-time imaging (Fig. 14) or in terms of a multi-dimensional NMR experiment where the spectroscopic dimension is recorded in the detection period without an applied gradient. Although the signal intensity decreases for higher multiple-quantum coherence orders, the amplification of the gradient strength is clearly demonstrated in Fig. 14 for coherence orders as high as 12.

The same principle has been applied to the double-quantum coherences of ^2H in deuterated solids. Because the ^2H has spin $I = 1$, the double-quantum transition is independent of the orientation of the quadrupole coupling tensor

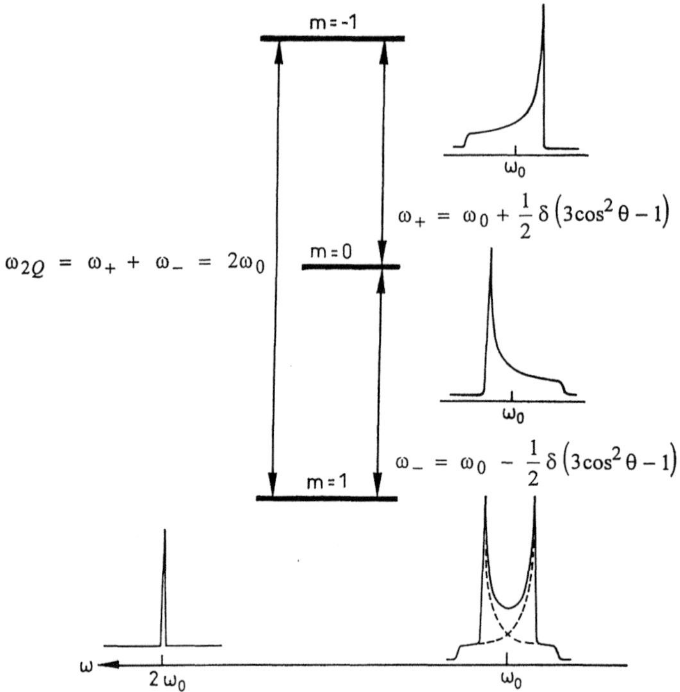

Fig. 15. Schematic presentation of energy levels, transitions and line shapes for single ($|p| \equiv |\Delta m = 1$) and double-quantum transitions ($|p| = 2$) in the powder limit for spins $I = 1$ and axially symmetric quadrupole coupling. (from Günther et al. [288])

in the magnetic field (Fig. 15) [287, 288]. This transition corresponds to a narrow line, which is advantageous for space encoding and also effects a gradient amplification by a factor of $p = 2$ providing an extra benefit. While spatial information must be detected indirectly via phase modulation of single-quantum coherences, the spectroscopic dimension is detected directly. Therefore, each point in space is characterized by a deuteron wide-line spectrum which provides information about molecular order and slow molecular reorientation [216, 217, 288].

Application of ^2H double-quantum imaging for the detection of molecular order and mobility are depicted in Fig. 16. The line shape in the spatially resolved ^2H spectra changes considerably with the orientation of a highly oriented polyethylene sample in the magnetic field (a). Similarly, different line shapes are observed for phenylene-deuterated polycarbonate with and without (anti-) plasticizer additive, indicating different mobilities of the chain segments. The high information potential of this method is offset by the fact that the samples need to be deuterated and that T_1 relaxation times may be long requiring long recycle delays and thus acquisition times of up to one day. In general, quadrupolar nuclei ($I > \frac{1}{2}$), for instance ^2H for polymers or ^{27}Al for ceramics

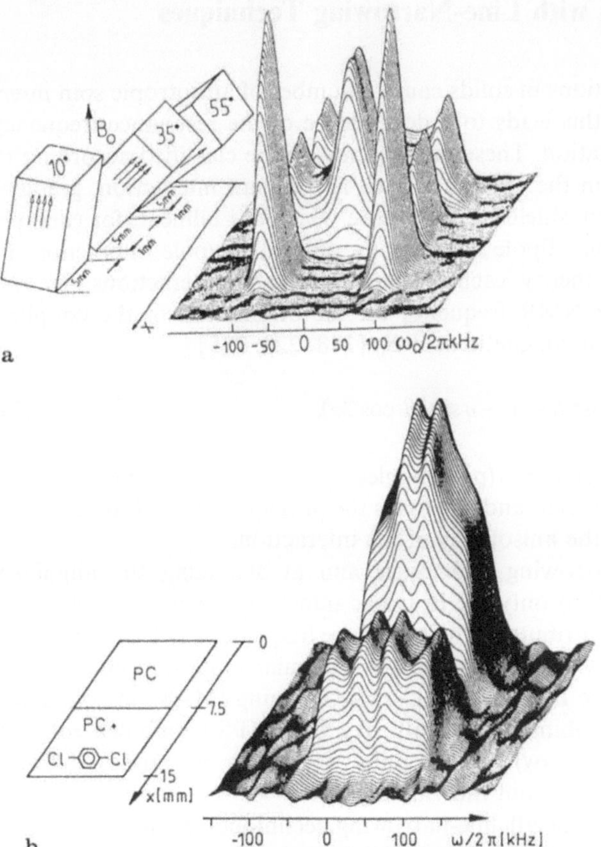

Fig. 16a,b. Spectroscopic double-quantum images of ^2H demonstrating the sensitivity of ^2H line shapes to molecular order and mobility ($\omega_Q = \omega - \omega_0$). **a** A phantom made of three parts of highly oriented, perdeuterated polyethylene. Different angles of the orientation axis relative to $\boldsymbol{B}_0$ result in different line shapes. **b** A phantom made of two parts of polycarbonate deuterated at the aromatic positions. One part contained a low-molecular weight additive which is suppressing the mechanical γ relaxation. The other part was pure bisphenol-A polycarbonate. In the spectrum of the first part, the motionally narrowed region in the center of the ^2H wideline spectrum is suppressed. (from Günther et al. [197–199, 288])

[133, 281], are promising candidates for this concept of spectroscopic multiple-quantum imaging, because their spectra can provide information about the local electron symmetry and therefore about molecular mobility and orientation [289, 290].

If the gradient is not applied during but after the multiple-quantum evolution, a multiple-quantum filter (cf. Sect. 2.5) is obtained for the magnetization which forms the image. In this way, single-quantum magnetization like that of water can be suppressed manipulating contrast in the final image or in the localized spectrum [291]. This concept was applied for the discrimination of different intra- and extra-cellular sodium and potassium in in vivo experiments [292–294].

5 Imaging of Solids with Line-Narrowing Techniques

The slow molecular motions in solids cause a number of anisotropic spin interactions to be effective; this leads to a dependence of the resonance frequency on the molecular orientation. These interactions can be classified according to those which are linear in the spin part, such as Zeeman interaction, gradient interaction, and magnetic shielding, and those which are bilinear, for example direct and indirect dipole–dipole interactions, and quadrupole interaction. In first order perturbation theory, each of these anisotropic interactions give rise to a dependence of the NMR frequency on the orientation of the coupling tensor with respect to the magnetic field $\boldsymbol{B}_0$ [218–220, 295],

$$\omega = \omega_0 + \frac{\delta}{2}(3\cos^2\beta - 1 - \eta\sin^2\beta\cos 2\alpha), \tag{24}$$

where α and β are the Euler angles (polar angles) specifying the tensor orientation, η is the asymmetry parameter, and δ denotes the principal value of the coupling tensor which measures the anisotropy of the interaction.

The various line-narrowing techniques aim at averaging the angularly dependent terms in (24), so only the isotropic interactions (isotropic chemical shift, scalar coupling) contribute to the resonance frequency [1, 218, 295]. Line-narrowing is achieved by mimicking the fast molecular motion in liquids. This can be done either in the LCF by magic-angle spinning (MAS), in spin space by rf pulses, or by a combination of both like CRAMPS (*c*ombined *r*otation *a*nd *m*ulti-*p*ulse *s*pectroscopy). When pulse sequences are used, different interactions can be averaged out individually.

The common feature of all line-narrowing techniques is the use echoes, which can be generated in different ways. In general, echoes are advantageous in NMR imaging for many reason: signal excitation and detection are separated to overcome receiver dead time and gradient-switching times. To achieve line narrowing, echoes must be formed in a repetitive fashion. Stroboscopic sampling of each echo maximum yields a signal with prolonged decay and thus a narrower line after Fourier transformation. The corresponding liquid-state experiment is the CPMG echo experiment [296, 297].

Many different methods for generation of echoes in solids are known [219]: amongst them are the solid-echo technique, the Jeener-Broekaert or alignment-echo technique for $I = 1$ nuclei, multi-pulse sequences (cf. Sect. 5.1), the magic-echo technique (cf. Sect. 5.2), mechanical sample spinning at the magic-angle (MAS, cf. Sect. 5.3), and spin-locking of the magnetization at the magic-angle (cf. Sect. 5.4).

5.1 Multi-Pulse Imaging

A number of reviews have been published about this particular approach to solid-state NMR imaging [10, 189, 298, 299]. The one by Cory is recommended in particular [10]. In the following, an overview of these techniques is given for spin-$\frac{1}{2}$ nuclei.

In high-resolution solid-state spectroscopy, the usual aim is to eliminate the orientation-dependent parts of bilinear interactions while the orientation-independent ones are preserved. Linear interactions are averaged out, for instance, by the CPMG sequence [296, 297], bilinear interactions by multi-pulse sequences like WAHUHA [300], MREV-8 [301, 302], BR-24 [303], and others. These pulse sequences also reduce the orientation-dependent parts of the interaction but do not eliminate them. For space encoding, ideally the frequency shifts of all interactions are reduced to zero, in order to prevent evolution of the magnetization. Such sequences have been designed for use in imaging and are called *time-suspension sequences* [61, 62, 304–306]. The signal evolution in the presence of such sequences is only determined by the relaxation $T_{1\rho}$ in the effective field of the pulse sequence which imposes a general limit upon the success of these techniques.

5.1.1 Multi-Pulse Imaging with Static Gradients

The ideal case of averaging out all line-broadening interactions by a *time-suspension sequence* in the presence of a static gradient would have no success for imaging because even the linear interaction of the gradient would be cancelled. Therefore, the simplest approach to solid-state imaging with line narrowing uses the MREV-8 sequence in the presence of a static field gradient. This reduces the homonuclear dipole–dipole coupling to zero and only diminishes the chemical shift including the frequency spread introduced by the magnetic field gradients. If the gradient is strong enough to dominate the other interactions, images of solids can be obtained with relatively high resolution [307–309]. The pulse sequences for that purpose are shown in Fig. 17. Two- and three-dimensional images can be measured by back-projection (Fig. 17a) and by the Fourier technique (Fig. 17b). In the first case, slow gradient switching can be afforded so the gradient settling time does not interfere with the pulse sequence. In the second case, additional evolution periods have to be implemented for phase encoding and gradient switching prior to detection. The latter can be realized preferably by storing the magnetization along the z-axis [226] or by spin locking [308] with a constant "protection" time of the order of T_1 or $T_{1\rho}$, respectively. However, only one quadrature component of the signal can be stored at a time requiring two experiments.

The disadvantage of the method is the well-known offest dependence of the line-narrowing efficiency of MREV-8 [310] and other multi-pulse cycles. As long as the gradients are turned on during the pulse sequence, this will be an inherent problem [311] and the averaging efficiency and thus the spatial resolution will decrease for off-resonance magnetization components. From experience, efficient line-narrowing is only achieved for magnetization with offset frequencies being much smaller than the inverse pulse separation in the sequence

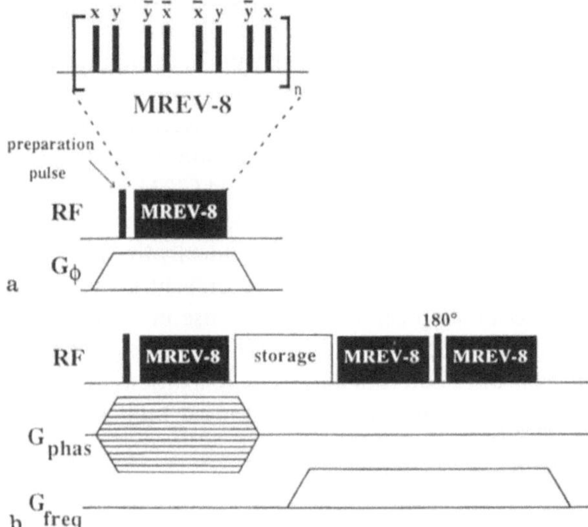

Fig. 17a,b. Multi-pulse imaging schemes [308] for slow switching of $\boldsymbol{B}_0$-field gradients: two-dimensional space encoding by **a**. frequency encoding (BP) and **b** phase and frequency encoding (2DFI) with intermediate storage of the magnetization components for gradient switching

[312]. This is because such multi-pulse sequences are constructed from solid-echo pulse pairs. The required 90° phase difference of the excitation pulses is only fulfilled for thoses magnetization components that do not precess in the delays between pulses. Therefore, the multi-pulse techniques discussed above have to carefully balance gradient strength and offset dependence for optimizing spatial resolution.

5.1.2 Multi-Pulse Imaging with Oscillating Gradients

An improvement is obtained when the field gradient is applied in a time-dependent fashion, rather than in a continuous. This even allows the application of *time-suspension sequences* [304]. Because the events in all multi-pulse sequences are applied periodically with stroboscopic signal sampling in specific windows, rapidly oscillating gradients can be employed instead of static gradients. Miller and Garroway [221] demonstrated that the effect of such a time-dependent gradient does not vanish during averaging of linear interactions by a CPMG sequence.

In comparison to static gradients, oscillating gradients offer some practical advantages [10]: (1) a sinusoidal current applied to a coil will always have the same steady-state frequency independent of changes in amplitude and phase; (2) the gradient strength can be improved, if the gradient coil is tuned to the desired frequency; (3) the gradient periodically becomes states zero, and these can be made to coincide with the rf events in the pulse cycle to avoid off-resonance effects of the rf pulses. However, it was demonstrated that the last point is of secondary importance, because the rf field strength should be much higher than the strength of the gradient field [313]. NMR images obtained with oscillating gradients show improved spatial resolution [276, 314–318] but still suffer from spatially varying resolution due to the non-compensated off-resonance dependence. This can be overcome if the phase shifts in the *r*otating *c*oordinate *f*rame (RCF), which cause the dependence of the line-narrowing efficiency on the resonance offset, are eliminated by additional toggling of the rf pulse phases during the multi-pulse sequence (cf. Fig. 18). This technique is called *second averaging* [319, 320].

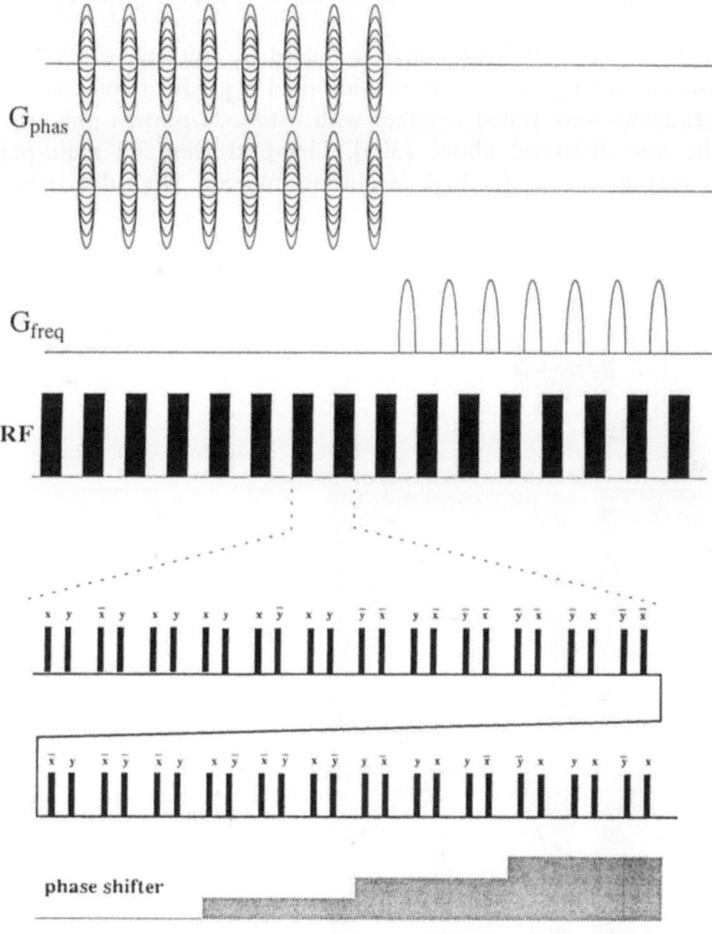

Fig. 18. The CMG-48 sequence for imaging with multi-pulse line narrowing [189, 322]. G_{phase} represents the phase encoding gradients and G_{freq} the frequency encoding during acquisition. For second averaging the phases of the rf pulse have to be shifted additionally

5.1.3 Multi-Pulse Imaging with Pulsed Gradients

The largest experimental flexibility can be gained at the expense of demanding gradient accessories [321] when using short (2-5 μs) gradient pulses in the windows of a multi-pulse cycle [311]. Optimum spatial resolution is obtained by a combination of *time-suspension* multi-pulse excitation with second averaging and pulsed gradients, because 1) rf pulses are applied in the absence of gradients and 2) by removing the resonance offset, the variation of spatial resolution over the sample is eliminated [304]. The most promising pulse cycle for this purpose consists of 48 pulses (Fig. 18) [304]. It can be applied either with oscillating or pulsed gradients. Together with second averaging it yields extremely narrow residual line widths for solids [304, 10]. The resulting high resolution is demonstrated in Fig. 19 with slice-selective images of a polyacetal phantom [322].

Alternatively to pulsed $\boldsymbol{B}_0$-field gradients, pulsed $\boldsymbol{B}_1$ gradients can be used. They are much easier to generate but an additional rf gradient coil has to be used. Such gradients were tested together with a *time-suspension* pulse cycle similar to the one discussed above [323]. Line-narrowing by multi-pulse techniques is very promising for high resolution imaging. High demands on

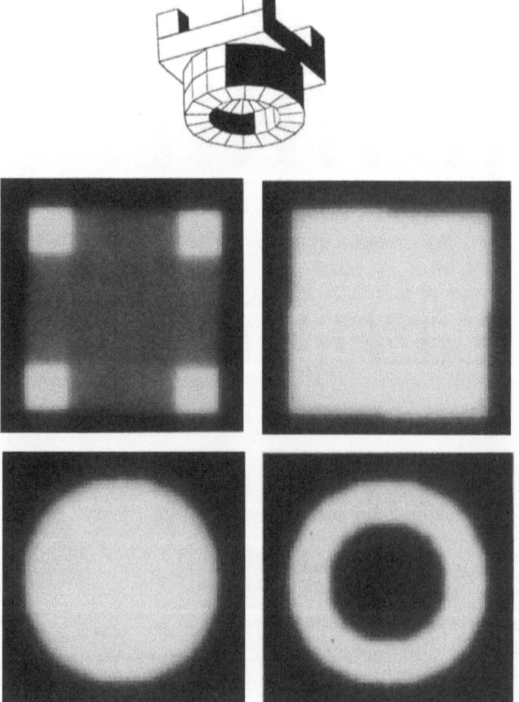

Fig. 19. Multi-pulse image of a phantom made from polyacetal (Delrin): sample geometry (*top*) and 4 slices taken from a 64 × 64 × 16-voxel measured with the sequence shown in Fig. 18. The voxel size is 92 μm × 92 μm × 625 μm and the spectrometer frequency was 400 MHz. (Reproduced with permission from D. G. Cory [322])

experimental setup and equipment, however, seem to hamper widespread use. Though some progress has been recently made in the implementation of such experiments on standard spectrometers [322, 324].

5.2 Magic-Echo Imaging

Prior to the discovery of the magic-echo [219, 325–330] it was believed that the decay of transverse magnetization as a result of homonuclear dipole–dipole interaction is an irreversible loss of signal. However, the magnetization can be refocused by changing the sign of the dipole–dipole interaction Hamiltonian under the action of a suitable rf excitation. This is equivalent to a time inversion in the spin evolution and produces the *magic-echo*.

The dipole–dipole interaction scales with the second Legendre polynomial $P_2(\cos\theta) = \frac{1}{2}(3\cos^2\theta - 1)$ where θ is the angle between the static magnetic field $\boldsymbol{B}_0$ and the effective field in the rotating coordinate frame (RCF). Sign and magnitude of the Hamiltonian can be changed by manipulating θ, so $P_2 = 1$ for $\theta = 0°$ and $P_2 = -\frac{1}{2}$ for $\theta = 90°$. For a spin system which evolves for a time 2τ under an angle $\theta = 0°$ and for a time 4τ under $\theta = 90°$, all interactions scaling with P_2 are refocused after 6τ. One pulse sequence by which this can be achieved is shown in Fig. 20a [331]. Other pulse sequences work as well [331–336].

The first pulse generates transverse magnetization which evolves during τ under a dipolar Hamiltonian $\mathbf{H}_D$. The second pulse then tips the magnetization in the direction of the effective field given by the following rf irradiation which has to be stronger than the couplings giving rise to $\mathbf{H}_D$. The effective field is adjusted to form an angle of 90° to the one in the period before the pulse by locking the magnetization in the rf field. Then the spin system evolves under a Hamiltonian $-\frac{1}{2}\mathbf{H}_D$. The interactions are refocused for a first time in a rotary echo which forms at 3τ after the first pulse. It can be observed stroboscopically if the lock field is replaced by a so-called "burst" field consisting of many short pulses [327]. A second echo forms if the evolution under $-\frac{1}{2}\mathbf{H}_D$ is 4 times longer than the first one. The magnetization is returned by the last pulse to precess around the static field for another time of duration τ. The interactions are then refocused at 6τ to form the magic-echo in the absence of any rf field. The refocusing pulse sequence $90_y - \alpha_x - \alpha_{-x} - 90_{-y}$ is referred to as *magic-sandwich pulse*. The phase inversion of the lock field is introduced to refocus effects of magnetic-field inhomogeneities and chemical shift dispersion. The nominal flip angle of the spin-lock pulses should be a multiple of 360° [331].

The gradients necessary for spatial resolution can be applied either in the windows before and after the magic-sandwich pulse as $\boldsymbol{B}_0$ gradients (G_0 in Fig. 20a) [332] or over the whole sequence [334–336] with nearly the same effect as the previous method. As long as the rf irradiation is much stronger than the additional offset of the gradients, only small artifacts are produced and the gradient switching is further simplified. Another possibility is the use of $\boldsymbol{B}_1$ gradients (G_1) during the magic-sandwich pulse [336] (cf. Sect. 5.4). This method

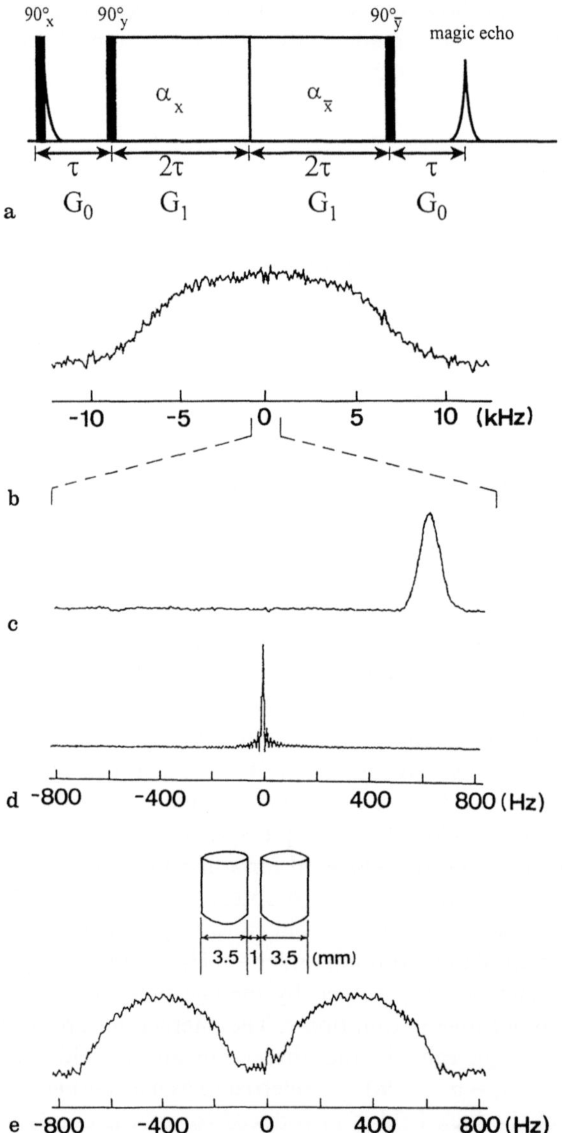

Fig. 20a–e. Illustration of magic-echo imaging on an adamantane phantom. **a** Basic pulse sequence without gradient pulses which can be applied in the windows marked with G (G_0 marks DC gradients and G_1 rf gradients). The magic-echo appears after a time of 6τ. **b** Dipolarly broadened spectrum without line-narrowing. **c** Line-narrowed spectrum by application of the pulse sequence in **a**. **d** Further line-narrowing obtained by refocused chemical shift interaction. The resonance offset retained in **c** is removed in **d**. **e** One-dimensional projection of the phantom with the geometry shown in the inset. The spatial resolution is about 70 μm with a gradient strength of 16 mT/m. The spectrometer frequency was 60 MHz and $\tau = 50\,\mu s$ corresponding to T_2 of adamantane. No signal accumulation was made except for **e** where two signals were added. (Reproduced with permission from S. Matsui [332])

offers the advantage that the evolution period under the gradient is twice as long resulting in better spatial resolution.

Under ideal conditions, the dipole–dipole interaction is completely refocused, so that even for $\tau \gg T_2$ a magic-echo can be observed (typical values are $\tau = 300\,\mu s$ for $T_2 = 50\,\mu s$). At the echo maximum, the bilinear interactions are refocused but not the linear ones. They can be refocused in addition by generating multiple magic-echoes from repetitive application of magic-echo sandwiches and adding the resulting sequence to a CPMG sequence [332]. The net result is a replacement of the last pulse (90°_{-y}) by a 90°_y pulse ($90^\circ_y = 90^\circ_{-y} + 180^\circ_y$), and the effect of the resultant pulse sequence is identical to that of the *time-suspension sequences* discussed in Sect. 5.1.

The different steps leading to the *time-suspension magic-echo* imaging sequence are illustrated in Fig. 20b-d with spectra of adamantane [332]. The original, dipolarly broadened signal exhibits a line width of about 13 kHz (b). The magic-echo sequence causes a reduction to 88 Hz (c). However, the spectrum is still broadened due to chemical shift dispersion. Additional refocusing of the chemical shift results in a line width of just 8 Hz at zero frequency in the RCF (d) which can readily be used for imaging at high resolution. Combining the magic-echo with the 180° pulse requires that the gradient applied in the first window has to be inverted in the second one to avoid cancellation of gradient evolution. Thus the signal can be only sampled after multiples of 12 τ. The quality of results achievable by this technique is illustrated in (e) by a projection of a phantom consisting of two parallel cylinders [332]. A signal was acquired every 600 μs ($\tau = 5\,\mu s$) corresponding to a bandwidth of 1.76 kHz.

The bandwidth limitation can be overcome by constant evolution time phase encoding (cf. Sect. 4.2). This is illustrated in Fig. 21 by a 2D image of a phantom of hexamethylbenzene [335, 336]. However, the spectral information accessible

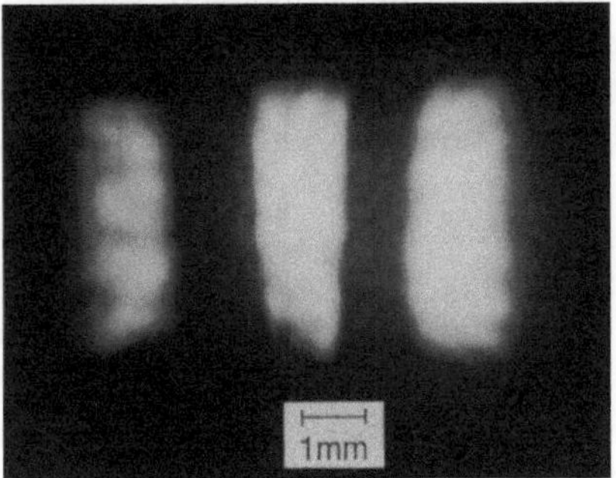

Fig. 21. Magic-echo image of a phantom (length 6 mm, diameter 4.2 mm) made from hexamethylbenzene (original line width about 12 kHz) with intercalated Teflon spacers of different size. The rugged contours of the image result from the hexamethylbenzene's powdery form. The different size of the spacers (one thinner than 1 mm and the other about 1.3 mm thick) is easily distinguished. A maximum field gradient strength of 150 mT/m was applied for $2\tau = 400\,\mu s$. The spectrometer frequency was 200 MHz. (Reproduced with permission of Kimmich et al. [335, 336])

in this way was not used. None of those techniques has been adapted for introduction of parameter contrast which can be easily realized by using such a magic-echo sequence in a preparation period to select spins with different coupling strengths for contrast improvement [337].

Compared to multi-pulse line narrowing, the magic-echo imaging technique offers significant advantages. The method is more forgiving with respect to experimental imperfections, less demanding on NMR hardware, and provides larger windows for the gradient pulses. Thus imaging can be performed under line-narrowing conditions with stronger gradients because switching times can be slower. Similarly the receiver dead time can be increased to enable the use of narrower receiver filters leading to improved sensitivity. Furthermore, the off-resonance dependence seems to be superior to most multi-pulse sequences [337]. Because the decay of the magic-echo corresponds to the FID, it can be sampled for exploitation of spectroscopic information [338]. The spatial information can be encoded in phase and frequency for use with BP or Fourier imaging.

5.3 Imaging with Magic-Angle Spinning

MAS is a standard technique in high-resolution solid-state NMR spectroscopy [1, 2, 339]. Essentially all anisotropic spin interactions described by second rank coupling tensors are averaged by MAS. However, spinning sidebands are observed if the spinning speed is smaller than the anisotropy of the interaction. Thus typical applications are restricted to averaging of the chemical-shift anisotropy. In combination with multipulse line narrowing (CRAMPS), MAS is also applied to the observation of high-resolution spectra of abundant nuclei like ^{1}H [2, 340–342]. Whenever homonuclear multi-pulse excitation is used, however, the receiver bandwidth has to be large enough to avoid filter ringing, leading to a decrease in S/N compared to the use of MAS alone. This benefit in sensitivity is one of the incentives to develop MAS imaging techniques. Another one is the lure of high-resolution ^{13}C spectroscopic imaging [222, 275].

However, the non-invasive character of NMR imaging is sacrificed in at least three ways. 1) As in most solid-state imaging techniques, the sample size is restricted to dimensions smaller than 10 mm due to the sizes of MAS rotors and the limitations of rf power around 1 kW for short 90° pulse. 2) The sample must be balanced for fast spinning (typical rotor frequency $\omega_r < 20$ kHz). Preferably, samples should exhibit axial symmetry. 3) Accelerations of the order of some 100 000 g are generated at 10 kHz spinning frequency. The corresponding centrifugal forces can cause deformation, fracture, and separation of materials. Different spectroscopic effects that depend on the rotor frequency are reported. In ion-containing materials, line shifts were observed due to changes in temperature, conformation, and ion mobility [279, 343–345]. More obvious is the change in efficiency of averaging the dipole–dipole interaction with varying spinning speeds.

Various experiments have been proposed for MAS imaging [145, 222, 275, 279, 346–351], but only experiments with rotor synchronized, rotating gradients have been realized yet. Thus only this class of experiments is reviewed. Here the gradients are rotating synchronously with the spinner, so that the gradients appear stationary in the sample frame. The hardware is technically demanding [345, 351] but once available most imaging techniques can be performed with MAS, in particular the 2DFI and backprojection methods. As an alternative to synchronized gradient-rotation gradient pulses synchronized to the rotor phase can be used [347].

If the gradient coils are aligned coaxially to the rotor axis including the magic angle (54.7°) with respect to $\boldsymbol{B}_0$, the full gradient tensor has to be considered (cf. Sect. 2.1) as well as the constraints imposed by Maxwell's equations [145], because a current in one of the gradient coils will lead to a projection which is tilted with respect to the apparent gradient direction. The effect can be explained in terms of a perturbation of the desired shape of the magnetic field by additional fields which leads to distortions of the projections (Fig. 22, left). On the other hand, under suitable conditions the gradient coils can be driven such that undistorted projections are obtained (Fig. 22, right) [145]. These demand that one of the gradient coils in the tilted gradient coordinate frame (GCF: x', y', z') is parallel to an axis in the laboratory coordinate frame (LCF: x, y, z) and also that the gradient coils are driven simultaneously with well defined currents. If the y- and the y'-axes are aligned, sinusoidal gradient currents $\boldsymbol{I}_\mathrm{i}$ oscillating at the spinner frequency ω_r are applied to the gradient coils in order to generate a gradient that rotates with the sample and appears static in the rotating sample coordinate frame (SCF: x'', y'', z''). These currents are given by

$$\begin{aligned} \boldsymbol{I}_{\mathrm{x}'} &= \tfrac{1}{2}\boldsymbol{I}_0 \cos(\omega_\mathrm{r} t + \varphi), \\ \boldsymbol{I}_{\mathrm{y}'} &= \boldsymbol{I}_0 \sin(\omega_\mathrm{r} t + \varphi), \\ \boldsymbol{I}_{\mathrm{z}'} &= \frac{1}{\sqrt{2}}\boldsymbol{I}_0 \cos(\omega_\mathrm{r} t + \varphi). \end{aligned} \tag{25}$$

This allows imaging in the $x''y''$ plane, where φ denotes the relative phase between the gradient direction and the x''-axis in the SCF. If projections are acquired for small increments of φ, an image can be reconstructed by conventional filtered back-projection (cf. Sect. 2.3) [279, 345]. On the other hand, the gradient orientation can be changed into an orthogonal direction by setting $\Delta\varphi = 90°$ for use of 2DFI techniques [10, 348].

Figure 23 demonstrates the method by two images of a polybutadiene/polystyrene blend. To avoid slice selection [265, 345], 750 μm thick discs were investigated [279]. The samples showed no obvious voids. Molecular mobility of the polybutadiene fraction is high enough to use MAS for averaging the residual line broadening by the ^{1}H homonuclear dipole–dipole interaction, while this

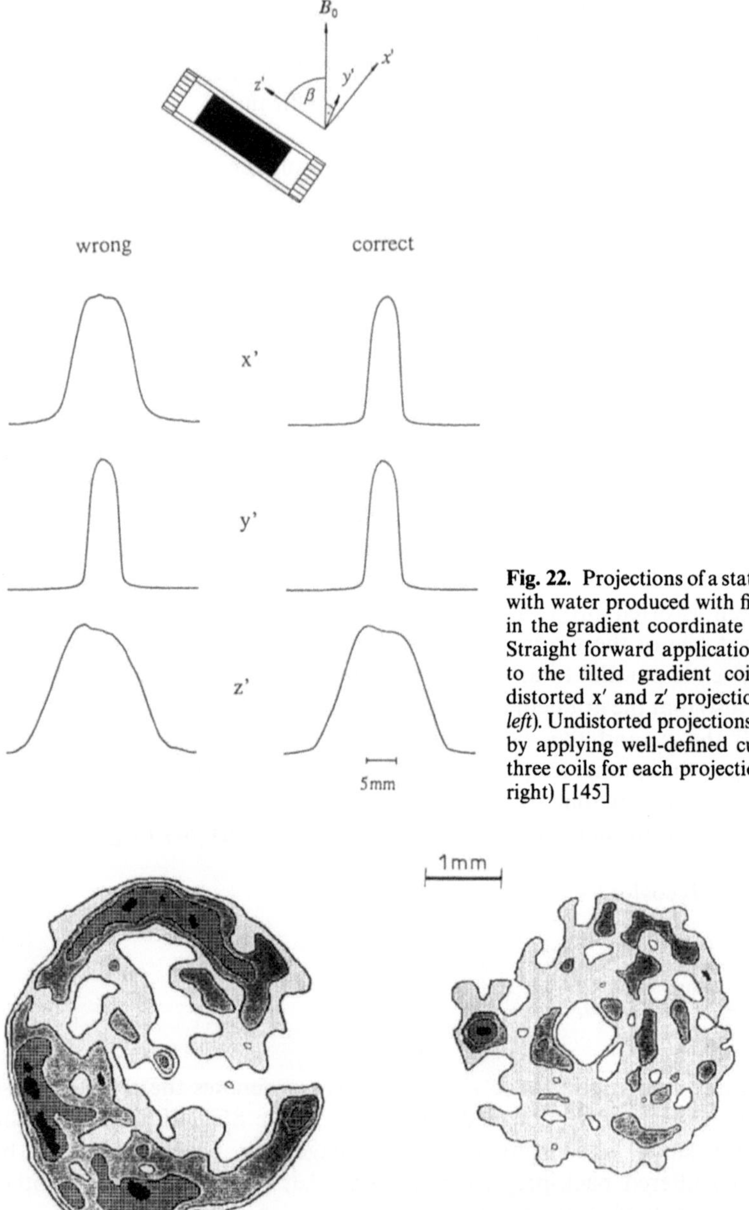

Fig. 22. Projections of a static rotor filled with water produced with field gradients in the gradient coordinate frame (*inset*). Straight forward application of currents to the tilted gradient coils results in distorted x′ and z′ projections ("wrong", *left*). Undistorted projections are obtained by applying well-defined currents to all three coils for each projection ("correct", right) [145]

Fig. 23a, b. ^{1}H-MAS images of solid discs (750 μm thick and 3.5 mm diameter) of polybutadiene/polystyrene blends. **a** Mechanical mixture of both components. **b** Blend cast from toluene. The image contrast is caused by the difference in homonuclear, dipolar coupling strength of the two components. The signals from polystyrene are filtered out (*white*) so only polybutadiene protons contribute to the image contrast (*dark*). The images were reconstructed by back projection from a set of 64 projections. The spatial resolution is better than 50 μm at a spinning speed of 5 kHz. (Reproduced with permission from Veeman et al. [279])

interaction is too strong for the rigid polystyrene to be averaged to zero. Therefore, the signal of the polystyrene is suppressed and the image contrast is formed only by polybutadiene protons. Differences in phase separation of the blend components resulting from different preparation techniques of the samples can be identified by the homogeneity of the image contrast.

Most experimental imperfections give rise to error signals which oscillate with frequencies ω_r and $2\omega_r$. These signals are transformed into spinning sidebands which can be shifted outside the image if the rotor period $2\pi/\omega_r$ is set twice the spectral width of the image. Furthermore, they can be eliminated by small gradient misadjustments [10, 350, 351] or averaged to zero in multiple accumulations with random rotor phases [348]. Aside such artifacts, sidebands occurring for spinnings speeds smaller that the observed couplings can be suppressed by TOSS (*to*tal *s*uppression of *s*pinning-sidebands) [352] similarly to spectroscopy experiments [353].

For imaging of protons in rigid solids, the strong homonuclear dipole–dipole interaction has to be narrowed by multi-pulse excitation in addition to MAS, that is, by CRAMPS techniques. This concept was demonstrated using synchronized rotating gradients and MREV-8 line-narrowing [349, 351]. Due to the offset dependence of the multi-pulse technique, the spatial resolution in the image varied over the sample. To eliminate this effect, the multi-pulse techniques reviewed in Sect. 5.1 must be adapted to CRAMPS. The most general case would be a CRAMPS experiment with MAS-synchronized gradient pulses [10]. The MAS-imaging technique was also applied to spectroscopic imaging of ^{2}H using a constant-evolution-time pulse sequence (cf. Sect. 4.2) [280]. Related techniques employ rotating gradients for the creation of new trajectories in $\boldsymbol{k}$-space [173–178] but not for line narrowing. Those methods could be combined with MAS and may lead to interesting variations of the experiments described above.

5.4 *Magic-Angle Rotating-Frame Imaging*

Another line-narrowing approach similar to MAS, with continuous rotation at the magic-angle in spin space for a static sample, is spin locking in an effective field tilted at the magic-angle [354–357]. If the angle between the applied static field $\boldsymbol{B}_0$ and the effective field in the rotating frame is denoted by θ, the isotropic chemical shift and the gradient interaction are scaled by $\cos\theta$ and the interaction anisotropies by $(3\cos^2\theta - 1)/2$. For θ adjusted to the magic-angle, the anisotropic parts of the interactions are averaged to zero.

To manipulate a precession of magnetization around the effective field $\boldsymbol{B}_e$ at the magic-angle in the rotating coordinate frame (RCF), the magnetization the spin-locked in a field with [358]

$$\omega_e = -\gamma B_e = \sqrt{\omega_1^2 + \Omega^2}, \tag{26}$$

where $\omega_1 = -\gamma B_1$ is the rf field amplitude and $\Omega = \omega_0 - \omega_{RF}$ is the resonance frequency in the rotating frame, so ω_e is an audio frequency. In this way, pulse sequences can be applied in the *m*agic-*a*ngle *r*otating *f*rame (MARF) in analogy to those in the rotating coordinate frame [359]. A further constraint concerning the magic-angle ($\theta = 54.7°$) condition is given by

$$\theta = \arctan \frac{\omega_1}{\Omega}. \tag{27}$$

MARF experiments were originally proposed for line narrowing in high-resolution solid-state NMR [354, 355] but the magic-angle condition can only be satisfied for one chemical shift at a time. Thus, the efficiency of MARF line narrowing varies with resonance offset similar to that of multi-pulse experiments. As there seem to be no other advantages, multi-pulse experiments are preferred for MARF line narrowing because they are easier to implement.

For imaging experiments, the resonance offset Ω is determined by the applied field gradient. De Luca, Maraviglia, and coworkers [360–368] have shown that an effective field, which matches the magic-angle condition over the whole sample, can be produced. This requires an effective gradient $\boldsymbol{G}_e$ being a superposition of a $\boldsymbol{B}_0$ gradient $\boldsymbol{G}_0$ and an rf gradient $\boldsymbol{G}_1$ oscillating with ω_{RF},

$$G_0 = (G_x, G_y, G_y), \qquad \text{where } G_r = \frac{\partial B_z}{\partial r}, \tag{28a}$$

$$G_1 = (G'_x, G'_y, G'_z), \qquad \text{where } G'_r = \frac{\partial B_1}{\partial r} \cos(\omega_{RF} t), \tag{28b}$$

and $r = x, y, z$. While $\boldsymbol{G}_0$ determines the spatial resonance offset Ω, $\boldsymbol{G}_1$ influences the $\boldsymbol{B}_1$ field strength, that is, ω_1 in Eq. (26). To maintain the orientation of the effective field at the magic-angle, the gradient amplitudes must be adjusted according to Eq. (27),

$$\frac{G'_r}{G_r} = \frac{\gamma B_1}{\Omega} = \tan 54.7° = \sqrt{2}. \tag{29}$$

If Eq. (29) is fulfilled, the angle θ of the effective field is independent of the position within the sample as long as frequency shifts from other spin interactions can be neglected. Then all bilinear interactions are cancelled over the whole sample and interactions of higher orders may be eliminated, for example, by double rotation in the MARF [356, 357].

The experimental realization (Fig. 24) requires at least three different types of signals: a) the rf transmitter signal (TX_{RF} with frequency ω_{RF}) for the initial 90° pulse which aligns the magnetization along the magic-angle in the RCF for subsequent spin-locking; b) the transmission of audio-frequency pulses (TX_{AF} with frequency ω_e) for manipulations of the magnetization in the MARF; c) an effective gradient of amplitude $G_e = \sqrt{G_0^2 + G_1^2}$ for spatial resolution. The same coil is used for excitation in the MARF by audio-frequency pulses and for detection of the system response. It can be aligned along the $\boldsymbol{B}_0$-axis and must be decoupled from the rf coil.

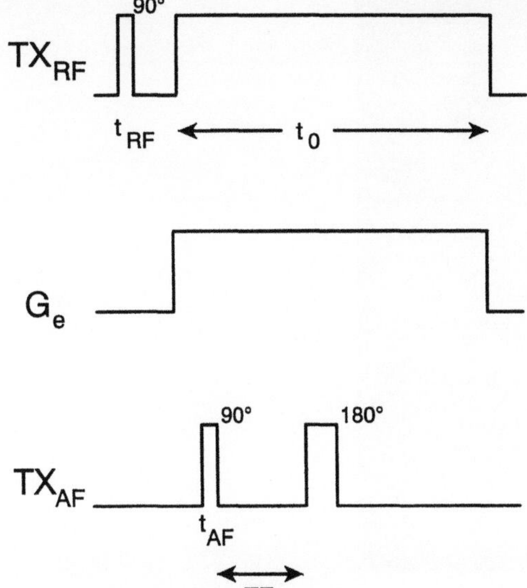

Fig. 24. Full pulse sequence for MARF-echo imaging [363]. See text for details. The gradient G_e is a combination of $\boldsymbol{B}_0$ and $\boldsymbol{B}_1$ gradients

Pulse sequences realized in this fashion behave analogously to those in the more common RCF, and a 90°–τ–180° sequence results in a spin echo at a time τ following the 180° pulse. In this way, the experiment can be modified for multi-pulse excitation on the TX_{AF} channel, which can be used for reducing the residual line broadening by a solid echo-based sequence (cf. Sect. 5.1) or for removing chemical shifts by a CPMG echo train. Furthermore, ultra-slow molecular motion can be detected by measuring the relaxation rates $T_{1\rho}$ and $T_{2\rho}$, for example, in the MARF [360]. The hardware for the experiment consists of an rf and an af excitation coil and two spatially matched gradient coils. Because of the demanding apparatus, experiments have been only implemented in one dimension so far, and the images were reconstructed from projections acquired by sample rotation [363]. Experimental examples are shown in Fig. 25 [363] with phantoms of adamantane.

The advantage of this method is the better line-narrowing efficiency for rigid solids compared to MAS; because the "spinning" speed in the MARF cannot be achieved by MAS. Furthermore, the sample remains static. The disadvantages are the demanding equipment, which in fact may be less complicated than commercial MAS probes, and the low S/N of the acquired signal because of the low excitation frequency $\omega_e/2\pi \approx 10^{-3}\gamma B_0$ [360, 362]. This problem can be alleviated to some degree by audio-frequency coils with a large number of turns (about 1000 [360]) which, however, result in prolonged receiver ringing and thus longer dead times.

The experiment can be simplified by discarding the TX_{AF} channel if the signal is acquired point by point using a dwell time Δt_0 (cf. Fig. 24). Therefore, the evolution time in the MARF [365, 366] must be stepwise varied by $n\ \Delta t_0$.

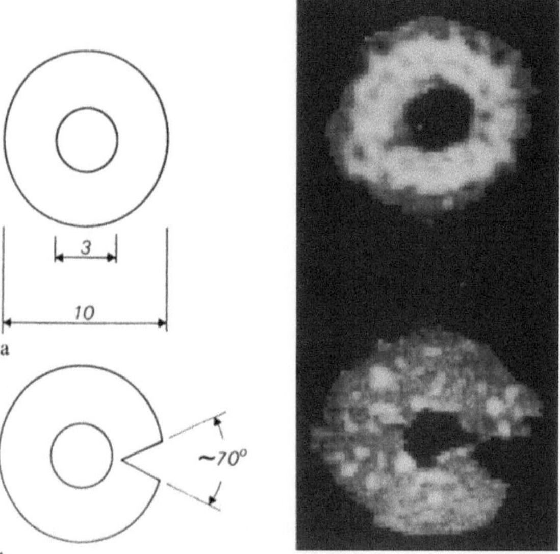

Fig. 25a, b. Phantoms (*left*) and MARF images (*right*) of compressed adamantane. The dimensions are given in mm. The sample thickness was 8 mm for sample **a** and 6 mm for sample **b**. The adamantane line width was about 20 kHz. The images were reconstructed from 16 projections measured with manual sample rotation. The experimental parameters with reference to Fig. 24 were: $TE = 1.2$ ms, $TR = 3$s, $G_1 = 14$ mT/m, $G_0 = 10$ mT/m, $G_e = 17$ mT/m, $\Omega = 31.2$ kHz, $\omega_e/2\pi =$ 54 kHz, $\omega_0/2\pi = 20$ MHz, $t_{RF} = 3.4$ µs, $t_{AF} = 40$ µs, and the total acquisition time was 1.3 h. The spatial resolution obtained was about 1 mm. The mottled appearance of the images is due to the poor definition by only a few numbers of projections. (Reproduced with permission from Maraviglia et al. [363])

Then the magnetization can be acquired in the RCF after application of a 90° pulse on the TX_{RF} channel. This can be considered as a variation of the early Lee-Goldburg experiment [369]. The method is more time-consuming and provides the same line narrowing but with increased S/N because the signal is detected in the RCF.

6 Selective Methods for Spectroscopy and Imaging

The combination of 1D spectroscopy and 3D imaging can be achieved by a 4D experiment which incorporates three spatial dimensions and a spectroscopic one. This type of imaging is most informative because a variety of image contrast parameters can be calculated from the NMR spectra. On the other hand, the experiment is extremely time consuming, because the dimensionality of the imaging scheme has to be augmented by one to include all dimensions. Selective-spectroscopy and localization methods acquire only a subset of the accessible spectral and spatial information of interest, so they becomes less expensive in terms of data acquisition time than imaging of all involved dimensions on the full scale.

6.1 Chemical Shift Selective Imaging

Often the complete NMR spectrum is not needed for generation of the desired image contrast, but only the spectral intensities at selected chemical shifts which vary over the sample volume. Such situations are more typical for liquid than for solid-state imaging because separate lines are observed. Typical examples are the acquisition of water and fat images in clinical applications [16, 19, 370] and in studies of plants [371, 372]. For the acquisition of images at two chemical shifts, a spin echo difference technique has been developed [373]. There two data sets are acquired: one with coinciding Hahn and gradient echoes; the other one with both echoes being delayed by the inverse frequency separation of both resonances. Sum and difference of both data sets produce images at either chemical shift.

If the image at only one chemical shift is to be measured, the magnetization in the frequency range of interest can be excited or the unwanted signals from frequencies outside are saturated by use of selective pulses. Selective pulses may be shaped in amplitude, phase, and frequency, or composed of hard pulses, so-called composite pulses. These are reviewed elsewhere [4, 8, 16, 20, 374–380].

In general, most selective excitation sequences are comparatively long. Therefore, they cannot be applied to solids where the signal decays rapidly, often within times of the order of a few microseconds. The few applications of chemical-shift selective imaging reported for the characterization of solids are restricted to differentiation of liquids in solids [120, 121, 123, 126–128] and chemical shift variations in rubber-like samples [139, 140, 144, 381, 382].

For chemical shift-selective excitation in solids, techniques based on DANTE (*d*elays *a*lternating with *n*utations for *t*ailored *e*xcitation) [155, 383, 384] or related multi-pulse techniques [385, 386] seem to be more successful. Maciel and Davis [227] used such a technique for ^{13}C-selective excitation in an imaging experiment, taking advantage of the relatively narrow and separated lines of the rare isotope (cf. Sect. 3.4). Other possibilities of spectroscopic selection may be spin locking (cf. next section) or the combination of selective pulses together with line-narrowing techniques.

6.2 Slice Selection and Localized Spectroscopy

Usually the dimensionality of the selected region is reduced in subsequent steps by repetitive application of different forms of selective excitation applied in orthogonal field gradients [4, 8, 16, 20, 387–392]. For instance, a narrow-bandwidth selective pulse (Fig. 26a right) excites a slice of transverse magnetiza-

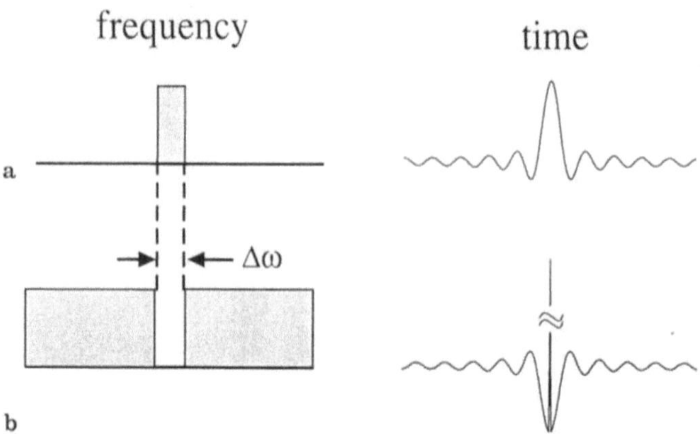

Fig. 26a, b. Frequency profiles (*left*) for different selective-excitation pulses (*right*): **a** excitation of magnetization within a given frequency window; **b** excitation of magnetization outside a given frequency window

tion (Fig. 26a left) perpendicular to the direction of the gradient. To saturate all magnetization except for the slice, a pulse with the inverse excitation profile, that is, broadband excitation with a narrow gap at the position of the line, is needed (Fig. 26b left). In general, selective excitation can be designed with either excitation profile for longitudinal or transverse magnetization.

Most selective-excitation techniques were developed for liquids, where T_2 is long, allowing the use of long lasting (narrow band) excitation. Solids, however, are characterized by long spin–lattice relaxation times T_1, which are usually longer than 1 s, sometimes even for ^{1}H NMR, and by short spin–spin relaxation times T_2 which are typically shorter than 1 ms. Thus the FID of rigid solids generally decays in times shorter than the necessary duration of a selective pulse. Therefore, techniques are needed that artificially slow down the decay of the transverse magnetization amplitude or avoid the generation of transverse magnetization altogether in order to select slices in short times. This can be achieved by use of spin lock pulses [157–160, 393], sequences of hard pulses [155, 384], and saturation and dephasing of unwanted longitudinal magnetization. A trivial but time-consuming solution is to measure a 3D image by the back-projection method, for instance, and compute slices through it at any angle [157]. Of course, methods designed for the liquid state can be used together with line-narrowing techniques (e.g., MAS) [345] for selective excitation, saturation, or localized spectroscopy.

6.2.1 *Selective Saturation of z-Magnetization*

By using a selective pulse with a wide excitation profile showing a hole at the frequency of the slice (Fig. 26b left), the longitudinal magnetization outside the slice in question can be saturated or converted to transverse magnetization for subsequent dephasing. The use of such pulses is favored by the large differences in T_1 and T_2 encountered in solids. In particular, there will be no phase twist and no intensity loss by dephasing of the selected magnetization. For good results, the pulse can be applied in a repetitive fashion. Pulses of this type are shaped pulses (Fig. 26b right) [394–396] or certain noise pulses [397–400]. Problems arise with the large signal bandwidths of solids requiring wide-band saturation and thus high rf power. Therefore, another approach of generating the desired pulse profile (Fig. 26b right) is a composite pulse consisting of two selective 45° pulses with a nonselective 90° pulse in the middle [401, 402]. By successive application of such pulses in orthogonal field gradients, longitudinal magnetization can be selected within a volume element for localized spectroscopy.

6.2.2 Spin-Lock Slice Selection

For solids, the most widely studied slice-selection technique uses spin locking in a magnetic field gradient [157–160, 393, 403]. It is applicable to solids as well as liquids [393, 404]. The *s*pin-*l*ock *i*nduced *s*lice *e*xcitation (SLISE) consists of a hard 90° pulse followed by a period during which the transverse magnetization is locked in the direction of the $\boldsymbol{B}_1$ field. During the lock pulse the locked magnetization relaxes with the relaxation time $T_{1\rho}$ in the rotating frame. The other magnetization dephases with a time constant T_2^* which is characteristic of the strength of the field gradient. If the gradient is strong enough, T_2^* is short compared to $T_{1\rho}$. Provided that the length τ_{SL} of the spin-lock period is longer than T_2^* and shorter than $T_{1\rho}$,

$$T_2^* < \tau_{SL} < T_{1\rho}, \tag{30}$$

the magnetization that survives the spin-lock period is the magnetization of the selected slice. Given a gradient G_x in x-direction, the slice profile is determined by the transverse component [393, 405]

$$M_{SL}(x) = M_0 \cos^2\left[\arctan \frac{xG_x}{B_1}\right], \tag{31}$$

with a width at half height of

$$\Delta x_{1/2} = \frac{2B_1}{G_x}. \tag{32}$$

The slice thickness is proportional to the amplitude B_1 of the spin-lock pulse and inversely proportional to the strength of the field gradient. Thus the thickness can be varied by changing $\boldsymbol{B}_1$ without changing the shape of a pulse, and it is independent of the length of the pulse which can be adjusted to yield $T_{1\rho}$ weighted image contrast. However, to guarantee the spin-lock effect, the lock field must be larger than the local fields in the sample which arise, for instance, from the homonuclear dipole–dipole coupling among ^{1}H.

The transverse magnetization selected by this SLISE excitation can be refocused by a nonselective pulse for subsequent gradient switching and imaging [157] or stored as longitudinal magnetization [157, 158]. Application of three such spin-lock sandwiches can be employed to select a volume element of longitudinal magnetization for localized spectroscopy [158]. This type of spatially resolved spectroscopy has been called *lo*ck pulse *s*elective spectroscop*y* (LOSY) [159, 160]. The method has been demonstrated to work on a phantom yielding localized ^{1}H spectra of hexamethylbenzene and polyethylene with about 3 mm spatial resolution [158] as well with cross polarization for localized ^{13}C spectroscopy [406]. The sensitivity of this technique can be further improved when the lock pulse is replaced by two adiabatic half passages [407, 408]. The concept of an adiabatic half passage can also be used for polarization transfer in heteronuclear localized spectroscopy [403, 406].

6.2.3 Sensitive Slice Selection

A rather simple technique for imaging and slice selection was derived from the early sensitive-point method [156, 409–412]. Free-induction decay signals are acquired in static field gradients which are stepped through a range of values with increasing scan number. The acquired signals are then added. In the signal

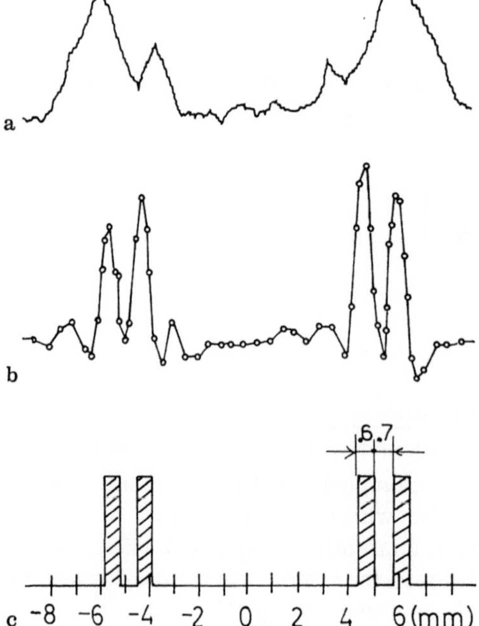

Fig. 27a–c. Images of a phantom of four sheets of PMMA acquired with a sensitive-point method. **a** Without gradient modulation. **b** With gradient modulation of a maximum strength of 650 mT/m. **c** Sample geometry. (Reproduced with permission from Rigamonti et al. [156])

sum, the only parts of the signal that add coherently are those for which the applied magnetic field is constant for all scans, that is, at the zero crossing of the gradient field. The other signal contributions are averaged out. For a single gradient, a slice is selected in this way.

With this technique the spatial resolution is not limited by the requirement that $\boldsymbol{B}_1$ must be larger than the largest frequency offset. In fact, selective excitation can improve the spatial selectivity. Thus the gradients can be increased to large values and slices down to tenth of millimeters in width can be obtained. The effectiveness of this method is demonstrated in Fig. 27 with a PMMA sample [156].

Given enough time for data acquisition, the method can be also used for point and line scanning in imaging similar to the different variants of the sensitive-point method [4]. This is a very simple realization of localized spectroscopy [412, 413] because it can be combined with various spectroscopic methods. In principle, there is no reason why only the gradient-independent signal average should be extracted from the acquired total signal. Because the gradient dependence is known for each voxel, the signal can be extracted for any voxel by suitable data processing. This introduces a multiplex advantage for all three space coordinates and the point method becomes a volume method [414].

6.2.4 Multi-Pulse Slice Selection

Multi-pulse sequences are used for line-narrowing by selective averaging of the homonuclear dipole–dipole interaction (cf. Sect. 5.1). Thus in their presence the length of the FID is effectively prolonged, and therefore time-extended selective excitation can be applied. Different schemes based on the MREV-8 sequence were designed for use with [384] and without [155] magic-angle spinning. A slice-selective train of pulses is either intergrated or interleaved with the multi-pulse sequence leading to analoga of the DANTE sequence used for selective excitation in high resolution NMR.

The original DANTE [383] sequence consists of a train of n small flip angle pulses, which are separated by free precession intervals Δt. The corresponding spectrum of the excitation sequence can be explained in a simplified way by linear response theory as an array of sinc functions with a width of $(n\Delta t)^{-1}$ spaced by Δt^{-1}. In these intervals, the magnetization precesses about the z-axis with a frequency which is determined by the resonance offset from the rf excitation. For slice selection, the dependence of this precession frequency on the chemical shift needs to be replaced by a dependence on the applied gradient field. To achieve this for rigid solids, the dominant dipole–dipole interaction needs to be overcome by replacing each segment in the original sequence, which consists of a pulse followed by a free precession interval, by a combination of two MREV-8 based multi-pulse sequences [155]. One sequence effects a precession around the z-axis which is independent of the chemical shift for all

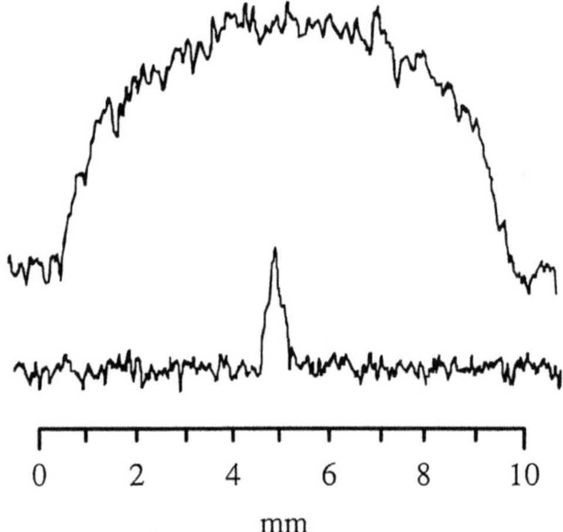

Fig. 28. Example for DANTE slice selection using a ferrocene cylinder. The upper 1D projection shows a cross-section of the sample without slice selection detected with an MREV-8 sequence. The lower slice was obtained by a DANTE sequence consisting of 50 single MREV-8 trains. The average gradient strength was about 55 mT/m. (Reproduced with permission from Cory et al. [155])

spins. In intervals between the rf pulses magnetic field gradients are applied so the precession frequency becomes dependent on the space coordinate. This requires short pulses of a few microseconds and eliminates excess line broadening induced by the offset dependence of the multi-pulse sequence (cf. Sect. 5.1). The other sequence causes a rotation around the y-axis which is identical for all spins. Second averaging by phase modulation of the rf pulses is used to cancel effects of chemical shift in addition to the dipole–dipole interaction.

The effectiveness of slice selection by the dipole-narrowed DANTE sequence is illustrated in Fig. 28 with 1D images of a ferrocene cylinder without and with slice selection [155]. The same MREV-8 sequence used for inducing the gradient-dependent z-precession in the DANTE sequence was applied for detection of the spatial response after slice selection. The technique is based solely on the use of hard pulses. The slice selection process does not interfere with the decoupling efficiency of the MREV-8 sequence, and distortions from chemical shift and susceptibility differences are absent.

Furthermore, the strong off-resonance dependence of multi-pulse experiments can be used for such selections. Because the offsets are caused by gradients, the failure of line-narrowing and corresponding rapid dephasing for off-resonance magnetization results in a slice selection for the on-resonance condition. Yannoni et al. [415, 416] used a certain multi-pulse sequence for that purpose where the offset dependence was increased by field or frequency modulation. The same idea was adopted for the offset dependence of MARF experiments (cf. Sect. 5.4) by Matsui [417].

6.2.5 *Localized Spectroscopy*

In many cases the morphology of the sample is known a priori by conventional NMR imaging or some other technique and the interest focuses on measuring the NMR spectrum in a selected volume for identification of the local chemistry. Then only the magnetization from the volume of interest needs to be excited for a subsequent acquisition of its spectrum by *localized spectroscopy* or *volume-selective spectroscopy*. Different techniques were designed to this end for liquid samples [4, 8, 16, 20, 418–428]. Techniques applied up to now for localized spectroscopy in solid-state NMR are the SLISE excitation with spin-lock pulse sandwiches applied in orthogonal field gradients, and the generation of stimulated and alignment echoes where each pulse is applied in a different field gradient for selective excitation [158].

6.3 *Surface Coils and Surface Methods*

NMR imaging of solids is restricted to small samples because the rf power scales with the fifth power of the sample dimensions and with the second power of the field strength [14]. Then sample diameters are restricted to less than 10 mm or the excitation is restricted to selected regions of the sample by the use of surface coils or short coils with large diameters which surround the sample.

Surface coils [429–436] are widely used in medical imaging to improve the signal-to-noise ratio for localized imaging and spectroscopy [436–441]. In most medical applications the rf excitation is applied with a large "whole-body" coil while the signal is detected with the surface coil (decoupled from the excitation coil) during an imaging or a spectroscopic experiment [442, 443]. The spatial localization in such an experiment is determined by the coil diameter and the decay in the receptivity of the surface coil. As a rule of thumb, the sensitive volume in such an experiment is given by half the sphere defined by the coil diameter.

A similar arrangement can be used in solids for slice selective spectroscopic imaging without a gradient. The sample is homogeneously excited with a transmitter coil which is much bigger than the receiver coil [253]. Then only the magnetization of a thin slice defined by the location of the sample relative to the receiver coil is detected. By repositioning the sample, 1D projections can be measured which must be deconvoluted (cf. Sect. 3.5) from the "blurring" by the space-dependent sensitivity of the receiver coil. The required calibration can be done, for instance, with a phantom of known geometry. The technique is very effective for passive slice selection and 1D imaging in combination with line-narrowing techniques. Because no gradients are employed, the residual band-width is small preventing loss of multipulse decoupling efficiency at large resonance offsets (cf. Sect. 5.1)

If a single surface coil is used for transmitting and receiving, the space dependence of the $\boldsymbol{B}_1$ contributes quadratically the signal amplitude. In this case, the space coordinate normal to the coil can be encoded without additional gradients, because the gradient introduced by $\boldsymbol{B}_1$ itself can be exploited. The majority of experiments, which make use of this $\boldsymbol{B}_1$ decay for spectroscopic imaging in liquids, can be classified into three groups: (a) nutation or rotating-frame techniques where the nominal excitation flip angle is varied systematically in a 2D fashion. Fourier transformation produces an axis which scales nonlinearly with space [444–451]; (b) selection of a localized volume by depth-selective pulse sequences [390, 429–432, 452–460]; (c) combinations of (a) and (b) [461–463]. Different geometries and technical modifications of surface coils were proposed for these purposes in single- [42, 436, 438, 442, 443, 462, 464–470] and double-resonance modes [471–476]. The methodical and technical aspects for liquid-state applications have been reviewed by M. R. Bendall [441].

The concept of depth-selective pulses was applied to solids by Miller and Garroway [477, 478]. They used a train of nominal 180° pulses to excite a thin slice by inversion of the magnetization in the inhomogeneous $\boldsymbol{B}_1$ field of a surface coil. For all other locations, transverse magnetization components remain and dephase if the duration between the pulses is longer than T_2. The position of the slice can than be moved through the sample by changing the rf power and thus the nutation angle. This concept can be used for slice selection in a preparation period followed by an MREV-8 sequence to average the homonuclear dipole–dipole coupling to zero [477] and for direct acquisition if the 180° pulses are used as part of a dipole decoupling sequence [478]. The first method is illustrated in Fig. 29 by depth-selective spectra of an adamantane sandwich. The fact that all three adamantane disks of the phantom are resolved, illustrates that certain multi-pulse sequence like the MREV-8 sequence can be used even in strongly inhomogeneous $\boldsymbol{B}_1$ fields.

The rf field inhomogeneity of surface coils can be overcome by composite [376] or adiabatic pulses [407, 408]. The latter approach allows the spins to align along local fields independent of the strength and offset of the excitation field.

An approach related to surface coil imaging is *s*kin *e*ffect *e*nhanced *i*magi*ng* (SEEING) [429, 480], which makes use of $\boldsymbol{B}_1$ attenuation by the skin effect in conducting materials [481]. The absorption of the excitation field results in a $\boldsymbol{B}_1$ gradient strength of about 2 T/m, but which in addition is a function of the sample geometry. This complicates investigations of inhomogeneous layers. By variation of the nutation angle of a single excitation pulse, the depth can be encoded and separated from the chemical shift but the phase of the recorded signal is a complex function of depth. Because the response to even a single pulse already consists of a superposition of signals with different phases from different depths [480, 482, 483], all slices in a 2D nutation experiment have to be phased by hand. In principle, there is no signal-to-noise limitation so the spatial resolution can be better than 1 μm (for Ag or Au) depending on the conductivity of the sample. Interesting applications of this method are conceivable, for example, the detection of chemical change in the layers of metallic composites [479] or catalytic converters [480].

For NMR investigations of objects too large to fit into any magnet, surface coil techniques can be combined with planar or shaped $\boldsymbol{B}_0$ magnets remote

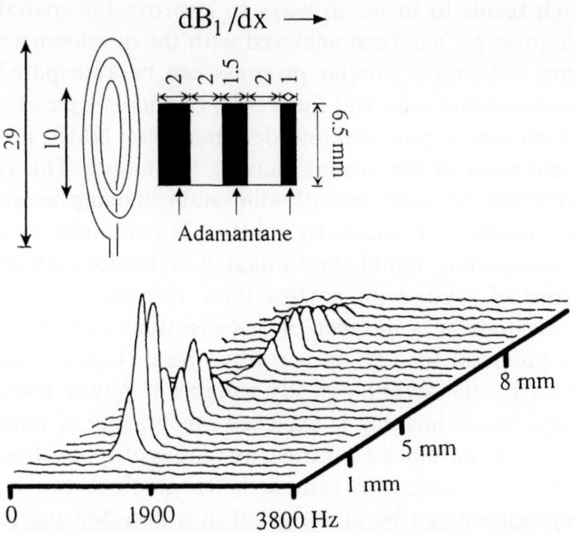

Fig. 29. Surface coil NMR spectra of an adamantane sandwich with teflon spacers. Sample and surface coil geometry are shown at the top. The spectra were measured with MREV-8 line narrowing after preparation with a sequence of 180° pulses for slice selection. The depth was scanned by changing the rf power of preparation and detection pulses. All three adamantane disks are well resolved with good line narrowing of the original 10 kHz line width. Note that the spatial axis is not linear. (Reproduced with permission from Miller et al. [477])

from the object of interest. Such a configuration permits the measurement of depth profiles of arbitrarily large objects which can also be remote from the spectroscopist [484–487]. Although the depth is restricted to a few centimeters, this approach can be very useful for scanning large objects consisting of thin shells or sheets, for example, airplane wings. The concept of combined surface magnet and surface coil for creating a sensitive plane with variable depth was used to detect moisture in soil and highway bridge decks by a nutation technique [488] as well as for in situ analysis of bore holes in oil prospecting [484–486]. It is conceivable that this approach can be adapted to NMR investigations of large solid objects.

7 Summary and Outlook

An overview of imaging techniques and applications to solids has been given. The field of solid-state imaging is currently enjoying vigorous development, but genuine applications are scarce. For this reason, several references to liquid-state imaging techniques have been included, which are believed to mark developments of interest to solid-state applications.

Most methodical research seems to focus on ways to improve the spatial resolution, and considerable progress has been achieved with the development of multi-pulse line-narrowing techniques. Similar progress can be anticipated for the development of surface-coil methods. But the wealth of microscopic and macroscopic parameters, which can be provided and determined by NMR, has not yet been incorporated into most of the current imaging techniques. This is believed to be the major obstacle to wider use of solid state imaging as an analytical research tool. Optimization of parameter contrast is well under way for the experimentally less demanding liquid-state imaging techniques which can be used for investigations of solids with narrow lines. For this class of materials, in particular for elastomers, a number of investigations have been reported which show great potential for use on a routine basis. Despite the limits in sample size imposed by the technologically suitable rf power level, a development similar to liquid-state imaging is expected for imaging of rigid solids. By proper exploitation of the parameter contrast, new areas of application can be identified where previously unknown sample heterogeneities can be localized and their time dependence can be investigated in a non-destructive fashion.

Acknowledgments. The authors would like to thank T. A. Carpenter, D. G. Cory, A. N. Garroway, L. D. Hall, R. Kimmich, R. A. Komoroski, B. Maraviglia, S. Matsui, J. B. Miller, A. Rigamonti, J. H. Strange, B. H. Suits, W. S. Veeman, and K. Zick for permission to use their results in this review. A special note of thanks is attributed to Gunther Schauß for imaging our drained heads (Fig. 1). Furthermore, we thank Carsten Fülber, Ewald Günther, Jan Jansen, Helge Nilgens, Jürgen Paff, Ulrich Scheler, and Frank Weigand for checking the manuscript and their excellent collaboration in the imaging group. We owe our particular gratitude to Michael Hansen, Manfred Wilhelm, and the Titman-Geen family for checking the manuscript and for many helpful discussions. Last but not least, the enlightening discussions, the continuous support, and the interest of H. W. Spiess in this work are gratefully acknowledged.

8 References

1. Mehring M (1983) Principles of high resolution NMR in solids. Springer, Berlin Heidelberg, New York
2. Gerstein BC, Dybowski CR (1985) Transient techniques in NMR of solids. Academic, London
3. Fyfe CA (1983) Solid state NMR for chemists. CFC, Guelph Canada
4. Mansfield P, Morris PG (1982) NMR imaging in biomedicine. In: Waugh JS (ed) Advances in Magnetic Resonance Suppl. 2. Academic
5. Mansfield P (1988) J. Phys. E: Sci. Instrum. 21: 18
6. Mansfield P in Ref. [18], p 345
7. Mansfield P, Pykett IL (1978) J. Magn. Reson. 29: 355
8. Callaghan PT (1991) Principles of nuclear magnetic resonance microscopy. Clarendon Press, Oxford

9. Andrew ER, Fitzsimmons JR (1987) Bull. Magn. Reson. 9(3): 53
10. Cory DG (1992) Ann. Reports on NMR 24: 87
11. Blümich B, Kuhn W (eds) Magnetic resonance microscopy. Verlag Chemie, Weinheim
12. Miller JB, Cory DG, Garroway AN (1990) Phil. Trans. R. Soc. Lond. A 333: 413
13. Kuhn W (1990) Angew. Chem. Int. Ed. Engl. 29: 1
14. Krestel E (1990) Imaging systems for medical diagnosis. Siemens AG, Berlin
15. Chen CN, Hoult (1989) DI biomedical magnetic resonance. Adam Hilger, New York
16. Partain CL, Price RR, Patton JA, Kulkarni MV, James AE (eds) (1988) Magnetic resonance imaging, vol 2. Saunders WB, Philadelphia
17. Talagala SL, Lowe IJ (1991) Concepts Magn. Reson. 3: 145
18. Maraviglia B (ed) (1988) Physics of NMR spectroscopy in biology and medicine. Elsevier, Amsterdam
19. Partin CL, Price RR, Patton JA, Kulkarni MV, James AE (eds) (1988) Magnetic resonance imaging, vol. 1. Saunders WB, Philadelphia
20. Morris PG (1986) Nuclear magnetic resonance imaging in medicine and biology. Clerendon, Oxford
21. Wehrli FW, Shaw D, Kneeland JB (1988) Biomedical magnetic resonance imaging. VCH, Weinheim
22. Newhouse VL (ed) (1988) Progress in medical imaging. Springer, Berlin Heidelberg, New York
23. Higgins CB Hricak H (eds) (1987) Magnetic resonance of the body. Raven, New York
24. Gordon R (ed) (1984) Industrial applications of computed tomography and NMR imaging. Book of abstracts from the topical meeting. Hecla island, Manitoba Canada
25. Jezzard P, Wiggins CJ, Carpenter TA, Hall LD, Jackson P, Clayden NJ, Walton NJ (1992) Adv. Mater. 4(2): 82
26. Listerud JM, Sinton SW, Drobny GP (1989) Anal. Chem. 61: 23A
27. Ackerman JL, Ellingson WA (eds) (1991) Advanced tomographic imaging methods for the analysis of materials. Materials research society, symposium proceedings vol 217
28. Kose K (1991) Phys. Rev. A 44(4): 2495
29. Singer JR (1978) J. Phys. E: Sci. Instrum. 11: 281
30. Caprihan A, Fukushima E (1990) Phys. Rep. 198: 195
31. Ikeya M (1992) New applications of electron spin resonance. World scientific, London
32. Eaton GR, Eaton SS, Ohno K (eds) (1991) EPR imaging and In Vivo EPR. CRC, London
33. Berliner LJ in Ref. [11], p 151
34. Zommerfelds W, Hoch MJR (1986) J. Magn. Reson. 67: 177
35. Keven L, Bowman MK (1990) Modern pulsed and continuous wave electron spin resonance. John Wiley, New York
36. Eaton GR, Eaton SS (1988) Bull. Magn. Reson. 10 (1, 2): 22
37. Matsui S, Kose K, Inouye T (1990) J. Magn. Reson. 88: 186
38. Rommel E, Nickel P, Kimmich R, Pusiol D (1991) J. Magn. Reson. 91: 630
39. Nickel P, Rommel E, Kimmich R, Pusiol D (1991) Chem. Phys. Lett. 183 (3, 4): 183
40. Rommel E, Kimmich R, Robert H, Pusiol D (1992) Meas. Sci. & Techn. 3: 446
41. Turner R (1986) J. Phys. D: Appl. Phys. 19: L147
42. Styles P in Ref. [18], p 412
43. Henkelman RM, Bronskill MJ (1987) Rev. Magn. Reson. Med. 2(1): 1
44. Gonzales RC, Wintz P (1987) Digital image processing. Addison-Wesley, New York
45. Sabatier PC (ed) (1987) Basic methods of tomography and inverse problems. Adam Hilger, Bristol
46. Braddick OJ, Sleigh AC (1983) Physical and biological processing of images. Springer, Berlin Heidelberg New York
47. Pratt WK (1978) Digital image processing. Wiley, New York
48. Blackledge MJ (1989) Quantitative coherent imaging. Academic, London
49. Baum J, Pines A (1976) J. Am. Chem. Soc. 108: 7447
50. Munowitz M, Pines A, Mehring M (1987) J. Chem. Phys. 86: 3172
51. Caravatti P, Neuenschwander P, Ernst RR (1986) Macromolecules 19: 1889
52. VanderHart DL (1987) J. Magn. Reson. 72: 13
53. Kärger J, Pfeifer H, Heink W (1988) Adv. Magn. Reson. 12: 1
54. Stilbs P (1987) Prog. NMR Spectrosc. 19: 1
55. Cory DG, Garroway AN (1990) Magn. Reson. Med. 14: 435
56. Fujara F, Fleischer D, in this issue.

57. Chmurny G, Hoult DI (1990) Concepts Magn. Reson. 2: 131
58. Lauterbur PC (1972) Bull. Am. Phys. Soc. 18: 86
59. Lauterbur PC (1973) Nature 242: 190
60. Lauterbur PC (1974) Pure Appl. Chem. 40: 149
61. Mansfield P, Grannell PK (1973) J. Phys. C: Solid state phys. 6: L422
62. Mansfield P, Grannell PK (1975) Phys. Rev. B 12: 3618
63. Eccles CD, Callaghan PT (1986) J. Magn. Reson. 68: 393
64. Callaghan PT, Eccles CD (1987) J. Magn. Reson. 71: 426
65. Hay GA (1978) J. Phys. E: Sci. Instrum. 11: 377
66. Rugar D, Yannoni CS, Sidles JA (1992) Nature 360: 563
67. Sidles JA (1991) Appl. Phys. Lett. 58: 2854
68. Sidles JA (1991) Phys. Rev. Lett. 68: 1124
69. Dieckman SL, Rizo P, Gopalsami N, Botto RE in Ref. [27], p 169
70. Banci L, Bertini I, Luchinat C (1991) Nuclear and electron relaxation. VCH, New York
71. Blümler P, Blümich B (1991) Macromolecules 24: 2183
72. Blümler P, Dumler H, Blümich B (1992) Kautschuk + Gummi, Kunststoffe 45: 699
73. Blümler P, Blümich B (1992) Mag. Reson. Imaging 10(5): 779
74. Kuhn W, Koeller E, Theis I in Ref. [11], p 217
75. Kuhn W, Theis I, Koeller E in Ref. [27], p 33
76. Guilfoyle DN, Mansfield P (1992) J. Magn. Reson. 97: 342
77. Gaigalas AK, Chidester A van Orden, Robertson B, Mareci TH, Lewis LA (1989) Nuclear technology 88: 113
78. Sinton SW, Iwamiya JH, Chow AW in Ref. [27], p 73
79. Xia Y, Callaghan PT (1991) Macromolecules 24: 4777
80. Icenogle MV, Caprihan A, Fukushima E (1992) J. Magn. Reson. 100: 376
81. Cory DG, Garroway AN, Miller JB (1990) Polymer preprints 31(1):149
82. Ilg M, Maier-Rosenkranz J, Müller W, Albert K, Bayer E, Höpfel D (1992) J. Magn. Reson. 96: 335
83. Ilg M (1990) PhD Thesis, University of Tübingen, Germany
84. Rothwell WP, Gentempo PP (1985) Bruker Report 1: 46
85. Blackband S, Mansfield P (1986) J. Phys. C: Solid state phys. 19: L49
86. Webb AG, Hall LD (1990) Polym. Commun. 31: 422
87. Webb AG, Hall LD (1990) Polym. Commun. 31: 425
88. Pütz B, Barsky D, Schulten K (1992) J. Magn. Reson. 97: 27
89. Weisenberger LA, Koenig JL (1989) J. Polym. Sci. C: Polym. Lett. 27: 55
90. Weisenberger LA, Koenig JL (1989) Appl. Spectroscopy 43: 1117
91. Smith SR, Koenig JL (1991) Macromolecules 24: 3496
92. Koening JL in Ref. [11], p 187
93. Asakura T, Demura M, Ogawa H, Matsushita K, Imanari M (1991) Macromolecules 24: 620
94. Ilg M, Albert K, Rapp W, Bayer E (1990) J. Magn. Reson. 90: 370
95. Ilg M, Pfleiderer B, Albert K, Rapp W, Bayer E in Ref. [27], p 27
96. Majors PD, Smith DM, Caprihan A in Ref. [27], p 43
97. Tabak F, Corti M (1990) J. Chem. Phys. 92:, 2673
98. Mareci TH, Donstrup S, Rigamonti A (1988) J. Mol. Liquids 38: 185
99. Weisenberger LA, Koenig JL (1990) Macromolecules 23: 2445
100. Weisenberger LA, Koenig JL (1990) Macromolecules 23: 2454
101. Clough RS, Koenig JL (1989) J. Polym. Sci. C: Polym. Lett. 27: 451
102. Fyfe CA, Randall LH, Burlinson NE (1992) Poster WP 168 at 33rd ENC. Asilomar, CA
103. Randall LH, Fyfe CA, Mei Z, Whitworth S (1992) Bull. Magn. Reson. 14: 108
104. Chingas GC, Milliken J, Resing HA, Tsang T (1985) Synthetic metals 12: 131
105. Perry BC, Koenig JL (1989) J. Polym. Sci. A: Polym. Chem. 27: 3429
106. de Crespigny AJS, Carpenter TA, Hall LD, Webb AG (1991) Polym. Commun. 32(2):36
107. Jackson P, Barnes JA, Clayden NJ, Carpenter TA, Hall LD, Jezzard P (1990) J Mater. Sci. Lett. 9: 1165
108. Pickap S, Blum FD (1987) Spectroscopy 2: 53
109. Hoh K, Perry B, Rotter G, Ishida H, Koenig JL (1989) J. Adhesion 27: 245
110. Samoilenko AA, Artemov DYu, Sibeldina LA (1987) Bruker Report 2: 30
111. Rothwell WP, Holecek DR, Kershaw JA (1984) J. Polym. Sci: Polym. Lett. 22: 241
112. Schuff N, Hornung PA, Williams EH in Ref. [27], p 55

113. Horsfield MA, Fordham EJ, Hall C, Hall LD (1989) J. Magn. Reson. 81: 593
114. Hall LD, Rajanayagam V (1986) Wood Sci. Technol. 20: 329
115. Rogers H, Bottomley PA (1987) Agronomy Journal 79: 957
116. Link J, Seelig J (1990) J. Magn. Reson. 89: 310
117. Bottomley PA, Rogers H, Foster TH (1986) Proc. Natl. Acad. Sci. USA 83: 87
118. Cofer GP, Brown JM, Johnson GA (1989) J. Magn. Reson. 83: 608
119. Attard JJ, Doran SJ, Herrod NJ, Carpenter TA, Hall LD (1992) J. Magn. Reson. 96: 514
120. Nesbitt GJ, Fens TW, van den JS Brink, Roberts N, in Ref [11], p 285
121. Dechter JJ, Komoroski RA, Ramaprasad S (1991) J. Magn. Reson. 93: 142
122. Komoroski RA, Sarkar SN (1991) Mat. Res. Soc. Symp. Proc. 217: 3
123. Horsfield MA, Hall C, Hall LD (1990) J. Magn. Reson. 87: 319
124. Dereppe JM, Schenker KV (1989) Bruker Report 2: 9
125. Dereppe JM, Moreaux C, Schenker KV (1991) J. Magn. Reson. 91: 596
126. Majors PD, Smith JL, Kovarik FS, Fukushima E (1990) J. Magn. Reson. 89: 470
127. Hall LD, Rajanayagam V, Hall C (1986) J. Magn. Reson. 68: 185
128. Hall LD, Rajanayagam V (1987) J. Magn. Reson. 74: 139
129. Randall LH, Sedgwick GE, Fyfe CA (1992) Bull. Magn. Reson. 14: 186
130. Osment PA, Packer KJ, Taylor MJ, Attard JJ, Carpenter TA, Hall LD, Herrod NJ, Doran SJ (1990) Phil. Trans. R. Soc. Lond. A 333:441
131. Ellingson WA, Ackerman JL, Garrido L, Wey JD, DiMilia RÁ (1987) Ceram. Eng. Sci. Proc. 8: 503
132. Ackerman JL, Garrido L, Ellingson WA, Weyand JD (1988) in: Nondestructive testing of high-performance ceramics. American Chemical Society, Westerville OH, p 88
133. Ackerman JL, Garrido L, Moore JR, Pfleiderer B, Wu Y in Ref. [11], p 237
134. Hayashi K, Kawashima K, Kose K, Inouye T (1988) J. Phys. D: Appl. Phys. 21: 1037
135. Garrido L, Ackerman JL, Ellingson WA (1990) J. Magn. Reson. 88: 340
136. Carpenter TA, Hall LD, Jezzard P, Wiggins CJ, Clayden NJ, Jackson P, Walton N in Ref. [11], p 267
137. Jackson P, Clayden NJ, Walton NJ, Carpenter TA, Hall LD, Jezzard P (1991) Polym. Int. 24: 139
138. Jezzard P, Wiggins CJ, Carpenter TA, Hall LD, Jackson P, Clayden NJ, Walton NJ, Polym. Int. (in press)
139. Günther U, Albert K (1992) J. Magn. Reson. 98: 593
140. Albert K, Günther U, Ilg M, Bayer E, Grossa M in Ref. [11], p 277
141. Nieminen AOK, Koenig JL (1988) J. Adhesion Sci. Technol. 2: 407 (1988)
142. Nieminen AOK, Koenig JL (1989) J. Adhesion 30: 47
143. Nieminen AOK, Liu J, Koenig JL (1989) J. Adhes. Sci. Technol. 3: 445
144. Nieminen AOK, Koenig JL (1989) Appl. Spectroscopy 43: 1358
145. Schauß G, Blümich B, Spiess HW (1991) J. Magn. Reson. 95: 437
146. McFarland EW (1992) Magn. Reson. Imag. 10: 269
147. Metz CE, Doi K (1979) Phys. Med. Biol. 24: 1079
148. Herman GT (1979) Image reconstruction from projections implementation and applications, Springer, Berlin
149. Kumar A, Welti D, Ernst RR (1975) J. Magn. Reson. 18: 69
150. Edelstein WA, Hutchinson JMS, Johnson G, Redpath TW (1980) Phys. Med. Biol. 25: 751
151. Ljunggren S (1983) J. Magn. Reson. 54: 338
152. Feiner LF, Lochner PR (1980) Appl. Phys. 22: 257
153. Lochner PR (1984) Bull. Magn. Reson. 6: 140
154. Carpenter TA, Hall LD, Jezzard P (1989) J. Magn. Reson. 84: 383
155. Cory DG, Miller JB, Garroway AN (1990) J. Magn. Reson. 90: 544
156. Corti M, Borsa F, Rigamonti A (1988) J. Magn. Reson. 79: 21
157. Cottrell SP, Halse MR, Ibbett DA, Boda-Novy BL, Strange JH (1991) Meas. Sci. Technol. 2: 860
158. Hafner S, Rommel E, Kimmich R (1990) J. Magn. Reson. 88: 449
159. Rommel E, Kimmich R (1989) J. Magn. Reson. 83: 299
160. Rommel E, Kimmich R (1989) Bull. Magn. Reson. 11(3, 4): 169
161. Ernst RR, Bodenhausen G, Wokaun A (1987) Principles of NMR in one and two dimensions. Clarendon, Oxford
162. Henning J (1991) Concepts Magn. Reson. 3: 125
163. Henning J (1991) Concepts Magn. Reson. 3: 179
164. Dasch CJ (1992) Appl. Optics 31(8): 1146

165. Haase A, Frahm J, Matthaei D, Hänicke W, Merboldt KD (1986) J. Magn. Reson. 67: 258
166. Haase A, Matthaei D, Barthkowski R, Dümke E, Leibfritz D (1989) J. Comp. Ass. Tom. 13: 1036
167. Haase A (1990) Magn. Reson. Med. 13: 77
168. Gyngell ML (1988) Magn. Reson. Imag. 6: 415
169. Cottrell SP, Halse MR, Strange JH (1990) Meas. Sci. Techol. 1: 624
170. Mat Daud Y, Halse MR (1992) Physica B 176: 167
171. Mat Daud Y, Halse MR, Strange JH (1992) In: Anagnostopoulos A, Milia F, Simopoulos A (eds) Extendend Abstracts of 26th Congress Ampere, Athens
172. Strange JH (1990) Phil. Trans. R. Soc. Lond. A 333: 427
173. Ogura Y, Sekihara K (1989) J. Magn. Reson. 83: 177
174. Ogura Y, Sekihara K (1990) J. Magn. Reson. 88: 359
175. Ogura Y, Sekihara K (1991) J. Magn. Reson. 92: 490
176. Matsui S, Kohno H (1986) J. Magn. Reson. 70: 157
177. Matsui S, Sekihara K, Shiono H, Kohno H, (1988) J. Magn. Reson. 77: 182
178. Cory DG (1989) J. Magn. Reson. 82: 337
179. Hall LD, Sukumur S (1984) J. Magn. Reson. 56: 314
180. Lauterbur PC, Levin DN, Marr RB (1984) J. Magn. Reson. 58: 536
181. Bernardo ML, Lauterbur PC, Hedges LK (1985) J. Magn. Reson. 61: 168
182. Jansen J, Blümich B (1992) J. Magn. Reson. 99: 525
183. Cory DG, Miller JB, Garroway AN, Veeman WS (1989) J. Magn. Reson. 85: 219
184. Callaghan PT (1984) Aust. J. Phys. 37: 369
185. Frydman L, Barrall G, Harwood J, Chingas GC in Ref. [11], p 373
186. Callaghan PT (1989) Bull. Magn. Reson. 11(3,4): 216
187. Cho ZH, Yi JH, Friedenberg RM (1992) Rev. Magn. Reson. Med. 4: 221
188. Callaghan PT (1990) J. Magn. Reson. 87: 304
189. Cory DG in Ref. [11], p 49
190. Moonen CTW, Soberling G, van Zijl PCM, Gillen J, von Kienlin M, Bizzi A (1992) J. Magn. Reson. 98: 556
191. Spiess HW (1985) Adv. Polym. Science 66: 23
192. Spiess HW (1982) In: Ward IM (ed) Development in oriented polymers 1. Applied Science, Barking
193. Suits BH, White D (1986) J. Appl. Phys. 60(10): 3772
194. Jezzard P, Carpenter TA, Hall LD, Jackson P, Clayden NJ (1991) Polym. Commun. 32(3): 74
195. Waller M, Fiddy MA, Leeman S in Ref. [24]. p MD2
196. Jackson P (1992) J. Mater. Sci. 27: 1302
197. Blümich B, Blümer P, Günther E, Schauß G, Spiess HW (1991) Makromol. Chem. Macromol. Symp. 44: 37
198. Blümich B, Blümler P, Günther E, Schauß G (1990) Bruker Report 2: 22
199. Blümich B, Blümler P, Günther E, Jansen J, Schauß G, Spiess HW in Ref. [11], p 167
200. Blümler P, Blümich B (1993) Acta Polymerica 44: 125
201. Campanella R, Casieri C, de Luca F, de Simone BC, Maraviglia B in Ref. [18], p 451
202. Chang C, Komoroski RA (1989) Macromolecules 22: 600 (1989)
203. Chang C, Komoroski RA (1991) In: Mathias L (ed) Solid state NMR of polymers. Plenum, New York, p 363
204. Sarkar SN, Komoroski RA (1992) Macromolecules 25: 1420
205. Komoroski RA, Sarkar SN in Ref. [11], p 203
206. Komoroski RA, Sarkar SN in Ref. [27], p 3
207. Garrido L, Ackerman JL, Mark JE (1990) Mat. Res. Soc. Symp. Proc. 171: 65
208. Garrido L, Mark JE, Sun CC, Ackerman JL, Chang C (1991) Macromolecules 24: 4067
209. Sinton SW, Iwamiya JH, Ewing B, Drobny GP (1991) Spectroscopy 6(3): 42
210. Webb AG, Jezzard P, Hall LD, Ng S (1989) Polym. Commun. 30: 363
211. Garrido L, Pfleiderer B, Ackerman JL, Moore J in Ref. [27], p 49
212. Chudek JA, Hunter G (1992) J. Mat. Sci. Lett. 11: 222
213. Suits BH, White D (1984) Solid state commun. 50(4): 291
214. Suits BH, Lutz JL (1989) J. Appl. Phys. 65(9): 3728
215. Suits BH in Ref. [27], p 67
216. Günther E, Blümich B, Spiess HW (1992) Macromolecules (1992) 25: 3315
217. Günther E, Raich H, Blümich B (1991/92) Bruker Report p 15
218. Haeberlen U (1976) High resolution NMR in solids: Selective averaging. Adv. Magn. Reson. Suppl. 1, Academic, New York

219. Slichter CP (1990) Principles of magnetic resonance. Springer, Berlin Heidelberg New York
220. Spiess HW (1978) NMR Basic principles and progress 15: 55
221. Miller JB, Garroway AN (1986) J. Magn. Reson. 67: 575
222. Cory DG, Veeman WS (1989) J. Phys. E: Sci. Instrum. 22: 180
223. Schenker KV, Suter D, Pines A (1987) J. Magn. Reson. 73: 99
224. Pines A, Gibby MG, Waugh JS (1973) J. Chem. Phys. 59(2): 569
225. Fry CG, Lind AC, Davis MF, Duff DW, Maciel GE (1989) J. Magn. Reson. 83: 656
226. Szeverenyi NM, Maciel GE (1984) J. Magn. Reson. 60: 460
227. Davis MF, Maciel GE (1991) J. Magn. Reson. 94: 617
228. Maciel GE, Davis MF (1985) J. Magn. Reson. 64: 356
229. Ackerman JL, Garrido L (1988) Poster at 7th Annual Meeting of the Society of Magnetic Resonance in Medicine, San Francisco
230. Kinchesh P, Randall EW, Williams SCR (1992) J. Magn. Reson. 98: 458
231. Bendel P, Davis M, Berman E, Kabalka GW (1990) J. Magn. Reson. 88: 369
232. Kabalka GW, Cheng GQ, Bendel P in Ref. [11], p 533
233. Liang ZP, Boada FE, Constable RT, Haacke EM, Lauterbur PC, Smith MR (1992) Rev. Magn. Reson. Med. 4: 67
234. Vaccaro R (1991) SVD signal processing II. Elsevier, New York
235. de Beer R, van Ormondt D (1992) NMR Basic principles and progress 26: 201
236. Delsuc MA, Levy GC (1988) J. Magn. Reson. 76: 306
237. Mazzero AR, Delsuc MA, Kumar A, Levy GC (1989) J. Magn. Reson. 81: 512
238. Belton PS, Wright KM (1986) J. Magn. Reson. 68: 564
239. Busse LJ, Pratt RG, Thomas SR (1988) J. Comput. Assist. Tomogr. 12(5): 824
240. Spielman D, Webb P, Macovski A (1988) J. Magn. Reson. 79: 66
241. van der Veen JWC, de Beer R, Luyten PR, van Ormondt D (1988) Magn. Reson. Med. 6: 92
242. Chi CY (1991) IEEE Transactions on signal processing 39(9): 2082
243. Barkhuysen H, de Beer R, Drogendijk AC, van Ormondt D, van der Veen JWC in Ref. [18], p 313
244. Bretthorst GL (1990) J. Magn. Reson. 88: 533
245. Bretthorst GL (1990) J. Magn. Reson. 88: 552
246. Bretthorst GL (1990) J. Magn. Reson. 88: 571
247. Brosnan T, Wright G, Nishimura D, Cao Q, Macovski A, Sommer FG (1988) Magn. Reson. Med. 8: 394
248. Macovski A, Spielman D (1986) Magn. Reson. Med. 3: 97
249. Hoch MJR, Day AR (1979) Solid state commun. 30: 211
250. Dahm J, Macho V, Maresch GG, Spiess HW (1991) Mol. Phys. 74: 591
251. Dahm J (1991) PhD Thesis, University of Mainz, Germany
252. Morris GA (1988) J. Magn. Reson. 80: 547
253. Cory DG, Miller JB, Garroway AN (1990) Meas. Sci. Technol. 1: 1338
254. Liu J, Nieminen AOK, Koenig JL (1989) Appl. spectroscopy 43(7): 1260
255. Cory DG, Reichwein AM, Veeman WS (1988) J. Magn. Reson. 80: 259
256. Veeman WS in Ref. [11] p 29
257. Majors PD, Caprihan A (1991) J. Magn. Reson. 94: 225
258. Hansen EW, Lan P, in Ref. [24], p TuC4
259. Pearlman JD, Leavitt M, Newell JB (1990) IEEE Transactions on medical imaging 9: 461
260. Haacke EM, Liang Z, Izen SH (1989) IEEE Transactions on acoustics, speech, and signal processing 37: 592
261. Sato T, Norton SJ, Linzer M, Ikeda O, Hirama M (1981) Applied optics 20: 395
262. Sati T, Saski K, Nakamura Y, Linzer M, Norton SJ (1981) Applied optics 20: 3073
263. Dhawan AP, Rangayyan RM, Gordon R, in Ref. [24], p TuA5
264. Blümler P, Greferath M, Blümich B, Spiess HW (1993) J. Magn. Reson. A103: 142
265. Damadian R, Minkoff L, Goldsmith M, Stanford M, Koutcher (1976) Physiol. Chem. Phys. 8: 61
266. Damadian R (1976) Science 194: 1430
267. Ewert W, Thiessenhusen KU in Ref. [32], p 119
268. Emid S, Creyghton JHN (1985) Physica B 128: 81
269. Garroway AN, Baum J, Munowitz MG, Pines A (1984) J. Magn. Reson. 60: 337
270. Samoilenko AA, Artemov DYu, Sibeldina LA (1988) JETP Lett. 47: 348
271. Samoilenko AA, Zick K (1990) Bruker Report 1: 40
272. Kinchesh P, Randall EW, Zick K (1992) J. Magn. Reson. 100: 411

273. Samoilenko AA, Zick K (1992) In: Anagnosopoulos A, Milia F, Smiopoulos A (eds) Extended Abstracts of 26th Congress Ampere, Athens
274. Emid S (1985) Physica B 128: 79
275. Cory DG, Veeman WS (1989) J. Magn. Reson. 84: 392
276. McDonald PJ, Attard JJ, Taylor DG (1987) J. Magn. Reson. 72: 224
277. Rommel E, Hafner S, Kimmich R (1990) J. Magn. Reson. 86: 264
278. Demco DE, Hafner S, Kimmich R (1991) J. Magn. Reson. 94: 333
279. Cory DG, de Boer JWC, Veeman WS (1989) Macromolecules, 22: 1618
280. Günther E, Blümich B, Spiess HW (1991) Chem. Phys. Let. 184: 251
281. Conradi MS (1991) J. Magn. Reson. 93: 419
282. Zax D, Pines A (1983) J. Chem. Phys. 78: 6333
283. Bodenhausen G (1981) Prog. Nucl. Magn. Spectrosc. 14: 137
284. Weitekamp DP (1983) Adv. Magn. Reson. 11: 111
285. Emid S, Smidt J, Pines A (1980) Bull. Magn. Reson. 2: 85
286. Emid S, (1982/83) Bull. Magn. Reson. 4: 99
287. Pines A in Ref. [18], p 43
288. Günther E, Blümich B, Spiess HW (1990) Molec. Phys. 71: 477
289. Samoson A, Lippmaa E (1983) Phys. Rev. B 28: 6567
290. Blümich B, Spiess HW (1988) Angew. Chem. 27: 1655
291. Freeman D, Hurd R (1992) NMR Basic principles and progress 27: 199
292. Lyon RC, Pekar J, Moonen CTW, McLaughlin AC (1991) Magn. Reson. Med. 18: 80
293. Cockman MD, Jelinski LW, Katz J, Sorce DJ, Boxt LM, Cannon PJ (1990) J. Magn. Reson. 90: 9
294. Wimperis S, Cole P, Styles P (1992) J. Magn. Reson. 98: 628
295. Haeberlen U, Waugh JS (1968) Phys. Rev. 175: 453
296. Carr HY, Purcell EM (1954) Phys. Rev. 94: 630
297. Meiboom S, Gill D (1958) Rev. Sci. Instrum. 29: 688
298. Miller JB, Garroway AN (1988) In: Thompson DO Chimenti DE (eds) Review of progress in quantitative nondestructive evaluation 7a. Plenum, p 609
299. Miller JB, Cory DG, Garroway AN (1989) Bull. Magn. Reson. 11(3, 4): 197
300. Waugh JS, Huber LM, Haeberlen U (1968) Phys. Rev. Lett. 220: 180
301. Mansfield P (1971) J. Phys. C: Solid state phys. 4: 1444
302. Rhim WK, Elleman DD, Vaughan RW (1973) J. Chem. Phys. 58: 1772
303. Burum DP, Rhim WK (1979) J. Chem. Phys. 71: 944
304. Con DG, Miller JB, Garroway AN (1990) J. Magn. Reson. 90: 205
305. Weitekamp DP, Garbow JR, Pines A (1982) J. Chem. Phys. 77: 2870
306. Weitekamp DP, Garbow JR, Pines A (1984) J. Chem. Phys. 80: 1372
307. Dieckman SL, Gopalsami N, Botto RE (1990) Energy and Fuels 4: 417
308. Chingas GC, Miller JB, Garroway AN (1986) J. Magn. Reson. 66: 530
309. Sinton SW, Iwamiya JH, Ewing B, Drobny GP (1991) Spectroscopy 6: 42
310. Garroway AN, Mansfield P, Stalker DC (1975) Phys. Rev. 11: 121
311. Miller JB, Cory DC, Garroway AN (1989) Chem. Phys. Lett. 164: 1
312. Pines A, Waugh JS (1972) J. Magn. Reson. 77: 187
313. Miller JB, Cory DC, Garroway AN (1989) Chem. Phys. Lett. 164: 1
314. Miller JB, Garroway AN (1989) J. Magn. Reson. 82: 529
315. Miller JB, Garroway AN (1989) J. Magn. Reson. 85: 255
316. McDonald, PJ Tokarczuk PF (1989) J. Phys. E: Sci. Instrum 22: 948
317. McDonlad, PJ Tokarczuk PF (1992) J. Magn. Reson. 99: 225
318. Cottrell SP, Halse MR, McDonald PJ, Strange JH, Tokarczuk PF (1989) Bull. Magn. Reson. 11: 310
319. Pines A, Waugh JS (1975) J. Magn. Reson. 8: 50
320. Cory DG, Miller JB, Garroway AN (1990) Mol. Phys. 70: 331
321. Conradi MS, Garroway AN, Cory DG, Miller JB (1991) J. Magn. Reson. 94: 370
322. Cory DG (1992) Talk at 33rd ENC. Asilomar, CA
323. Cho HM, Lee CJ, Shykind DN, Weitekamp DP (1985) Phys. Rev. Lett. 55: 1923
324. Cory DG, Gravina SJ in Ref. [27], 0 p 15
325. Rhim WK, Pines A, Waugh JS (1970) Phys. Rev. Lett. 25: 218
326. Rhim WK, Pines A, Waugh JS (1971) Phys. Rev. B 3: 684
327. Rhim WK, Pines A, Waugh JS (1971) Phys. Rev. B 3: 684
328. Schneider H Schmiedel H (1969) Phys. Lett. 30A: 298 (1969)

329. Schneider H Schmiedel H (1975) Ann. Physik 32: 249
330. Pines A, Rhim WK, Waugh JS (1972) J. Magn. Reson. 6: 457
331. Takegoshi K, McDowell CA (1985) Chem. Phys. Lett. 116: 100
332. Matsui S (1991) Chem. Phys. Lett. 179: 187
333. Matsui S (1992) J. Magn. Reson. 98: 618
334. Hafner S (1992) PhD Thesis, University of Ulm, Germany
335. Hafner S, Demco DE, Kimmich R (1991) Meas. Sci. Technol. 2: 882
336. Demco DE, Hafner S, Kimmich R (1992) J. Magn. Reson. 96: 307
337. Weigand F (1993) Diploma Thesis, University of Mainz
338. Matsui S (1991) J. Magn. Reson. 95: 149
339. Griffin RG, Aue WP, Haberkorn RA, Harbison GS, Herzfield J, Menger EM, Munowitz MG, Olejniczak ET, Raleigh DP, Roberts JE, Ruben DJ, Schmidt A, Smith SO, Vega S in Ref. [18], p 203
340. Maciel GE, Bronnimann CE, Hawkins BL (1990) Adv. Magn. Reson. 14: 125
341. Burum DP (1990) Concepts Magn. Reson. 2: 213
342. Ryan LM, Taylor RE, Paff AJ, Gerstein BC (1980) J. Chem. Phys. 77: 508
343. Hayashi S Hayamizu K (1990) J. Chem. Phys. 92: 2818
344. Hayashi S Hayamizu K (1991) Bull. Chem. Soc. Japan 64: 1386
345. Schauß G (1992) PhD Thesis, University of Mainz, Germany
346. Wind RA, Yannoni CS (17, Nov. 1981) US-Patent 4. 301. 410
347. Ackerman JL (31, Mar. 1987) US-Patent 4.654.593
348. Cory DG, van Os JWM, Veeman VS (1988) J. Magn. Reson. 76: 543
349. Cory DG, Reichwein AM, van Os JWM, Veeman WS (1988) Chem. Phys. Lett. 143: 467
350. Cory DG, Veeman WS (1989) J. Magn. Reson. 82: 374
351. Veeman WS, Cory DG (1989) Adv. Magn. Reson. 13: 43
352. Dixon WT (1982) J. Chem. Phys. 77: 1800
353. Scheler U, Blümich B, Spiess HW (to be published)
354. Mefed AE, Atsarkin VA (1977) JETP Lett. 25: 215
355. Mefed AE, Atsarkin VA (1978) Sov. Phys. JETP 47: 378
356. Atsarkin VA, Mefed AE, Rodak I (1979) Sov. Phys. Solid State 21: 1537
357. Mefed AE (1984) Sov. Phys. JETP 60: 162
358. Redfield AG (1955) Phys. Rev. 98: 1787
359. de Luca F, Nucceteli C, de Simone BC, Maraviglia B (1989) Solid state comm.70: 797
360. de Luca F, de Simone BC, Maraviglia B, Nuccetelli C (1986) Bull. Magn. Reson. 8: 102
361. de Luca F, Nuccetelli C, de Simone BC, Maraviglia B (1986) J. Magn. Reson. 69: 496
362. de Luca F, Maravilia B (1986) J. Magn. Reson. 67: 169
363. de Luca F, de Simmone BC, Lugeri N, Maraviglia B, Nuccetelli C (1990) J. Magn. Reson. 90: 124
364. de Luca F, Fattibene P, Lugeri N, Campanella R, Maraviglia B (1991) Appl. Magn. Reson. 2: 93
365. de Luca F, Lugeri N, de Simone BC, Maraviglia B (1992) Solid state comm. 82: 151
366. de Luca F, de Simone BC, Lugeri N, Maraviglia B (1993) J. Magn. Reson., in press
367. de Luca F, de Simone BC, Maraviglia B, Nuccetelli C in Ref. [18], p 382
368. Lugeri N, de Luca F, Maraviglia B in Ref. [11], p 101
369. Lee M, Goldburg WI (1955) Phys. Rev. 98: 1787
370. Kaldoudi E, Williams SCR (1992) Concepts Magn. Reson. 4: 53
371. Rumpel H, Pope JM, Concepts Magn. Reson., (in press)
372. Pope JM in Ref. [11], p 441
373. Dixon WT (1984) Radiology 153: 189
374. Freeman R (1991) Chem. Rev. 91: 1397
375. Shaka AJ, Freeman R (1984) J. Magn. Reson. 59: 169
376. Levitt MH (1986) Prog. NMR Spectros. 18: 61
377. Warren WS, Silver MS, (1988) Adv. Magn. Reson. 12: 247
378. Kessler H, Mronga S, Gemmecker G (1991) Magn. Reson. Chem. 29: 527
379. Morris PG (1992) NMR Basic priniciple and progress 26: 149
380. Cho ZH, Park HW (1989) Adv. Magn. Reson. Imag. 1: 1
381. Garrido L, Mark JE (1989) Polym. Prepr. 30: 217
382. Pfleiderer B, Ackerman JL, Garrido L (1992) Poster WP 156 33rd ENC. Asilomar, CA
383. Morris GA, Freeman R (1978) J. Magn. Reson. 29: 433

384. Caravatti P, Levitt MH, Ernst RR (1986) J. Magn. Reson. 68: 323
385. Kuhns PL, Conradi MS (1982) J. Chem. Phys. 77(4): 1771
386. Tekely P, Brondeau J, Elbayed K, Retourned A, Canet D (1988) J. Magn. Reson. 80: 509
387. Garroway AN, Grannell PK, Mansfield P (1974) J. Phys. C: Solid state phys. 7: L575
388. Mansfield P (1977) J. Phys. C: Solid State Phys. 10: L55
389. Mansfield P, Maudsley AA, Baines T (1976) J. Phys. E: Sci. Instrum. 9: 271
390. Hoult DI (1979) J. Magn. Reson. 35: 69
391. Mulkern R (1992) Concepts Magn. Reson. 4: 307
392. Briand J, Hall LD (1989) J. Magn. Reson. 83: 418
393. Wind RA, Creyghton JHN, Ligthelm DJ, Smidt J (1978) J. Phys. C: Solid state phys. 11: L223
394. Doddrell DM, Brooks WM, Bulsing JM, Field J, Irving MG, Baddeley H (1986) J. Magn. Reson. 68: 367
395. Doddrell DM, Bulsing JM, Galloway GJ, Brooks WM, Field J, Irving M, Baddeley H (1986) J. Magn. Reson. 70: 319
396. Luyten PR, Marien AJH, Sijtsma B, Hollander JA (1986) J. Magn. Reson. 67: 148
397. de Crespigny AJS, Carpenter TA, Hall LD (1989) J. Magn. Reson. 85: 595
398. de Crespigny AJS, Carpenter TA, Hall LD (1990) J. Magn. Reson. 88: 406
399. Ordidge RJ (1987) Magn. Reson. Med. 5: 93
400. Roberts TPL, Carpenter TA, Hall LD (1991) J. Magn. Reson. 91: 204
401. Aue WP, Müller S, Cross TA, Seelig J (1984) J. Magn. Reson. 56: 350
402. Müller S, Aue WP, Seelig J (1985) J. Magn. Reson. 63: 530
403. Kimmich R, Demco DE Hafner S, in Ref. [11], p 57
404. Rommel E, Kimmich R (1989) Magn. Reson. Med. 12: 390
405. Demco DE, Kimmich R, Hafner S, Weber HW (1991) J. Magn. Reson. 94: 317
406. Hafner S, Demco DE, Kimmich R (1991) Chem. Phys. Lett. 187: 53
407. Johnson AJ, Garwood M, Ugurbil K (1989) J. Magn. Reson. 81: 653
408. Garwood M, Ugurbil K (1992) NMR Basic principles and progess 26: 109
409. Hinshaw WS (1974) Phys. Lett. 48A(2): 87
410. Hinshaw WS (1976) J. Appl. Phys. 47(8): 3709
411. Bottomley PA (1982) Rev. Sci. Instrum. 53(9): 1319
412. Bottomley PA (1982) J. Magn. Reson. 50: 335
413. Scott KN, Brooker HR, Fitzsimmons JR, Bennett HF, Mick RC (1982) J. Magn. Reson. 50: 339
414. Macovski A (1985) Magn. Reson. Med. 2: 29
415. Yannoni CS, Vieth HM (1976) Phys. Rev. Lett. 37: 1230
416. Wind RA, Yannoni CS (1979) J. Magn. Reson. 36: 273
417. Matsui S (1992) J. Magn. Reson. 97: 335
418. Ordidge RJ, Connelly A, Lohman JAB (1986) J. Magn. Reson. 66: 283
419. Kimmich R, Hoepfel D (1987) J. Magn. Reson. 72: 379
420. Kimmich R, Schnur G, Hoepfel D, Ratzel D (1987) Phys. Med. Biol. 32: 1335
421. Knüttel A, Kimmich R (1988) J. Magn. Reson. 78: 205
422. Knüttel A, Rommel E, Clausen M, Kimmich R (1988) Magn. Reson. Med. 8: 70
423. Kimmich R, Rommel E, Knüttel A (1989) J. Magn. Reson. 81: 333
424. Knüttel A, Kimmich R (1989) Magn. Reson. Med. 9: 254
425. Knüttel A, Kimmich R, Spohn KH (1989) J. Magn. Reson. 81: 570
426. Knüttel A, Kimmich R, Spohn KH (1990) J. Magn. Reson. 86: 526
427. van Zijl PCM, Moonen CTW, Alger JR, Cohen JS, Chesnick SA (1989) Magn. Reson. Med. 10: 256
428. Decorps M, Bourgenois D (1992) NMR Basic principal and progress 27: 119
429. Bendall MR, Pegg DT (1984) J. Magn. Reson. 57: 337
430. Bendall MR, Pegg DT (1985) J. Magn. Reson. 63: 494
431. Bendall MR, Pegg DT (1985) Magn. Reson. Med. 3: 91
432. Bendall MR, Pegg DT (1986) J. Magn. Reson. 68: 252
433. Bosch CS, Ackerman JJH (1992) NMR Basic Principles and Progress 27: 3
434. Bendall MR (1986) Bull. Magn. Reson. 8(1, 2): 17
435. Béné GJ, Borcard B, Hiltbrand E, Magnin P (1981) NMR Basic principles and progress 19: 81
436. Ackerman JJH (1990) Concepts Magn. Reson. 2: 33
437. Ackerman JJH, Grove TH, Wong GG, Gadian DG, Radda GK (1980) Nature 283: 167
438. Evelhoch JL, Crowley MG, Ackermann JJH (1984) J. Magn. Reson. 56: 110

439. Harpen MD (1987) Med. Phys. 14: 616
440. Lawry TJ, Weiner MW, Matson GB (1990) Magn. Reson. Med. 16: 294
441. Bendall MR in Ref. [16], p 1201
442. Crowley MG, Evelhoch JL, Ackerman JJH (1985) J. Magn. Reson. 64: 20
443. Haase A, Hänicke W, Frahm J (1984) J. Magn. Reson. 56: 401
444. Hoult DI (1979) J. Magn. Reson. 33: 183
445. Mitsumori F, Bolas NM (1992) J. Magn. Reson. 97: 282
446. Bolton PH (1985) J. Magn. Reson. 63: 620
447. Haase A, Malloy C, Radda GK (1983) J. Magn. Reson. 55: 164
448. Garwood M, Schleich T, Ross BD, Matson GB, Winters WD (1985) J. Magn. Reson. 65: 239
449. Cox SJ, Styles P (1980) J. Magn. Reson. 40: 209
450. Karczmar GS, Weiner MW, Matson GB (1987) J. Magn. Reson. 71: 360
451. Styles P (1992) NMR Basic principles and progress 27: 45
452. Bendall MR, Gordon RE (1983) J. Magn. Reson. 53: 365
453. Bendall MR (1984) J. Magn. Reson. 59: 406
454. Pegg MR, Bendall DT, Tyburn GM, Brevard C (1983) J. Magn. Reson. 55: 322
455. Pegg DT, Bendall MR (1985) J. Magn. Reson. 63: 556
456. Hetherington HP, Wishart D, Fitzpatrick SM, Cole P, Shulman RG (1986) J. Magn. Reson. 66: 313
457. Tycko R, Pines A (1984) J. Magn. Reson. 60: 156
458. Ng TC, Glickson JD, Bendall MR (1984) Magn. Reson. Med. 1: 450
459. Pope JM, Eberl S (1985) Magn. Reson. Imag. 3: 389
460. Bottomley PA (1992) NMR Basic principles and progress 27: 67
461. Garwood M, Schleich T, Matson GB, Acosta G (1984) J. Magn. Reson. 60: 268
462. Garwood M, Robitaille PM, Ugurbil K (1987) J. Magn. Reson. 75: 244
463. Blackledge MJ, Styles P, Radda GK (1987) J. Magn. Reson. 71: 246
464. Bechinger B, Opella SJ (1991) J. Magn. Reson. 95: 585
465. Buess ML, Garroway AN, Miller JB (1991) J. Magn. Reson. 92: 348
466. Letcher JH (1989) Magn. Reson. Imag. 7: 581
467. Hyde JS, Jesmanowicz A, Grist TM, Froncisz W, Kneeland B (1987) Magn. Reson. Med. 4: 179
468. Hyde JS, Jesmanowicz A, Grist TM, Froncisz W, Kneeland B (1988) Magn. Reson. Med. 6: 253
469. Rath AR, Roeder SBW, Fukushima E (1988) J. Magn. Reson. 79: 461
470. Hayes CE, Roemer PB (1990) Magn. Reson. Med. 16: 181
471. Schnall M (1992) NMR Basic principles and progress 26: 33
472. Grist TM, Jesmanowicz A, Kneeland JB, Froncisz W, Hyde JS (1988) Magn. Reson. Med. 6: 253
473. Gonnella NC, Silverman RF (1989) J. Magn. Reson. 85: 24
474. Li L, Sotak CH (1991) J. Magn. Reson. 93: 207
475. Edelstein WA, Hardy CJ, Mueller OM (1986) J. Magn. Reson. 67: 156
476. den Hollander JA, Behar KL, Shulman RG (1984) J. Magn. Reson. 57: 311
477. Miller JB, Garroway AN (1988) J. Magn. Reson. 77: 187
478. Miller JB, Garroway AN (1989) J. Magn. Reson. 85: 432
479. Neue G (1990) Bruker Report 2: 25
480. Skibbe U, Neue G (1990) Colloids and Surfaces 45: 235
481. Gadian DG, Robinson FNH (1979) J. Magn. Reson. 34: 449
482. Mehring M, Kotzur D, Kanert O (1972) Phys. Status solidi B 53: 25
483. Kotzur D, Mehring M, Kanert O (1973) Z. Naturforsch. 28: 1607
484. Cooper RK, Jackson JA (1980) J. Magn. Reson. 41: 400 (1980)
485. Burnett LJ, Jackson JA (1980) J. Magn. Reson. 41: 406 (1980)
486. Jackson JA, Burnett LJ, Harmon JF (1980) J. Magn. Reson. 41: 411
487. Kleinberg RL, Sezginer A, Griffin DD, Fukuhara M (1992) J. Magn. Reson. 97: 466
488. King JD, Matzkanin GA, Rollwitz WL in Ref. [24], p MD4

Author Index Volumes 21–30

MIX
Papier aus verantwortungsvollen Quellen
Paper from responsible sources
FSC® C105338

If you have any concerns about our products, you can contact us on
ProductSafety@springernature.com

In case Publisher is established outside the EU, the EU authorized representative is:
Springer Nature Customer Service Center GmbH
Europaplatz 3, 69115 Heidelberg, Germany

Printed by Libri Plureos GmbH
in Hamburg, Germany